CONTENTS

KU-024-472

[Mathematically oriented sections are marked with an asterisk *; these may be skipped without the flow of the text being lost]

Contents

CAMBRIDGE S
IN MATHEMAT

Editors

C. CANNINGS
Department of Probability and Statistics, University of Sheffield, UK
F. C. HOPPENSTEADT
College of Natural Sciences, Michigan State University, East Lansing, USA
L. A. SEGEL
Weizmann Institute of Science, Rehovot, Israel

MODELLING BIOLOGICAL POPULATIONS IN SPACE AND TIME

CAMBRIDGE STUDIES
IN MATHEMATICAL BIOLOGY

ERIC RENSHAW

Department of Statistics and Modelling Science, University of Strathclyde

Modelling biological populations in space and time

CAMBRIDGE
UNIVERSITY PRESS

Published by the Press Syndicate of the University of Cambridge
The Pitt Building, Trumpington Street, Cambridge CB2 1RP
40 West 20th Street, New York, NY 10011–4211, USA
10 Stamford Road, Oakleigh, Melbourne 3166, Australia

First published 1991
First paperback edition 1993

Printed in Great Britain at the University Press, Cambridge

British Library cataloguing in publication data

Renshaw, Eric
  Modelling biological populations in space and time.
  1. Organisms. Population. Dynamics. Mathematical models
  I. Title
  574.5248011

Library of Congress cataloguing in publication data

Renshaw, Eric
  Modelling biological populations in space and time / Eric Renshaw.
    p. cm.
  Includes bibilographical references.
  ISBN 0-521-30388-5 (hbk)   ISBN 0-521-44855-7 (pbk)
  1. Population biology—Statistical methods. 2. Spatial analysis
(Statistics) I. Title.
  QH352.R46 1990
  574.5'248'072—dc20 90-33653 CIP

ISBN 0 521 30388 5 hardback
ISBN 0 521 44855 7 paperback

ETA

Contents

Contents

Contents

Contents

PREFACE

The remarkable variety of dynamic behaviour exhibited by many species of plants, insects and animals has stimulated great interest in the development of both biological experiments and mathematical models. From a relatively slow start in the 1920s and 1930s, the pace of research has quickened dramatically over the past few years. Unfortunately, however, ideas have polarized at the same rate. Theoreticians often model purely in terms of manipulating mathematical equations, throwing in the occasional biological reference merely to gain practical respectability; whilst biologists may develop vaguely plausible deterministic models which reflect mathematical hope rather than biological reality.

Many researchers still use one approach to the total exclusion of the other. The reasons are two-fold. First, pioneering biological studies were greatly influenced by deterministic mathematics, and reluctance to accept stochastic ideas is still ingrained. Second, too many mathematicians are taught in a practical vacuum, with the result that instead of using mathematics to interpret and understand biological phenomena they become transfixed by the models themselves.

In this book we develop a unifying approach. First, we show that both deterministic and stochastic models have important roles to play and should therefore be considered together; popular deterministic ideas of logistic, chaotic and predator–prey relationships can change markedly when viewed in a stochastic light.

Second, in biology we are often asked to infer the nature of population development from a single data set, yet different realizations of the same process can vary enormously. Since even stochastic solutions are only of limited help here, we shall construct simple computer simulation procedures which provide much needed insight into the underlying generating mechanisms. Indeed, such model-based simulations can highlight hitherto unforeseen features of a process and thereby suggest further profitable lines of biological investigation. Sample programs are provided to enable readers to perform their own simulation experiments.

The third approach we advocate is recognition that the environment has a spatial dimension, since individual population members rarely mix homogeneously over the territory available to them but develop instead within separate sub-regions. Subsequent migration between these sub-regions can vary from being purely local to involving extensive migration patterns covering the Earth's surface. Fortunately, fairly simple models can be developed which highlight the effects that geographic restrictions and species' mobility may have on population development. These models can provide vital knowledge about the dispersal and control of many natural populations, not just of animals, insects and plants but also of diseases such as malaria, rabies and AIDS.

Only relatively simple population models for single species and two interacting species are used, though these are extremely important in their own right. The resulting mathematical analyses are then sufficiently transparent to enable useful biological conclusions to be drawn from them. If elegant mathematics occurs as a result, then all well and good, but we shall not regard mathematical proof as being of primary interest. Readers with virtually no mathematical expertise can gain a great deal simply by skipping over the more theoretically oriented sections (marked with an asterisk *). Indeed, this is to be actively encouraged at a first reading; much can be understood without the basic flow of the text being lost. All the subject areas covered are accessible to undergraduate students in biology and mathematics alike.

This said, I hope that many people with theoretical interests in population dynamics will also find much to influence them. Since as well as discussing models of single-species population growth and two-species interaction, we also introduce chaos, fluctuating environments, spatial predator–prey systems and population dynamics, epidemics, and spatial branching processes. The possibilities for further development are endless.

It is a pleasure to record my grateful appreciation to Richard Cormack, John Meldrum and Byron Morgan who kindly read the final manuscript; I am especially indebted to Richard for all his perceptive comments.

A LIST OF SYMBOLS AND NOTATION

Whilst some of the conventions below, such as $\leqslant$, are strictly adhered to, others just record the usual meaning or meanings of a symbol. For example, N usually denotes deterministic population size, but i may denote species, be a general index, or (as i) represent $\sqrt{(-1)}$, depending on the particular context in use.

Symbol	Usual meaning
a	position of absorbing barrier
a_{ij}	interaction parameters
$B(N)$	general birth rate for a population of size N
c	velocity of propagation
c_k	autocovariance at lag k
$\cos(\theta)$	cosine of θ
$\cosh(\theta)$	$\frac{1}{2}(e^\theta + e^{-\theta})$
Cov, σ_{12}	covariance
$CV(t)$	coefficient of variation at time t, $\sqrt{(V(t))}/m(t)$
$D(N)$	general death rate for a population of size N
E	expectation, mean
e^θ, $\exp(\theta)$	exponential of θ
$e^{i\theta}$	$\cos(\theta) + i\sin(\theta)$ for $i = \sqrt{(-1)}$
f	frequency (but only in a time-series context)
i	$\sqrt{(-1)}$
$I_i(x)$	modified Bessel function of the first kind
h	small time length
$\log_e$	logarithm to the base e $[\log_e(e^x) = x]$
$\log_{10}$	logarithm to the base 10
K	logistic carrying capacity
$K_\nu(z)$	modified Bessel function of the second kind
max	maximum value
min	minimum value
$m(t)$	mean population size at time t

Symbols and notation

$n_0, N(0)$	initial population size
$N(t)$	deterministic population size at time t
$n(t), u(t)$	small random variable, e.g. $X(t) = N*[1 + n(t)]$
$N*$	deterministic equilibrium value
$n!$	$n \times (n - 1) \times (n - 2) \times \cdots \times 2 \times 1$
$\binom{n}{r}$	$(n!)/[(n - r)!r!]$
p.d.f.	probability density function
$p_0(t)$	probability that the population is extinct by time t
$p_0(\infty)$	probability of eventual extinction
$p_N(t)$	probability that the population is of size N at time t
$p_{n_0}(q_{n_0})$	probability of eventual absorption at $a(0)$ for an initial population of size n_0
$\Pr(A)$	probability that event A occurs
r	net growth rate in the absence of regulation
r_k	autocorrelation at lag k
s, S	inter-event time
$\text{sech}(\theta)$	$2/(e^{\theta} + e^{-\theta})$
$\sin(\theta)$	sine of θ
s_k	cross-correlation at lag k
t	time
T	period of oscillation
T_a	time to first occupation of state a
$\tanh(\theta)$	$(e^{\theta} - e^{-\theta})/(e^{\theta} + e^{-\theta})$
t_c	coherence time
t_D	reaction time-lag
t_G	reproductive time-lag
$T_E(n_0)$	mean time to extinction for an initial population of size n_0
$(t, t + h)$	small time interval from time t to $t + h$
$V(t), \text{Var}(t), \sigma^2(t)$	variance of population size at time t
$X(t), Y(t)$	stochastic population size at time t
$\{X(t)\}$	sequence of values of $X(t)$
Y	uniform pseudo-random number in the range $0 \leqslant Y \leqslant 1$
Z_N	length of time population is exactly of size N
$z(u)$	weighting function
α	immigration rate

β	coefficient of skewness, and rate of infection of susceptibles
γ	Euler's constant $(0.577\,216\ldots)$, and rate of removal of infectives
γ_3	measure of skewness
δt	small time increment
δZ	random 'noise'
λ	birth rate
μ	death rate
ν	migration rate
ζ	index of stability $(\log_e(T_e))$
π	pi $(3.141\,593\ldots)$
π_N	equilibrium probability that the population is of size N
$\pi_N^{(Q)}$	pseudo-equilibrium probability that the population is of size N (i.e. conditional on extinction not having occurred)
ω	angular frequency, $\omega = 2\pi f = 2\pi t/T$
$df(t)/dt$	total derivative of $f(t)$ with respect to t
$\partial f(s, t)/\partial t$	partial derivative of $f(s, t)$ with respect to t
NAG	Numerical Algorithms Group
$\displaystyle\sum_{i=1}^{n} a_i$	the sum $a_1 + a_2 + \cdots + a_n$
$\displaystyle\int_a^b f(x)\,dx$	integral of $f(x)$ between a and b
$\displaystyle\prod_{i=1}^{n} a_i$	the product $a_1 \times a_2 \times \cdots \times a_n$
$\Phi(.)$	standard Normal distribution function with mean 0 and variance 1
$\pm$	plus or minus
$\sqrt{}$	square root
$=$	equal to
$\equiv$	identically equal to
$\simeq$	approximately equal to
$\sim$	$a(t) \sim b(t)$ means that $a(t)/b(t) \to 1$ as $t \to \infty$
$\to$	tends to
$\neq$	not equal to
$> (<)$	strictly greater (less) than
$\geqslant (\leqslant)$	greater (less) than or equal to
$\gg (\ll)$	a lot greater (less) than
∞	infinity

1

Introductory remarks

Of all areas of ecology, population biology is perhaps the most mathematically developed, and has involved a long history of mathematicians fascinated by problems associated with the dynamics of population development. Interest was induced by early studies of small mammals and laboratory controlled organisms, since these easily lent themselves to a mathematical formulation. A great deal of more recent research is concerned with modelling multi-species and spatial population growth, though it is not clear just how effective these models are for predicting behaviour outside the laboratory. There is general uncertainty regarding whether populations in the natural environment are mostly regulated from within by density-dependent factors, or whether the main influence is due to external density-independent factors. Theoretical developments have generally followed the former route, primarily because there is much less information on external factors due to their complexity and variability (see Gross, 1986).

Throughout most of this text we shall therefore disregard the (generally unknown) external influences on population growth, and develop the ideas of density-dependence. Moreover, since even apparently minor modifications to simple biological models can lead to difficult, if not intractable, mathematics, we shall begin by investigating the simplest possible forms of model structure (Chapter 2). In these, members of a population are assumed to develop *independently* from each other, for then the resulting mathematical analyses are sufficiently transparent to enable useful biological conclusions to be drawn.

1.1 Deterministic or stochastic models?

Recent interest in population dynamics has polarized to an undesirable extent, though there are hopeful signs that this situation may soon improve. On the one hand, there are mathematicians whose idea of collaboration in 'applied' biological research is to spend just enough time with the biologist to be able to write down a set of probability equations which will keep himself amused over the next few months! Whilst conversely, some mathematically oriented biologists have gained considerable mileage out of developing supposedly plausible deterministic models with interesting mathematical features. As long as no one questions whether these features relate to biological reality or pure imagination, then not only will this approach remain unchallenged but it will also thrive on its own increasing aura of biological respectability.

The tragedy is that too few researchers realize that *both* deterministic and stochastic models have important roles to play in the analysis of any particular system. Slavish obedience to one specific approach can lead to disaster. Provided that population numbers never become too small then a deterministic model *may* enable sufficient biological understanding to be gained about the system; if at any time population numbers do become small then a stochastic analysis is vital. So pursuing both approaches simultaneously ensures that we do not become trapped either by deterministic fantasy or unnecessary mathematical detail.

As an example, suppose that all members of a population develop independently from each other, and reproduce at rate λ and die at rate μ. Then the deterministic number of individuals alive at time t, starting from an initial population of size $N(0)$ at time 0, is given by

$$N(t) = N(0) \exp\{(\lambda - \mu)t\}. \tag{1.1}$$

If births predominate over deaths then this result tells us that the population size will explode exponentially fast, whilst if deaths predominate then extinction is inevitable.

For large $N(0)$ and realistically small t this is indeed an excellent description, but let us reflect for a moment on what happens if $N(0)$ is small. Indeed, suppose that $N(0) = 1$ and $\lambda = 2\mu$ so that births are twice as likely to occur as deaths. Then result (1.1) predicts exponential growth with $N(t) = \exp(\mu t)$. But the first event to occur may be a death, with probability $\mu/(\mu + \lambda) = \frac{1}{3}$, and this results in the population immediately becoming extinct. Thus the probability of ultimate extinction is at least $\frac{1}{3}$ which is in direct contradiction to the deterministic prediction. The situation becomes even more contradictory when $\lambda = \mu$. Then $N(t)$ remains absolutely constant at $N(t) = 1$ in spite of the fact that the actual (i.e. stochastic) process specifically involves birth and death.

Far more sinister (because it is not so intuitively obvious) is the popular desire to infer the nature of population development from a single experimental run. Ask most people to infer the average weight of a Scotsman by weighing just one and they would (one hopes!) express at least some degree of reservation. But ask them to predict how a population develops through time when they are presented with a single sequence of observations and their response suddenly changes. An apparently hard lesson to learn is that different realizations from the same process can vary enormously. A graphic illustration of this is provided in Figure 1.1 which shows three simulations of a simple birth–death model with $\lambda = 1.0$, $\mu = 0.5$, and $N(0) = 3$. Whilst one simulation exhibits exponential population growth (in line with deterministic theory), the second grows roughly linearly, whilst the third dies out completely! Yet each is generated from exactly the same process; the only changes are the purely random effects.

Such simulations are easy to compute, and are especially valuable in showing the degree of variability that we might expect to observe in practice. If plots of say 20 simulations all lie reasonably close to the deterministic curve then we can be satisfied that a deterministic approach will provide an adequate description of population development. If they do not then a stochastic description clearly needs to be developed. Moreover, such model-based simulations can highlight hitherto unforeseen features of a process and thereby suggest further profitable lines of biological investigation. It is surely in this continued interplay between modelling and experimentation that population studies are at their most powerful.

Figure 1.1. Three from 20 simulations of a simple birth–death process with $\lambda = 1.0$, $\mu = 0.5$ and $n_0 = 3$, showing $N(t)$ for values of $N(6)$ that are smallest (———), tenth smallest (– – –) and largest (— — —).

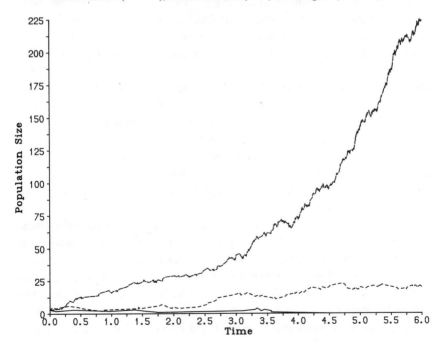

Such interplay is vital in epidemiology. For example, the current AIDS epidemic presents one of the major health problems of the twentieth century (see the report in the *Journal of the Royal Statistical Society*, 1988, A, 151, 3–136). Yet effective data are both sparse and patchy, and so their efficient use demands extremely close collaboration between modellers and medical practitioners.

The modelling of epidemics (Chapter 10) exposes the problem that deterministic prediction can be substantially different from what is expected under a stochastic model (i.e. the average of a large number of simulated realizations). For example, the expected rate at which new infectives accrue is of considerable interest to public health officers in deciding the severity and likely length of duration of an epidemic. Yet deterministic and stochastic assumptions generate substantially different values for this rate, and so the two approaches might easily lead to different conclusions. Full conditions for agreement between deterministic and expected stochastic solutions are at present unknown.

One area in which determinists have become extremely enthusiastic is the application of chaos theory, a subject with considerable popular appeal. Apparently trivial non-linear models such as

$$N(t + 1) = N(t)[a - bN(t)] \qquad (t = 0, 1, 2, \ldots) \tag{1.2}$$

(Chapter 4) give rise to a surprisingly rich diversity of mathematical behaviour ranging from stable equilibrium points, to stable oscillations between several points, through to a completely chaotic regime in which cycles of any period (or even totally aperiodic fluctuations) can occur. Though such systems have been known to mathematicians for a long time their considerable relevance in helping to understand biological phenomena remained essentially hidden until May (1975) highlighted their importance. A remarkable feature is that simple, purely deterministic, models can give rise to apparently totally unpredictable behaviour.

The trap for the unwary lies in being carried away on a wave of *mathematical* enthusiasm, since there is absolutely no reason why model (1.2) should give rise to an equally diverse range of *biological* behaviour. Indeed, as nature is inherently stochastic we have to investigate what mathematical features remain when a random component is added to completely deterministic equations such as (1.2). We shall learn that random variation can obliterate multi-point cycles, just leaving a stochastic 2-point cycle, though in the chaotic regime 3-point cycles occur both with and without the presence of random variation. Where fantasy takes over is in the belief that the mathematically exquisite and delicate fine structure of deterministic chaotic solutions (see May, 1986) might be observed in biology. Any superimposed environmental noise, almost no matter how small, destroys it. So for practical ecological purposes the chaotic regime relates to apparently random dynamics.

Note that we are not being totally critical here, as the deterministic prediction of stable, 2-point cyclic, and chaotic regimes has been immensely beneficial to greater understanding of population growth. What has been at fault is that too few stochastic checks have occurred *en route*.

1.2   Single-species populations

As Maynard Smith (1974) remarks, theory has never played the role in ecology that it has in population genetics, perhaps because there is nothing in ecology comparable to Mendel's laws. So when considering the place of mathematics in biology we need to develop models whose design is strongly influenced by considerations of mathematical simplicity. Moreover, these models should be constructed with the aim of providing a conceptual framework for the discussion of broad classes of phenomena (May, 1974b).

A good example is provided by the deterministic logistic equation

$$dN(t)/dt = rN(t)[1 - N(t)/K] \qquad (1.3)$$

(Chapter 3). Although this model provides nothing more than a crude (and sometimes wrong!) representation of biological growth under limited re-sources, it nevertheless 'summarises' a wide variety of data sets extremely well. The temptation of deducing the logistic curve as a universal law of population growth is unfortunately not always avoided.

Once shortage of resources precludes further population growth, equation (1.3) predicts that the population level $N(t) = K$ (called the carrying capacity) will be maintained exactly, contrary to general observation. In practice, many populations will fluctuate around their carrying capacity, so we need to be able to construct a probability distribution $\{p_N(t)\}$ of population size. Though analysis of even this simple logistic process does not readily yield an expression for $p_N(t)$, if a biological process has been developing long enough to ensure that it has stabilized around its carrying capacity then general 'equilibrium' probabilities $\{\pi_N\}$ may be obtained. Their derivation assumes that both population explosion and extinction are unlikely during the observed time span, and applies to any single-species model which satisfies this condition. However, considerable care is required when handling such situations, for we shall later see that some natural populations which exhibit initial logistic growth do not subsequently stabilize. Thus the logistic model may cease to be valid once the carrying capacity has been attained.

A major deficiency of the probability approach is that it relates to behaviour over all possible realizations. Now performing a population experiment many times under identical initial and operating conditions can produce totally different data sets; we have noted earlier that some (simulation) experiments may lead to extinction and others to population explosion. Theoretical probabilities may therefore tell us little about a single experimental run. Conversely, fitting a model to a single run of data implicitly assumes that repeated experiments will lead to similar results, which may well be far from the truth.

Fortunately, information on the likely behaviour of individual realizations may be obtained by repeatedly simulating the underlying model on a computer. This not only enables us to see the extent of possible behavioural differences, but it also allows the possibility of uncovering features that have not been observed in the field or laboratory. Specific biological experiments may then be constructed to see if these model-based features can be substantiated. The resulting sequential interaction between modeller and biologist should be extremely rewarding.

Of major interest is the mean time to extinction (T_E) of a population for which eventual extinction is certain, since it can be used to determine an index of stability $\xi = \log_e(T_E)$. A population is defined as being ecologically stable if it persists for a large number of generations, and ecologically unstable if it persists for only a few. Thus large T_E implies that extinction is unlikely to occur within any sensible time-scale, whence the population may simply be described in terms of its equilibrium distribution.

Many species exhibit behaviour far wider than can be described by the simple birth–death and logistic models, and in Chapter 4 we explore ways of expanding our suite of model-types by studying systems which possess a time-delay. For example, the population model

$$dN(t)/dt = rN(t - t_G)[1 - N(t - t_D)/K] \tag{1.4}$$

allows for both a reproductive time-lag (t_G) and a logistic reaction time-lag (t_D). In general, if the duration of the delay is longer than the 'natural period' of the system (usually defined as $1/r$ where r denotes population growth rate in the absence of regulation) then large amplitude oscillations will result. Indeed, as the time-delay increases it can have an increasingly destabilizing effect on $N(t)$, giving rise first to damped oscillations and then possibly to divergent oscillations. Since time-delays feature extensively in biological growth and control mechanisms, studying how changing the size and type of time-lag affects the dynamics of population growth yields valuable insight into the underlying biological process.

1.3   Two-species populations

Organisms do not generally exist in isolated populations but they live alongside organisms from other species. Whilst many of these species will not affect each other, in some cases they will interact. This may take the form of predation, cultivation, or competition for a common resource such as food or space. Now even the basic extension of the single-species logistic equation (1.3) to two species, namely

$$dN_1/dt = N_1(r_1 + a_{11}N_1 + a_{12}N_2)$$
$$dN_2/dt = N_2(r_2 + a_{21}N_1 + a_{22}N_2), \tag{1.5}$$

has six parameters. Hence as each parameter can be either positive, zero or negative the total possible number of such models is $3^6 = 729$ – with more species it becomes mind-boggling! We shall therefore confine our attention to two-species populations, and consider the two interaction processes of greatest biological relevance, namely competition (Chapter 5) and predation (Chapter 6).

Considering the former first, under the deterministic model (1.5) with $r_i > 0$ and $a_{ij} < 0$ ($i, j = 1, 2$) one of the species will definitely become extinct. The winning species then subsequently develops as a single-species logistic process. Pioneering experiments by Birch (1953) showed not only that a slight change in environment could alter which of the two competing species became extinct, but that in all the cases he examined the species which held the advantage always won. Thus in this situation a deterministic description is perfectly in order, especially since the quadratic equations (1.5) can be replaced by extremely robust linear approximations which enable the prediction of the winning species for given parameter values and initial population sizes.

Note that it does not follow that any set of repeated 'identical' experiments must necessarily lead to the same outcome (Krebs, 1985, gives examples). Moreover, differences in some species' habitat or requirements need only be very slight for both species to coexist quite happily.

Analysis of the deterministic equations (1.5) shows the existence of an equilibrium point (N_1^*, N_2^*) at which the population sizes $\{N_1(t), N_2(t)\}$ remain permanently unchanged (i.e. $dN_1/dt = dN_2/dt = 0$). If $N_1(t)$ and $N_2(t)$ are slightly perturbed from this point, then either they will be attracted back towards it (stable equilibrium – corresponds to coexistence) or else they will move away towards one of the axes (unstable equilibrium – corresponds to early extinction). In the unstable case simulation of the stochastic competition model shows that the probability that a particular species loses depends on the initial values $(N_1(0), N_2(0))$.

A totally different kind of outcome can occur under predation, in which members of one species eat those of the other. One of the earliest representations of predator–prey behaviour, constructed independently by Lotka (1925) and Volterra (1926), involves a particular case of (1.5), namely

$$dN_1/dt = N_1(r_1 - b_1N_2) \quad \text{(prey)} \tag{1.6a}$$
$$dN_2/dt = N_2(-r_2 + b_2N_1) \quad \text{(predator)}, \tag{1.6b}$$

and has interesting dynamics. We shall see that the solution $\{N_1(t), N_2(t)\}$ follows a family of closed curves in which each curve corresponds to a different initial point $(N_1(0), N_2(0))$.

This appealing result led Gause (1934) to make an empirical test of the model by rearing the protozoans *Paramecium caudatum* (prey) and *Didinium nasutum* (predator) together in an oat medium. He found that no matter how he altered the circumstances of his experiment *Didinium* always defeated *Paramecium*. So instead of generating oscillations predicted by the deterministic model (1.6), the experiment always yielded divergent oscillations which resulted in extinction of the prey. Note the interesting approach here, in which *biology* attempts to mimic *mathematics*!

The problem lies not with the mathematics, nor with the biology, but with an over-riding preoccupation with deterministic models. We shall discover that simulations of the stochastic form of the Lotka–Volterra process (1.6) do indeed lead to extinction of one of the species before the first cycle can be completed. Although sustained stochastic oscillations can be generated by imposing a maximum prey population size, i.e. by changing equation (1.6a) to

$$dN_1/dt = N_1(r_1 - cN_1 - b_1N_2), \tag{1.7}$$

this new system gives rise to convergent deterministic cycles towards an equilibrium point (Volterra, 1926). This great difference between stochastic and deterministic behaviour highlights the dangers inherent in making biological judgements which are based solely on deterministic models.

Difficulties can also arise when considering the relationship between models (of mathematical imagination) and biology (reality). For example, the deterministic Lotka–Volterra model (1.6) and the stochastic Volterra model (based on (1.6b) and (1.7)) have similar qualitative features. Thus if all we require is a 'black-box' description of observed predator–prey cycles then either approach may be used. However, if we wish to use the observations to *infer* or *understand* the underlying biological process then we need to be more specific. If the prey population is at all limited by resource, as well as by predator attack, then the Lotka–Volterra approach is automatically precluded.

Finally, we note that modelling provides a powerful tool for investigating the consequence of environmental and behavioural change (especially useful when initiating ideas for population control). Obvious examples include: introducing a refuge so that some prey will always be safe from predator attack; and letting the predator growth rate take account of the *relative* sizes of the two populations.

1.4   The spatial effect

Huffaker (1958) questioned Gause's conclusions that a predator–prey system is inherently self-annihilating without some outside interference, such as immigration. He recognized that individuals rarely mix homogeneously over a whole site but develop instead within separate (though possibly ill-defined) sub-regions. Indeed, there is currently a growing body of direct and circumstantial evidence which points to the particular importance of dispersal in controlling both the total size and the local variations in density of many natural populations (Nisbet and Gurney, 1982). Examples include comparisons of island and mainland vole populations (Tamarin, 1977), and of fenced and unfenced populations (Krebs, Keller and Tamarin, 1969).

To investigate whether the inability of the stochastic Lotka–Volterra model to generate sustained cyclic behaviour is due to the fact that it does not contain a spatial component, Huffaker set out to construct a large and complex spatial experiment in which the system would not be self-exterminating (Chapter 7). Two species of mites were allowed to develop and migrate over a spatial array of oranges arranged on trays, with migration of predators across trays being prevented by vaseline barriers and migration of prey over them being helped by air currents from a fan (Figure 1.2). It is

Figure 1.2. 120 oranges, each with $\frac{1}{20}$ orange-area exposed, occupying all positions in a three-tray system with partial-barriers of vaseline and supplied with wooden posts. Trays are joined by paper bridges. (Reproduced from Huffaker, 1958, by permission of Agriculture and Natural Resources Publications.)

interesting to speculate whether he would have bothered had he known that Gause's results are in full accord with the stochastic Lotka–Volterra model! Fortunately for us he was unaware of this result, and so he left the legacy of a beautiful experiment the likes of which would almost certainly not be undertaken today.

Huffaker found that although a single tray of 40 oranges always led to extinction after just one cycle, a three-tray system with 120 oranges did produce three full cycles before extinction occurred. A later experiment with 252 oranges even gave rise to four cycles! Is this extreme difficulty in sustaining oscillations with such a heavily contrived laboratory experiment really a fact of life, or did Huffaker just choose the wrong set of biological parameters? Once again computer simulation can provide the answer. Indeed, although a full-scale analysis of his system requires the power of a supercomputer, by recognizing that we do not need to know the exact numbers of prey and predators, but only whether there are none, few or many, we can dramatically reduce the computing requirements to the level of a personal computer. We shall see that sustained oscillations may be achieved with as few as ten sites provided that the migration rates are carefully chosen; not only must predators be more mobile than prey, but their migration rate must be high enough to prevent the occurrence of localized prey population explosions. Who will follow in the footsteps of Gause and Huffaker by designing an experiment to examine whether this computer prediction may be verified biologically?

Whilst simulation reveals general features of population spread and development, it does not provide complete insight into the underlying process. The bad news is that introducing a spatial component into equations for population growth makes even approximate solutions difficult to obtain. The even worse news is that some modellers use this as an excuse to disregard spatial effects, and then claim that their resulting conclusions reflect reality. This is fair enough if all they are doing is to construct an artificial examination question, but the practice becomes extremely dangerous when analysing situations of such immense importance as the spread and control of malaria or AIDS.

Tackling the problem head-on is no good, since even the stochastic two-colony version of (1.1) yields intractable mathematics. A sensible approach is to retain the formulation of deterministic mathematics and stochastic simulation, and to assume a fairly simple migration strategy (Chapter 9). We can use either a 'stepping-stone' model in which individuals can migrate only to immediately neighbouring sites (introduced in a genetics context by Kimura, 1953), or else a diffusion model in which individuals move continuously over the whole region. This approach is especially useful in helping us to understand animal, insect and plant dispersal through time, as deterministic results can be constructed which highlight how the interplay of migration and growth parameters affects the speed at which populations spread.

Laboratory experiments are by nature confined to a restricted area, which poses the question of how behaviour at the perimeter affects population growth throughout the region. A classic example is provided by Neyman, Park and Scott (1956) who investigate the spatial development of a population of flour beetles living in a $10 \times 10 \times 10$-inch cube of flour. Both sexes show an increase in population density away from the centre of the cube towards the sides, and along the sides towards the edges. Though a number of different models have been applied to this data set without too much success, the birth–death–migration stepping-stone approach works remarkably well. Moreover, as the equivalent deterministic diffusion process yields the same results, exact model choice is relatively unimportant. Stochastic behavioural differences can be investigated via simulation.

Such differences may well become substantial when simple (i.e. linear) birth is replaced by (non-linear) logistic growth or a two-species interaction process. Simulation runs suggest that whilst population numbers in neighbouring stepping-stone sites are closely related, those in neighbouring diffusion sites are far more variable. Note that a spatial predator–prey process can give rise to the spontaneous emergence of a large single patch of prey, claimed to be like the real situation for plankton populations observed in the sea. Moreover, the addition of spatial mobility to the stochastic Lotka–Volterra model can stabilize the non-spatial process and give rise to sustained population cycles. This marked change in behaviour clearly highlights the danger in ignoring the spatial component of a process purely for mathematical convenience.

An obvious way of avoiding problems with edge-effects is to arrange for sites to lie on a circle (or in two dimensions to lie on a torus!). Turing (1952) developed mathematical solutions based on such a layout, and although his analyses relate to the diffusion of reacting chemical substances his approach is completely general in its applicability (Section 9.10). Two distinct situations arise: in one case a number of stationary waves develop around the circle; in the other wave trains progress in opposite directions. The former is of particular biological applicability, and may account for dappled colour patterns in certain animals and the development of tentacled sea-creatures.

1.5   Related topics

Although this text concentrates primarily on internal density-dependent factors affecting population growth (see the opening paragraph) we must not forget that real populations are also controlled by external factors, even if it may seem convenient to do so since far less is known about them. An environment which fluctuates to any significant extent within the lifetime of a given population will clearly exert a considerable effect upon population development (Chapter 8). Moreover, a fluctuating environment provides an additional source of variability on top of the natural variation already present in the population itself. For example, plant species are strongly susceptible to climatic variation, and this exhibits a whole spectrum of pattern ranging in scale from the age of the Earth, through inter-glacial time-spans, down to tens of years. Sunspot cycles are widely thought to affect the development of several animal populations, whilst most living organisms are controlled to some extent by seasonal variation at either the yearly or daily level.

Thus four types of model are now available since both event and environment can vary deterministically or stochastically. Deterministic environmental change might involve trend or cycles. Stochastic variation may comprise the addition of random, or possibly autocorrelated, noise; or perhaps the frequency of a periodic component may itself be subject to some type of variation. Practical assessment of the potentially large number of models that can be generated from such assumptions may be made by recourse to laboratory experiments. Jillson (1980), for example, investigated the response of populations of flour beetles cultured in a series of regularly fluctuating environments.

Whilst this laboratory situation exhibits obvious cause and effect, field observations do not generally afford such luxury and great care is needed when explanations are being proffered for particular types of population pattern. One of the classic scenarios is the famous Canadian lynx data which show a highly regular cycle with a period of just under 10 years. This has inspired the generation of many mathematically based models, some environmental and some not, and whilst most *describe* the data tolerably well any *explanation* of the cyclic phenomenon must involve at least a certain amount of biological evidence. Where inferring process from pattern is concerned, mathematics is by no means an end in itself.

Periodic behaviour can also be observed in the study of infectious diseases (Chapter 10). For example, the numbers of measles and chicken-pox cases in large towns often exhibit cyclic rhythms. Though epidemiology is a major field of study in its own right, we include it here since it (perhaps surprisingly) involves only minor variations on predator–prey behaviour. Placing $r_1 = 0$, $b_1 = b_2 = \beta$ and $r_2 = \gamma$ in equations (1.6) gives

$$\begin{cases} dN_1/dt = -\beta N_1 N_2 \quad \text{(susceptibles)} \\ dN_2/dt = N_2(-\gamma_2 + \beta N_1) \quad \text{(infectives)}, \end{cases} \tag{1.8}$$

where $\{N_1(t), N_2(t)\}$ now denote the number of people susceptible to, and infected by, a certain disease. Thus β denotes the rate of infection and γ the removal rate of infectives (either by isolation or death).

Health authorities are particularly interested in the total number of removals before the end of an outbreak, a feature first investigated in a pioneering paper by Kermack and McKendrick (1927). They showed that: (i) there is a population threshold of susceptibles below which an outbreak cannot occur; and (ii) the number of susceptibles is eventually reduced to a value as far below this threshold as it was initially above. Unfortunately their paper languished in obscurity for 20 years, and one cannot help but wonder whether the progress of epidemic theory would have been greatly advanced had the authors published in a more widely available journal.

By the time this paper surfaced, interest was swiftly mounting in the application of stochastic processes, primarily due to two outstanding papers read to the Royal Statistical Society Symposium in Stochastic Processes in 1949 (Bartlett and Kendall). Epidemic models became an obvious target for investigation, since they have both mathematical and medical appeal, and a stochastic version of the threshold result soon appeared. Contrary to deterministic prediction, stochastic analysis shows that both major and minor outbreaks can occur if the threshold value is exceeded. This result may be studied theoretically and by simulation, and illustrates just how restrictive reliance on a totally deterministic approach can be. Moreover, even the simple model (1.8) can generate quite different stochastic realizations, which has serious implications when we ascribe abnormal results to abnormal virulence or infectiousness. Such results might arise purely from chance fluctuations.

Allowing for immigration of new infectives, studying more complicated diseases which pass through several recognizable stages, and introducing the spatial dimension, follows naturally along lines laid down earlier in this text. However, before we depart one totally fresh idea must be introduced (Chapter 11). Our study of the population development and spread of individuals such as flour beetles, plankton and infectives has so far disregarded the routes taken by these individuals to reach their destination. This becomes a serious omission when we wish to extend our field of investigation to cover the development of whole plants, nerve cells, lung airways, rivers, etc., since we are then interested in the complete morphology of a system.

Morphological modelling opens up a whole new world before us, and all we can do here is to make its existence known to the community at large. Plant development is determined by the rules which govern the production of successive components in its branching structure. Fortunately, a wide variety of different forms can arise from varying just a few simple branching rules, and even the most basic deterministic process can give rise to surprisingly lifelike forms. Though people may naturally be frightened by the thought of analyzing such apparently highly complex structures as the spatial configuration of tree crowns and structural root systems, the underlying sequential rules for individual branch and root generation may well be relatively simple.

Since analysis of biological branching systems depends heavily on computer simulation, we now have to adopt a totally different modelling approach. Elsewhere in this text we keep our models fairly simple in order to expose the qualitative features of a process through a theoretical investigation. The fact that such models only partially mimic reality does not bother us unduly provided that the essence of the process under study has been captured. However, if we are to use computers to 'grow' realistic plants then a far more quantitative approach is needed. Since all the work is performed on the computer, there is no advantage in making *any* initial simplifying assumptions. With modern computing power the running of long and detailed simulation programs which accurately reflect complex biological mechanisms is no longer a problem. We therefore isolate all the separate (i.e. elementary) stages of plant development, gather as much data as is necessary to describe each stage in statistical terms, translate these into computer code, and then combine the resulting subroutines into a simulation program. Output should be generated in the form of graph plots, preferably onto a VDU screen in the first instance. Obviously 'unplantlike' simulations can then be immediately discarded, and hard-copy need only be taken of successful ones. Two profound benefits emerge. First, by using a 'stepping-down' procedure, in which we study the effect of simplifying each subroutine in turn, we can deduce the minimal set of rules consistent with producing visually and statistically convincing realizations of plant architecture. Second, construction of the subroutines requires a considerable amount of biological knowledge, and a surprising number of gaps may well emerge during the course of the exercise, so pointing the way towards new areas of biological investigation.

Perhaps the most important aspect of all this work is that to be successful it necessitates strong interaction between mathematician and biologist.

2

Simple birth–death processes

The practical consequence of independent development is that individual members of a population must be able to live in virtually unrestricted environments in which no intraspecific competition can occur. Obvious potential situations are where species have been introduced into, or have invaded, isolated areas. Unfortunately, very few such cases have been extensively studied, though one that has is the invasion of Great Britain by the collared dove, *Streptopelia decaocto* (Hutchinson, 1978; Hengeveld, 1989). The bird spread westwards in Europe and started to breed in Britain in 1954. For the next ten years reasonably accurate censuses exist, but then the dove became sufficiently common to reduce its appeal to bird watchers and so data collection became inadequate. The rise in population growth between 1955 and 1963 (Figure 2.1) shows an almost linear relationship between the logarithm of population size and time. As we shall soon see, this is associated with independent growth during the early years; breeding birds were pioneers with no immediate neighbours. However, by the mid-1960s the dove had colonized most of Britain and competition for resource was beginning to take effect, shown by the crude 1970 estimate which is quite out of line with the values for earlier years.

If we ignore the invasive spatial element, the development of the dove population involves just two features, birth and death. Since each of these is easiest to understand in isolation, we shall first study them separately (Sections 2.1 and 2.2) before bringing them together as the simple birth–death process (Section 2.3). Readers should not be too concerned about the highly restrictive assumptions involved. As biological complexity increases, theoretical progress usually becomes more difficult; so if we are to gain an understanding of population development it is necessary to make simplifying assumptions. Provided that fundamental biology has been captured in a mathematical model then useful *qualitative* information should result. Here it is the appreciation and prediction of general behaviour, and not the production of exact quantitative information, that is truly important.

2.1   The pure birth process

Suppose that a population of organisms develops over a short period of time (relative to their lifespan) in crowd-free conditions and with unlimited food resource. Then let us make the assumptions that: (i) organisms do not die; (ii) they develop without interacting with each other; and (iii) the birth rate (λ) is the same for all organisms, regardless of their age, and does not change with time. This last assumption is particularly appropriate to single-celled organisms that reproduce by dividing.

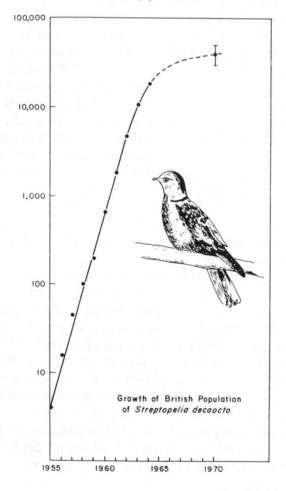

Figure 2.1. Population estimates of the collared dove (*Streptopelia decaocto*) in Great Britain since 1955, logarithmic scale. (Reproduced from Hutchinson, 1978, by permission of Yale University Press.)

Growth of British Population
of *Streptopelia decaocto*

2.1.1   *Deterministic model*
Considering first the classic deterministic approach, let $N(t)$ denote the population size at time t. Then in the subsequent *small* time interval of length h the increase in population size due to a single organism is $\lambda \times h$ (i.e. rate $\times$ time), so the increase in size due to all $N(t)$ organisms is $\lambda \times h \times N(t)$. Thus

$$N(t + h) = N(t) + \lambda h N(t), \tag{2.1}$$

which on dividing both sides by h gives

$$[N(t + h) - N(t)]/h = \lambda N(t).$$

Letting h approach zero then yields the differential equation

$$dN(t)/dt = \lambda N(t), \tag{2.2}$$

which integrates to give

$$N(t) = N(0) \exp(\lambda t), \tag{2.3}$$

where $N(0)$ denotes the initial population size at time $t = 0$. This form for $N(t)$ is known as the Malthusian expression for population development, and shows that the simple rules (i) to (iii) above give rise to exponential growth.

Taking logarithms of both sides of (2.3) gives

$$\log_e[N(t)] = \log_e[N(0)] + \lambda t, \tag{2.4}$$

and this relation provides a useful way of seeing whether a given data set exhibits exponential growth since a plot of $\log_e[N(t)]$ against t should be approximately linear.

2.1.2   *Stochastic model*
This model is purely deterministic since it assumes that each organism reproduces on a completely predictable basis at constant rate. In reality, however, population growth is 'stochastic' (i.e. random). Given a population of cells which grow by division one cannot say that a particular cell *will* divide in a specific time interval, only that there is a certain *probability* that it will do so. Moreover, whilst observed population sizes can take only the integer values $1, 2, 3, \ldots$, expression (2.3) gives rise to all real numbers (such as 1.375 and π) from $N(0)$ upwards which can be particularly unsatisfactory when $N(0)$ is small. Both of these objections can be overcome by studying the stochastic form of the pure birth process.

Suppose that in a short time interval of length h the probability that any particular cell will divide is λh. Then for the population to be of size N at time $t + h$, either it is of size N at time t and no birth occurs in the subsequent short time interval $(t, t + h)$, or else it is of size $N - 1$ at time t and exactly one birth occurs in $(t, t + h)$. By choosing h sufficiently small we may ensure that the probability of more than one birth occurring is negligible. Since the probability of N increasing to $N + 1$ in $(t, t + h)$ is $(\lambda h) \times N$, it follows that the probability of no increase in $(t, t + h)$ is $1 - \lambda N h$. Similarly, the probability of $N - 1$ increasing to N in $(t, t + h)$ is $\lambda (N - 1)h$. Thus on denoting

$$p_N(t) = \text{Pr(population is of size } N \text{ at time } t),$$

we have

$$
\begin{aligned}
p_N(t + h) = p_N(t) &\times \text{Pr\{no birth in } (t, t + h)\} \\
&+ p_{N-1}(t) \times \text{Pr\{one birth in } (t, t + h)\},
\end{aligned}
$$

i.e. $p_N(t + h) = p_N(t) \times (1 - \lambda N h) + p_{N-1}(t) \times \lambda (N - 1)h.$ (2.5)

On dividing both sides by h

$$[p_N(t + h) - p_N(t)]/h = -\lambda N p_N(t) + \lambda (N - 1)p_{N-1}(t),$$

and as h approaches zero this becomes

$$\mathrm{d}p_N(t)/\mathrm{d}t = -\lambda N p_N(t) + \lambda (N - 1)p_{N-1}(t) \qquad (2.6)$$

for $N = N(0), N(0) + 1, \ldots$.

Now whilst this differential equation can be solved by a variety of theoretical techniques, in this text we are concerned far more with the application of results than with the derivation of mathematical formulae. Suffice it to say that the solution to equation (2.6) is given by the negative binomial distribution

$$p_N(t) = \binom{N - 1}{n_0 - 1} \mathrm{e}^{-\lambda n_0 t}(1 - \mathrm{e}^{-\lambda t})^{N - n_0} \qquad (N = n_0, n_0 + 1, \ldots), \tag{2.7}$$

where for convenience we have written $N(0)$ as n_0. Readers interested in the proof should consult Bailey (1964). Note that λ and t occur only in the form of the product λt. Thus a high birth rate (λ) acting over a short time (t) will yield the same set of probabilities $\{p_N(t)\}$ as a lower rate over a longer time if λt is the same for both.

Though this process was first proposed by Yule (1925) to describe the rate of evolution of a new species within a genus, in most practical circumstances the pure birth assumption on which it is based is unrealistic and so the model's application in ecology is rather limited. Nevertheless, it does have the advantage of being the simplest stochastic process which involves reproduction by 'splitting'.

The probability structure $\{p_N(t)\}$ tells us the likely range of the population size at time t. As an illustration, Figure 2.2 shows four plots of solution (2.7) with $n_0 = 4$, $\lambda = 0.1$ and $t = 1$, 5, 10 and 20. We see that this distribution varies from being J-shaped at $t = 1$, just modal (maximum at $N = 5$) and extremely skew at $t = 5$, to being moderately skew with maxima at $N = 9$ and 23 when $t = 10$ and 20, respectively. As (2.7) is of standard negative binomial form the mean and variance are given by

$$m(t) = n_0\, e^{\lambda t} \quad \text{and} \quad V(t) = n_0\, e^{\lambda t}(e^{\lambda t} - 1), \tag{2.8}$$

respectively.

When $n_0 = 1$ the negative binomial probabilities (2.7) reduce to the geometric form

$$p_N(t) = e^{-\lambda t}(1 - e^{-\lambda t})^{N-1} \qquad (N = 1, 2, \ldots), \tag{2.9}$$

and as $1 - \exp(-\lambda t)$ lies between 0 and 1 for all positive values of λt this distribution is permanently J-shaped. When $n_0 > 1$ expression (2.7) is J-shaped for sufficiently small λt, but otherwise rises to a maximum value at $N = N_m$ (say). We evaluate N_m by noting that the ratio of successive terms

Figure 2.2. The probability distribution for a pure birth process with n_0 = 4, $\lambda = 0.1$ and (a) $t = 1$, (b) $t = 5$, (c) $t = 10$ and (d) $t = 20$.

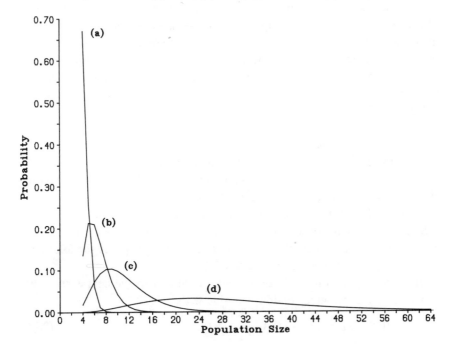

$$p_N(t)/p_{N-1}(t) = [(N-1)/(N-n_0)](1 - e^{-\lambda t}) \qquad (2.10)$$

will only equal 1 when $N = N_m$. Thus

$$(N_m - 1)(1 - e^{-\lambda t}) = N_m - n_0,$$

giving

$$N_m = (n_0 - 1)e^{\lambda t} + 1 \qquad (2.11)$$

which is slightly smaller than the mean $n_0 \exp(\lambda t)$.

It follows from (2.8) that as t increases, $V(t)/n_0$ increases virtually as the square of $m(t)/n_0$, and this result can create the misleading impression that the process $\{N(t)\}$ 'wanders' further and further away from the mean value $m(t)$. A clue to what really happens is given by the coefficient of variation

$$\sqrt{\{V(t)\}/m(t)} = \sqrt{\{n_0 e^{\lambda t}(e^{\lambda t} - 1)\}/n_0 e^{\lambda t}}$$
$$= \sqrt{(1 - e^{-\lambda t})}/\sqrt{(n_0)} \sim 1/\sqrt{(n_0)} \qquad (2.12)$$

for large t. This approaches zero as the initial population size n_0 increases, indicating that large variation about the mean $m(t) = n_0 \exp(\lambda t)$ is associated only with *small* values of n_0.

Although we know the probability distribution of population size through (2.7), the probabilities $\{p_N(t)\}$ do not explicitly tell us what a *particular realization* (i.e. an individual time history) of the process looks like, but give instead the distributional properties of a *large ensemble* of such realizations. Moreover, 'mean behaviour', represented by $m(t)$, may be very different from the behaviour of individual realizations. A further complication is that whilst the deterministic solution (2.3) agrees with the mean (2.8), such agreement does not hold for all population processes – a sufficient condition for agreement is that the individual members of the population develop independently of one another.

2.1.3 *Simulated model*

Information on the shape of individual realizations may be readily obtained by simulating the underlying process. Not only does this technique give considerable insight into the process, but behavioural features may well emerge that had not previously been anticipated.

Since each 'event' consists of the population size increasing by one, all we have to do is to construct the inter-event times. This requires the following two results, proved formally in Section 3.4.2.

(i) If the population is of size N, then the time S to the next event is an exponentially distributed random variable with

$$\Pr(S \geqslant s) = \exp(-\lambda N s) \qquad (s \geqslant 0). \qquad (2.13)$$

Table 2.1. Simulating a realization of a simple birth process with $n_0 = 1$ and $\lambda = 0.2$

Pop. size $N(t)$	Event time t	Random no. Y	Inter-event time s	Time of next event $t + s$
1	0.000	0.1431	9.721	9.721
2	9.721	0.8586	0.381	10.102
3	10.102	0.8637	0.244	10.346
4	10.346	0.4546	0.985	11.332
5	11.332	0.9548	0.046	11.378
6	11.378	0.6369	0.376	11.754
7	11.754	0.2284	1.055	12.809
$\vdots$	$\vdots$	$\vdots$	$\vdots$	$\vdots$

(ii) To simulate a value s we select a uniformly distributed random number Y in the range $0 \leqslant Y \leqslant 1$ and put

$$\exp(-\lambda N s) = Y \tag{2.14}$$

which, on taking logarithms, gives

$$s = -[\log_e(Y)]/(\lambda N). \tag{2.15}$$

Any reasonable pseudo-random number generator can be used for Y; the FORTRAN program 'BIRTH' listed in the Appendix to this chapter uses the NAG-routine G05CAF. Different realizations of the process are obtained by changing the starting point (called the 'seed') of the random number sequence $\{Y\}$.

Readers interested in learning about general aspects of simulation, from the generation and testing of pseudo-random numbers through to their use in a variety of applications, are recommended to consult the excellent introductory text of Morgan (1984). Ripley (1987) provides a comprehensive, mathematically oriented, guide to the current state of knowledge about simulation methods.

To simulate the pure birth process starting with $N(0) = n_0$ individuals at time $t = 0$, we therefore generate a sequence of pseudo-random numbers $Y_1, Y_2, Y_3, \ldots$ and use result (2.15) to obtain the corresponding inter-event times $s_1, s_2, s_3, \ldots$, whence

$$N = n_0 \qquad \text{at time } 0$$
$$N = n_0 + 1 \qquad \text{at time } s_1$$
$$N = n_0 + 2 \qquad \text{at time } s_1 + s_2, \text{ etc.}$$

This procedure is illustrated numerically in Table 2.1 for $n_0 = 1$ and $\lambda = 0.2$, and given the wide availability of random number tables it is clearly very easy to implement even on a pocket calculator.

Figure 2.3a shows the results from nine simulation runs with $n_0 = 1$ and $\lambda = 1$. Each took a different time to 'get going', and A, B and C show the quickest, median and slowest starters, respectively. Curve M shows the mean (Malthusian) population size $m(t) = \exp(t)$, and apart from a time shift the stochastic realizations follow M quite closely once $N(t)$ has grown a little. As already suggested by (2.12), variation between realizations is mainly due to the opening stages of the process; when N is small the first few inter-event times (2.15) have a wide range of possible values. As N increases, expression (2.15) decreases as $1/N$, and so the occasional very large value of $\log_e(Y)$ has less and less effect.

An even clearer picture emerges on plotting $\log_e\{N(t)\}$ against t (Figure 2.3b), since $\log_e\{m(t)\} = \log_e\{\exp(t)\} = t$ and so a plot (M) of $\log_e\{m(t)\}$ against t is a straight line of slope one through the origin. The corresponding

Figure 2.3. Three from nine simulations of a pure birth process showing (a) $N(t)$ and (b) $\log_e[N(t)]$ against t for $n_0 = 1$ and $\lambda = 1$; A and C show the two extreme realizations, B the median realization, and M the mean $m(t) = \exp(t)$.

(a)

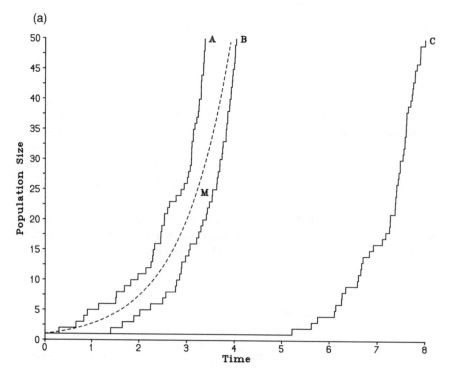

realizations A, B, C are parallel to M and show relatively little departure from linearity once N exceeds 7; only the intercepts on the time axis change.

Note that as an alternative to using the FORTRAN program BIRTH (or its equivalent in say BASIC), the pure birth process is also easily simulated by using the package MINITAB. For example, the commands

BASE 4688	sets the seed
URAN 50 C1	calculates 50 Y-values
LET C2 = $-$LOGE(C1)	forms $-\log_e(Y)$
GENE 1 50 C3	generates $N = 1, 2, ..., 50$
MULT C3 0.2 C4	forms λN
DIVI C2 C4 C5	forms $s = -[\log_e(Y)]/(\lambda N)$
PARSUM C5 C6	forms total time t
INSERT 0 1 C6 ⎫	
0.0 ⎬	balances column lengths
LET C3(51) = 51 ⎭	
PLOT C3 C6	plots $N(t)$ against t

simulate the first 50 event times for $n_0 = 1$, $\lambda = 0.2$ and seed 4688.

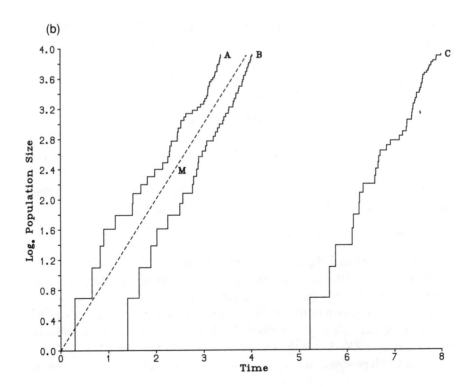

2.1.4   Time to a given state

The simulations of Figure 2.3b together with result (2.12) for the coefficient of variation, namely

$$CV(t) = \sqrt{\{(1 - e^{-\lambda t})/n_0\}} \quad \text{which tends to 0 as } n_0 \to \infty,$$

suggest that the larger $N(t)$ becomes the more indistinguishable are the realizations $\log_e\{N(t)\}$ from a series of straight lines parallel to $\log_e\{m(t)\}$. The variance of the time taken to reach any given large population size should therefore be constant. To prove this assertion let T_a denote the time at which 'state a' is first occupied, where $a > n_0$. Then if T_a is at most t, at time t the population size $N(t)$ must be at least a, whence we have the fundamental result that

$$\Pr(T_a \leqslant t) = \Pr(N(t) \geqslant a). \tag{2.16}$$

This expression relates time to numbers, and has interesting applications in the study of microbial infections (Morgan and Watts, 1980). A host is initially infected with a number of bacteria (n_0), and exhibits symptoms of disease only when the number of bacteria ($N(t)$) has grown to some threshold value (a). The time (T_a) from initial infection to first showing symptoms is called the incubation period.

Explicit results can be obtained directly from (2.7) and (2.16). We may immediately write down that

$$\Pr(T_a \leqslant t) = \sum_{N=a}^{\infty} \binom{N-1}{n_0 - 1} e^{-\lambda n_0 t}(1 - e^{-\lambda t})^{N - n_0}. \tag{2.17}$$

In particular, when $n_0 = 1$ the distribution function (2.17) reduces to

$$\Pr(T_a \leqslant t) = \sum_{N=a}^{\infty} e^{-\lambda t}(1 - e^{-\lambda t})^{N-1} = (1 - e^{-\lambda t})^{a-1}. \tag{2.18}$$

On differentiating this expression we see that the random variable T_a (for $n_0 = 1$) has the probability density function (p.d.f.) $g(t)$ given by

$$g(t) = \lambda(a - 1) e^{-\lambda t}(1 - e^{-\lambda t})^{a-2}. \tag{2.19}$$

Means, variances and higher-order moments may be derived directly from this p.d.f., but it is instructive to develop an alternative approach (for general n_0) which does not rely on first constructing $g(t)$.

Let Z_N be the length of time for which the population is exactly of size N. Then, as we have already seen from (2.13), Z_N is exponentially distributed with parameter λN. As all the successive times Z_N for $N = n_0, n_0 + 1, \ldots, a - 1$ are independent, and as T_a is the sum of these inter-event times, namely

$$T_a = Z_{n_0} + Z_{n_0+1} + \cdots + Z_{a-1}, \tag{2.20}$$

it follows that the expected (i.e. mean) value of T_a is

$$E(T_a) = \sum_{N=n_0}^{a-1} E(Z_N) = \sum_{N=n_0}^{a-1} (1/\lambda N). \tag{2.21}$$

Although this summation can be computed numerically fairly quickly, a more convenient algebraic form may be derived by noting that as $m \to \infty$

$$1 + \frac{1}{2} + \frac{1}{3} + \cdots + \frac{1}{m} \to \gamma + \log_e(m) \tag{2.22}$$

where $\gamma = 0.577\ 216$ (to six decimal places) is known as Euler's constant. On writing (2.21) in the form

$$E(T_a) = \frac{1}{\lambda}\left[\left(1 + \frac{1}{2} + \cdots + \frac{1}{a-1}\right) - \left(1 + \frac{1}{2} + \cdots + \frac{1}{n_0-1}\right)\right] \tag{2.23}$$

we see that for a and n_0 both large

$$E(T_a) \simeq (1/\lambda)[(\gamma + \log_e(a-1)) - (\gamma + \log_e(n_0-1))]$$

i.e.

$$\begin{aligned}E(T_a) &\simeq (1/\lambda)\log_e[(a-1)/(n_0-1)]\\ &\simeq (1/\lambda)\log_e(a/n_0).\end{aligned} \tag{2.24}$$

Note the agreement between this stochastic limit (2.24) and the deterministic value obtained from (2.3), namely

$$N(T_a) = a = n_0 \exp(\lambda T_a)$$

which yields the deterministic 'first passage time'

$$T_a = (1/\lambda)\log_e(a/n_0). \tag{2.25}$$

As the variance of Z_N is $(1/\lambda N)^2$, we can also use (2.20) to evaluate

$$\mathrm{Var}(T_a) = \sum_{N=n_0}^{a-1} \mathrm{Var}(Z_N) = \sum_{N=n_0}^{a-1} (1/\lambda N)^2. \tag{2.26}$$

For large a this approximates to

$$\mathrm{Var}(T_a) \simeq \sum_{N=1}^{\infty} (1/\lambda N)^2 - \sum_{N=1}^{n_0-1} (1/\lambda N)^2,$$

whence the standard result

$$\sum_{N=1}^{\infty} 1/N^2 = \pi^2/6 \tag{2.27}$$

leads to

$$\mathrm{Var}(T_a) \simeq (\pi^2/6\lambda^2) - \sum_{N=1}^{n_0-1} (1/\lambda N)^2. \tag{2.28}$$

Thus in spite of the considerable variation between different realizations of the simple birth process (see Figure 2.3), the variance of T_a is always less than $\pi^2/6\lambda^2$ no matter how large a is. Indeed, as n_0 increases, the variance of T_a tends to zero for all $a > n_0$, which demonstrates once again that variation between different realizations develops early in the process when population size is relatively small.

To illustrate this result, Figure 2.4 shows the frequencies of 1000 simulated first passage times (T_a) to state $a = 1000$ for $\lambda = 1$ and initial states $n_0 = 1, 5$ and 100. Not only is the marked decrease in variance as n_0 increases clearly

Figure 2.4. Frequencies of first passage time (T_a) to state $a = 1000$ constructed from 1000 simulations of a pure birth process with $\lambda = 1$ and initial states $n_0 = 1$ (———), 5 (– – –) and 100 (— — —).

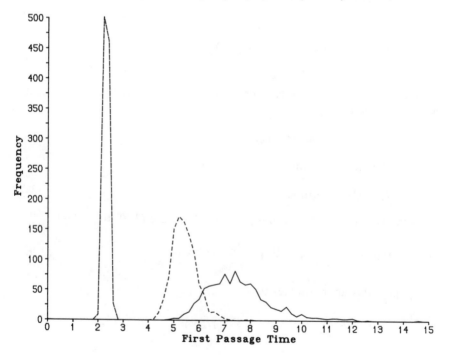

evident, but theoretical and simulated means and variances compare well, viz:

n_0	Sim. mean	Th. mean	Sim. var.	Th. var.
1	7.48	7.56	1.645	1.647
5	5.40	5.42	0.221	0.232
100	2.31	2.30	0.0105	0.0089

Note that when a is large but n_0 is small, the expression

$$E(T_a) = (1/\lambda)\left\{\gamma + \log_e(a-1) - \sum_{j=1}^{n_0-1}(1/j)\right\} \tag{2.29}$$

should be used instead of (2.24). The theoretical means were therefore calculated from (2.29) for $n_0 = 1$ and 5, and from (2.24) for $n_0 = 100$; whilst (2.28) was used to evaluate variances for $n = 1, 5$ and 100. The six largest simulated times were

12.01, 12.02, 12.05, 12.14, 12.52 and 14.68,

and their successive differences

0.01, 0.03, 0.09, 0.38 and 2.16

highlight the long tail of the first passage time distribution.

2.2 The pure death process

The longevity of organisms varies greatly, from a few hours for healthy bacteria to several millennia for the bristlecone pine *Pinus longaeva*. A question of considerable biological interest is to determine how the chance of death relates to the length of time an organism has been alive. We are therefore now considering the opposite to birth, namely death, and although this is less interesting from a mathematical viewpoint as no 'new' individuals are produced, it is nevertheless more widely applicable. For example, if the environment of an isolated population is polluted to such an extent that all future reproduction is prevented, and if the death rate of individual members is independent of their age, then a pure death process will result.

We shall therefore now assume that: (i) organisms do not give birth; (ii) they develop completely independently from each other; and (iii) the death rate (μ) is the same for all individuals and does not change with time. This last assumption is equivalent to saying that individuals do not 'age'; a feature thought to be true for many species of birds (Deevey, 1947) once they have become adult, and possibly true for some species of fish. In such situations the death rate is usually determined by the species' liability to accidental causes of death rather than to natural causes.

2.2.1   *Deterministic model*

Considering the deterministic approach first, suppose that in a small time interval of length h the decrease in population size due to a single organism is $\mu \times h$. Then the decrease in size due to all $N(t)$ organisms is $\mu \times h \times N(t)$. Thus

$$N(t + h) = N(t) - \mu h N(t),$$

which, on paralleling the pure birth argument, becomes

$$dN(t)/dt = -\mu N(t) \tag{2.30}$$

with the solution

$$N(t) = N(0)\exp(-\mu t). \tag{2.31}$$

Taking logarithms gives the linear relationship

$$\log_e\{N(t)\} = \log_e\{N(0)\} - \mu t, \tag{2.32}$$

which provides a quick graphical check on whether a given data set exhibits exponential decay.

For example, three ornithologists (Kraak, Rinkel and Hoogenheide, 1940) collected survival data on the lapwing, *Vanellus vanellus*. The birds were ringed as fledglings, and Figure 2.5 shows a logarithmic plot of the number of birds alive at a specific age t. Apart from a period of relatively high juvenile mortality during the first half-year of life (data not presented), the linear relationship (2.32) clearly holds true, so after this initial period the mortality rate is essentially constant from year to year. High senile mortality does not seem to occur. Experience with captive birds living under good conditions indicates that they can live much longer than wild birds (Hutchinson, 1978), which implies that very few wild birds actually reach 'old age'.

2.2.2   *Stochastic model*

The stochastic approach is intrinsically simpler to develop than that for the pure birth process since no new members are produced. Denote

$$q(t) = \text{Pr(a particular organism is alive at time } t).$$

Then

> $q(t + h) = \text{Pr}$(it is alive at time t and does not die in the
>   subsequent small time interval h)
>
> $= q(t) \times (1 - \mu h)$.

On letting h approach zero this yields

$$dq(t)/dt = -\mu q(t) \tag{2.33}$$

Figure 2.5. Composite age-specific survivorship curve for the lapwing, *Vanellus vanellus*, based on ringed birds found dead in Europe. Data are presented from the end of the first year of life onward. (Reproduced from Hutchinson, 1978, by permission of Yale University Press.)

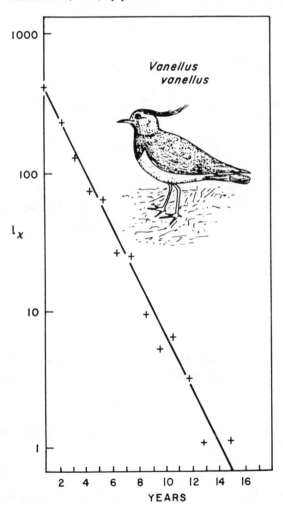

which integrates directly to give

$$q(t) = \exp(-\mu t). \tag{2.34}$$

Hence

$$p(t) = \text{Pr(the organism is dead by time } t)$$
$$= 1 - q(t) = 1 - e^{-\mu t}.$$

Thus if the initial population is of size $N(0) = n_0$, and if all organisms behave independently of each other, then $N(t)$ satisfies the conditions for the binomial distribution with probabilities

$$p_N(t) = \binom{n_0}{N} [q(t)]^N [p(t)]^{n_0 - N},$$

i.e.

$$p_N(t) = \binom{n_0}{N} e^{-N\mu t} (1 - e^{-\mu t})^{n_0 - N} \qquad (N = 0, 1, \ldots, n_0). \tag{2.35}$$

The associated mean and variance are therefore given by the standard results $n_0 q(t)$ and $n_0 q(t) p(t)$, i.e.

$$m(t) = n_0 e^{-\mu t} \quad \text{and} \quad V(t) = n_0 e^{-\mu t} (1 - e^{-\mu t}), \tag{2.36}$$

respectively.

Reading the lapwing population size directly off Figure 2.5 we see that it drops from 100 at time 4.1 years to 1 at time 14.2 years. We can therefore determine an estimate $(\hat{\mu})$ of the death rate μ from (2.36) by writing

$$1 = 100 \exp\{-\hat{\mu}(14.2 - 4.1)\},$$

i.e., on taking logarithms,

$$\hat{\mu} = [\log_e(100)]/(10.1) = 0.46.$$

*2.2.3   *Time to extinction*

Once all n_0 organisms have died the population is *extinct*, and we see from (2.35) that the time to extinction (T_0) has the distribution function

$$\text{Pr}(T_0 \leqslant t) = p_0(t) = (1 - e^{-\mu t})^{n_0} \qquad (t \geqslant 0). \tag{2.37}$$

Differentiating (2.37) then yields the corresponding p.d.f. of the random variable T_0, namely

$$f_{\text{ext}}(t) = n_0 \mu e^{-\mu t} (1 - e^{-\mu t})^{n_0 - 1}. \tag{2.38}$$

Although (2.38) may be used to find the expected time to extinction, an easier way is to parallel the technique already developed for the pure birth process by denoting Z_N to be the length of time for which the population is exactly of size N. Then

$$T_0 = Z_{n_0} + Z_{n_0 - 1} + \cdots + Z_1 \qquad (2.39)$$

is simply the sum of the times spent in each of the successive states n_0, $n_0 - 1, \ldots, 1$. As Z_N are independent exponentially distributed random variables with mean $(1/N\mu)$ and variance $(1/N\mu)^2$, we therefore have

$$E(T_0) = \sum_{N=1}^{n_0} (1/N\mu) \quad \text{and} \quad \text{Var}(T_0) = \sum_{N=1}^{n_0} (1/N\mu)^2. \qquad (2.40)$$

On using the asymptotic results (2.22) and (2.27) we see that for large n_0

$$E(T_0) \simeq (1/\mu)[\gamma + \log_e(n_0)] \quad \text{and} \quad \text{Var}(T_0) \simeq \pi^2/6\mu^2. \qquad (2.41)$$

Thus whilst the variance of the time to extinction remains finite irrespective of the value of n_0, the expected time to extinction increases with $\log_e(n_0)$. Hence under this model, if the lapwing population starts with 1000 birds then it has an expected time to extinction of 16.3 years, rising to 31.3 years if $n_0 = 10^6$ birds; whilst the variance is 7.8 years^2 in both cases.

2.2.4   *Simulation results*
Simulating the simple death process 1000 times with $\mu = 1$ for both $n_0 = 100$ and 1000 produced the following results.

		Exact	Asymptotic	Simulated
Mean	$n_0 = 100$	5.1874	5.1824 (-0.01%)	5.1936 ($+0.12\%$)
Variance		1.6350	1.6449 ($+0.61\%$)	1.5036 (-8.04%)
Mean	$n_0 = 1000$	7.4855	7.4850 (-0.01%)	7.5153 ($+0.40\%$)
Variance		1.6439	1.6449 ($+0.06\%$)	1.7096 ($+4.00\%$)

(We simply change the events $N \to N + 1$ to $N \to N - 1$ in the program BIRTH.) Exact and asymptotic values were computed from (2.40) (using MINITAB) and (2.41), respectively, and agreement between them is clearly excellent. However, whilst 1000 simulation runs are more than enough to obtain a simulated mean value with reasonable precision, the variance estimates are really too crude and a larger number of runs is needed to take account of the skewness of the extinction time distribution. With $n_0 = 100$, for example, the simulated T_0-values have a range of 2.92 to 11.00 yet the median value is only 4.98.

This exercise demonstrates the advantage of being able to derive exact, or at least asymptotically exact, theoretical results. However, if such results may not be obtained then simulation is clearly a powerful substitute. Indeed, not only may simulation be used to derive p.d.f.s and summary statistics empirically, but it also highlights the shape of individual realizations (see Figures 2.3a and 2.6). If possible, both theory and simulation should be applied in any given situation, as one complements the other.

Figure 2.6 illustrates the stochastic behaviour of the pure death process, showing six simulated realizations for $\mu = 1$ and initial value $n_0 = 100$. We see that for population sizes:

(i) from $N = 100$ to 40, time variation remains small across five of the six realizations;

(ii) from $N = 40$ to 10, it increases, though the range is still less than one time unit;

(iii) for N below 10, time variation increases rapidly due to $\text{Var}(Z_N) = 1/N^2$ rising quickly as N becomes small.

Figure 2.6. Six realizations of a pure death process with $\mu = 1$ and initial size $n_0 = 100$.

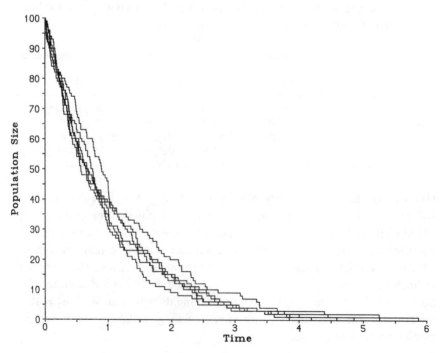

Hence the process is more or less deterministic until N drops to about 40, after which stochastic effects become increasingly important.

2.3 The simple linear birth and death process

Having studied birth and death separately, we shall now combine them into a single process. All members of the population are assumed capable of reproducing; either the organisms are asexual or, if the species is bisexual, we consider only the females and postulate that there is never a shortage of males.

The implicit assumption that individual birth and death rates remain constant, independent of both age and population size, will clearly be violated in virtually all natural situations involving population growth. However, whilst descriptive ecology is unconstrained by considerations of mathematical modelling, and is therefore free to study the complexities underlying real populations, the development of stochastic models is mathematically intractable in all but the simplest of cases. Models such as the simple linear birth–death process should therefore not be looked upon as providing a literal representation of reality, but should be treated instead as a means to obtaining greater understanding of the behavioural characteristics of the ecological processes which they attempt to mimic. More realistic (i.e. non-linear) birth and death rates can of course be tried, but the price paid in terms of added mathematical complexity is usually high.

2.3.1 *Deterministic model*

The deterministic approach proceeds in the same way as for the pure birth and pure death processes. The deterministic equation is now

$$dN(t)/dt = (\lambda - \mu)N(t),\tag{2.42}$$

with solution

$$N(t) = N(0)\exp\{(\lambda - \mu)t\}.\tag{2.43}$$

This expression is formally the same as (2.3) and (2.31), though the net rate of increase $(\lambda - \mu)$ may now be either positive or negative. For naturally occurring populations to grow in this exponential manner, numbers must be low enough relative to the available food resource to ensure that there is no competition between individuals.

An example of such growth is provided by the dove population (Figure 2.1). On taking the population to be of size 10 in 1956 and 1000 $4\frac{1}{2}$ years later on, we can obtain an estimate $(\hat{\lambda} - \hat{\mu})$ of the net annual growth rate $(\lambda - \mu)$ by writing (2.43) as

$$1000 = 10\exp\{(\hat{\lambda} - \hat{\mu})(9/2)\},$$

whence

$$\hat{\lambda} - \hat{\mu} = (2/9) \log_e(100) = 1.02.$$

2.3.2   *Stochastic model*

Analysis of stochastic behaviour follows along exactly the same lines as for the pure birth process, except that in the short time interval $(t, t + h)$ there is now a probability (λh) that a particular organism gives birth *and* a probability (μh) that it dies. With a population of size $N(t)$ at time t the probability that no event occurs is therefore $1 - \lambda Nh - \mu Nh$, since h is assumed to be sufficiently small to ensure that the probability of more than one event occurring in $(t, t + h)$ is negligible. Equation (2.5) is therefore extended to

$$p_N(t + h) = p_N(t)[1 - N(\lambda + \mu)h] + p_{N-1}(t)(N - 1)\lambda h$$
$$+ p_{N+1}(t)(N + 1)\mu h, \tag{2.44}$$

as state N can now be reached from states $N - 1$ (by a birth), $N + 1$ (by a death) or N (by neither). Dividing (2.44) by h and letting h approach zero then yields the set of equations

$$dp_N(t)/dt = \lambda(N-1)p_{N-1}(t) - (\lambda+\mu)Np_N(t)$$
$$+ \mu(N + 1)p_{N+1}(t) \tag{2.45}$$

over $N = 0, 1, 2, \ldots$ and $t \geqslant 0$.

Equations (2.45) may be solved using standard differential equation techniques (e.g. Cox and Miller, 1965, chapter 4); the solution for $p_N(t)$ is most easily expressed as being the coefficient of z^N in the expansion of

$$\left\{ \frac{\mu(1 - z) - (\mu - \lambda z) \exp[-(\lambda - \mu)t]}{\lambda(1 - z) - (\mu - \lambda z) \exp[-(\lambda - \mu)t]} \right\}^{n_0}. \tag{2.46}$$

Though an exact value for $p_N(t)$ may be found by expanding (2.46) as a power series in z, for $n_0 > 1$ the resulting expression (see Bailey, 1964) is really too messy to be of much practical use. However, as organisms are assumed to develop independently of one another, we may consider the development of a population of initial size n_0 as being equivalent to the development of n_0 *separate* populations each of initial size 1. Thus in order to understand the stochastic behaviour of the process it is sufficient just to consider $n_0 = 1$. In this case the population size $N(t)$ follows the geometric distribution

$$p_0(t) = \alpha(t)$$
$$p_N(t) = [1 - \alpha(t)][1 - \beta(t)][\beta(t)]^{N-1} \quad (N = 1, 2, \ldots) \tag{2.47}$$

where

$$\alpha(t) = [\mu(e^{(\lambda-\mu)t} - 1)]/[\lambda e^{(\lambda-\mu)t} - \mu]$$

and

$$\beta(t) = [\lambda(e^{(\lambda-\mu)t} - 1)]/[\lambda e^{(\lambda-\mu)t} - \mu]. \tag{2.48}$$

Standard results for the geometric distribution then yield the mean $m(t)$ and variance $V(t)$ of population size as

$$m(t) = n_0 e^{(\lambda-\mu)t}, \tag{2.49}$$

which is the same as the deterministic value (2.43), and

$$V(t) = n_0[(\lambda + \mu)/(\lambda - \mu)] e^{(\lambda-\mu)t}(e^{(\lambda-\mu)t} - 1) \tag{2.50}$$

(first derived in this context by Feller, 1939). Unlike $m(t)$, $V(t)$ depends not only on the difference between the birth and death rates, but also on their absolute magnitudes. This is what we should expect, because predictions about the future size of a population will be less precise if births and deaths occur in rapid succession than if they occur only occasionally. When birth and death rates are in balance (i.e. $\lambda = \mu$)

$$m(t) = n_0 \text{ (a constant)} \quad \text{and} \quad V(t) = 2n_0\lambda t, \tag{2.51}$$

the latter changing linearly, and not exponentially, with time.

The relative values of λ and μ affect the coefficient of variation $CV(t) = \sqrt{\{V(t)\}}/m(t)$ as follows:

(i) $\lambda > \mu$: $CV(t) \sim \sqrt{\{(\lambda + \mu)/n_0(\lambda - \mu)\}}$
(ii) $\lambda = \mu$: $CV(t) = \sqrt{\{2\lambda t/n_0\}}$ (2.52)
(iii) $\lambda < \mu$: $CV(t) \sim \sqrt{\{(\mu + \lambda)/n_0(\mu - \lambda)\}} \exp\{\tfrac{1}{2}(\mu - \lambda)t\}$.

Thus $CV(t)$ remains constant, or increases as $\sqrt{t}$, or grows exponentially at rate $\tfrac{1}{2}(\mu - \lambda)$, depending on whether $\lambda > \mu$, $\lambda = \mu$ or $\lambda < \mu$, respectively.

At this point it is worth reflecting on a criticism sometimes levelled against this model of exponential population growth, namely that if λ exceeds μ then it ultimately leads to populations so large that their existence is physically impossible. This argument is of course nonsensical, since for any given values of λ and μ we can always choose a 'sensible' time t_0 such that $m(t) = n_0 \exp\{(\lambda - \mu)t\}$ remains within realistic bounds for all t in the range $0 \leqslant t \leqslant t_0$ (Pielou, 1974).

A far more serious, but often neglected, question is 'How far into the future is ecological prediction, based on simple models, feasible?' (Pielou, 1977). The answer must depend on the situation being considered, since it is influenced to a large extent by biological factors such as over what length of time the (actual) birth and death rates can be expected to remain reasonably constant, and how large $N(t)$ may become before organisms can no longer be assumed to develop independently of each other. Clearly, whenever a biological process, no matter how innocently simple it may first appear, is being modelled, the underlying assumptions used in the construction of the model must be constantly questioned. The feasibly predictable future may well be disappointingly short.

2.3.3   *Probability of extinction*

We have already seen through expression (2.52) that three modes of behaviour are possible depending on the relative sizes of λ and μ. If $\lambda < \mu$ then it is intuitively reasonable to suppose that since deaths predominate the population will eventually die out, and since arrivals from outside the system (i.e. immigrants) do not occur this implies *extinction*. If $\lambda > \mu$ then births predominate, and either an initial downward surge causes the population to become extinct or else $N(t)$ avoids becoming zero and the population grows indefinitely.

That such differences in behaviour do indeed occur may be seen from (2.47), since with $N(0) = n_0$ the probability of extinction by time t is

$$p_0(t) = [\alpha(t)]^{n_0} = \left\{ \frac{\mu - \mu \exp\{-(\lambda - \mu)t\}}{\lambda - \mu \exp\{-(\lambda - \mu)t\}} \right\}^{n_0} \tag{2.53}$$

as n_0 separate lines of descent each have to die out. To find the probability of *ultimate extinction*, $p_0(\infty)$, we must now allow t to become large. If $\lambda < \mu$ then writing (2.53) as

$$p_0(t) = \left\{ \frac{\mu \exp\{(\lambda - \mu)t\} - \mu}{\lambda \exp\{(\lambda - \mu)t\} - \mu} \right\}^{n_0} \rightarrow \left(\frac{-\mu}{-\mu} \right)^{n_0} = 1 \quad \text{as } t \rightarrow \infty$$

shows that $p_0(\infty) = 1$. Ultimate extinction is therefore certain. Conversely, if $\lambda > \mu$ then the exponential terms vanish, so

$$p_0(\infty) = (\mu/\lambda)^{n_0}. \tag{2.54}$$

Note that if λ is greater than μ then extinction can still occur, though the probability that it does so decreases geometrically fast at rate μ/λ with increasing n_0. Thus if we wish to choose n_0 to ensure that $p_0(\infty) \leqslant 0.001$ (say), then we require

$$(\lambda/\mu)^{n_0} \geqslant 1000, \qquad \text{i.e. } n_0 \geqslant 3/\log_{10}(\lambda/\mu).$$

For example, if $\mu = 1$ and

$$\lambda = 1.01, \quad 1.1, \quad 2, \quad 10, \quad 100, \quad 1000,$$

then

$$n_0 \geqslant 695, \quad 73, \quad 10, \quad 3, \quad 2, \quad 1.$$

To determine $p_0(t)$ when $\lambda = \mu$ we first expand the exponential terms in (2.53) to obtain

$$p_0(t) = \left\{ \frac{\mu - \mu[1 - (\lambda - \mu)t + (\lambda - \mu)^2 t^2/2 - \cdots]}{\lambda - \mu[1 - (\lambda - \mu)t + (\lambda - \mu)^2 t^2/2 - \cdots]} \right\}^{n_0}.$$

This expression simplifies to give

$$p_0(t) = \left\{ \frac{\mu t + \text{terms in } (\lambda - \mu)}{1 + \mu t + \text{terms in } (\lambda - \mu)} \right\}^{n_0},$$

whence putting $\lambda = \mu$ yields

$$p_0(t) = [\mu t/(1 + \mu t)]^{n_0}. \tag{2.55}$$

As t increases, $p_0(t) \to [\mu t/\mu t]^{n_0} = 1$, and so ultimate extinction is certain even though the birth and death rates are equal. Thus although the expected size $m(t)$ of the population remains fixed at n_0, stochastic fluctuations about n_0 will inevitably lead to extinction if a sufficiently long period of time is allowed to elapse.

For the pure death process with a large initial population size n_0, the expected time to extinction $T_E(n_0)$ is (asymptotically) equal to $(1/\mu)[\gamma + \log_e(n_0)]$ (result (2.41)). Since γ is small $T_E(n_0)$ is therefore effectively equal to $(1/\mu)\log_e(n_0)$, which is the time required for the deterministic population to decline in size from n_0 to 1. This suggests that an approximation to the mean time to extinction for the pure birth–death process with $\lambda < \mu$ may be obtained by writing (2.43) as

$$1 \simeq n_0 \exp\{(\lambda - \mu)T_E(n_0)\},$$

whence taking logarithms gives

$$T_E(n_0) \simeq (\mu - \lambda)^{-1} \log_e(n_0). \tag{2.56}$$

[The exact result for $n_0 = 1$ is

$$T_E(1) = -(1/\lambda) \log_e[1 - (\lambda/\mu)], \tag{2.57}$$

and for $n_0 > 1$

$$T_E(n_0) = \frac{1}{\mu} \sum_{j=0}^{\infty} \left[\frac{\lambda}{\mu} \right]^j \sum_{i=0}^{n_0 - 1} \frac{1}{i + j + 1} \tag{2.58}$$

(Nisbet and Gurney, 1982, pp. 222–3).]

2.3.4   *Simulated model*

We have already seen through the pure birth process how computer simulation can provide considerable insight into the shape of single realizations. The program BIRTH is easily modified to allow individual events to be either births or deaths, and a program run is terminated when either a given time has elapsed (MAXTIME) or the population has become extinct (see Section 3.4 for full details). Individual realizations are constructed in the same manner as shown in Table 2.1, except that births and deaths now occur with probabilities $\lambda/(\lambda+\mu)$ and $\mu/(\lambda+\mu)$, respectively. Thus with $\{Y\}$ denoting a sequence of uniform pseudo-random numbers in the range 0 to 1,

if $0 \leqslant Y \leqslant \lambda/(\lambda + \mu)$ the next event is a birth,
otherwise   it is a death.

To illustrate this procedure consider $\lambda = 2$, $\mu = 1$ and $n_0 = 1$. The inter-event times $s = -\log_e(Y)/[N(\lambda + \mu)]$ (see expression (2.15)) and $\lambda/(\lambda + \mu)$ = 0.6667, whence one such realization is:

N	t	Y	s	t (new)=t+s	Y	Event	N (new)
1	0.0000	0.6237	0.1574	0.1574	0.3281	birth	2
2	0.1574	0.4735	0.1508	0.3082	0.0642	birth	3
3	0.3082	0.3967	0.1207	0.4109	0.8028	death	2
2	0.4109	etc.					

This simulation approach, initiated by Kendall (1950), is historically interesting since it preceded the availability of electronic computers.

Twenty simulations were performed for parameter values $n_0 = 3$, MAXTIME $= 6$, $\lambda = 1.0$ and $\mu = 0.5$, and Figure 1.1 shows the plots of $N(t)$ corresponding to values of $N(6)$ that are the smallest (only one led to extinction), tenth smallest, and largest. These plots clearly highlight the wide variety of stochastic behaviour possible from one specific process. The deterministic approach, with its dogmatic view of growth proceeding exactly as $3\exp(0.5t)$, can now be seen in its true perspective. Whilst the 'largest' run does indeed exhibit classic exponential form, the 'median' run remains relatively static between times 4.2 and 6.0, and the 'smallest' run leads to extinction by time 4.54. If these three sequences were observed in the laboratory, how many applied researchers would seriously consider the possibility that the same underlying process was responsible for each one?

With such widespread variation, the average behaviour of as few as 20 simulations may well not coincide with the theoretically determined mean behaviour. The 20 $N(6)$-values have a mean of 46.05 and a variance of 3141, considerably different from the theoretical values of 60.26 and 3450 derived from expressions (2.49) and (2.50), respectively. Whilst from (2.53) the probability of extinction by time $t = 6$ is $p_0(6) = 0.1157$, yet only 1 of the 20 runs became extinct. As would be expected, increasing the number of runs to 1000 (30 seconds of c.p.u. time on an ICL 2988 machine) and 10 000 (292 seconds of c.p.u. time) leads to much closer agreement, namely:

	Mean(6)	Var(6)	$p_0(6)$
20 runs	46.05	3141	0.0500
1000 runs	58.73	3197	0.1160
10000 runs	60.37	3540	0.1180
Theoretical values	60.26	3450	0.1157

If required, simulated estimates for the probabilities $\{p_N(t)\}$ may be obtained by dividing the number of runs that give rise to population size N at time t by the total number of runs used.

Since situations little more complex than this birth–death process give rise to intractable mathematics, simulation is clearly a powerful tool for studying population dynamics, especially as it demonstrates the individual behaviour of a process. In the above example 1000 runs provides a good perspective, and more general impressions may be achieved by using considerably fewer.

2.3.5   *Representation as a simple random walk*

We have seen that evaluating the first passage time T_a from n_0 to a for the pure birth process, and the extinction time T_0 from n_0 to 0 for the pure death process, is relatively easy as the population size either increases or decreases by one unit, respectively, at each step. However, for the simple birth–death process the population size can both increase and decrease and so the summations (2.20) and (2.39) no longer apply.

In principle, the mean and variance of the time to extinction T_0 may be obtained from the distribution function (2.53), namely

$$\Pr(T_0 \leqslant t) = p_0(t) = [\alpha(t)]^{n_0}. \tag{2.59}$$

However, the evaluation of T_a is a considerably harder problem to solve, and so instead of examining *time* let us simplify matters by just considering *number of events*.

Consider the population size at time $t = 0$ and at the instants just *after* the first, second, etc., changes of state. We then have a new process which evolves at the discrete 'time' points $1, 2, 3, \ldots$, and so is equivalent to a 'simple random walk' starting at n_0 (Figure 2.7). Once the population size reaches zero it stays there since no more births can take place, so state 0 (i.e. extinction) is called an 'absorbing barrier'. Each individual step may be upwards (corresponding to a birth) or downwards (corresponding to a death) with probability $p = \lambda/(\lambda + \mu)$ or $q = \mu/(\lambda + \mu)$, respectively.

To determine the probability that the population ever reaches size a we introduce an absorbing barrier at position $N = a$. The probability that position a is ever reached is then identical to the probability of absorption at a. Classic simple random walk results (see, for example, Feller, 1966; Bailey, 1964) show that

$$q_{n_0} = \text{Pr(eventual absorption at 0)}$$
$$= [(q/p)^a - (q/p)^{n_0}]/[(q/p)^a - 1]$$

$$p_{n_0} = \text{Pr(eventual absorption at } a)$$
$$= [(q/p)^{n_0} - 1]/[(q/p)^a - 1]. \tag{2.60}$$

Hence on substituting for $p = \lambda/(\lambda + \mu)$ and $q = \mu/(\lambda + \mu)$, we have

$$q_{n_0} = \text{Pr(extinction occurs before population reaches size } a)$$
$$= [(\mu/\lambda)^a - (\mu/\lambda)^{n_0}]/[(\mu/\lambda)^a - 1] \tag{2.61}$$

$$p_{n_0} = \text{Pr(population ever reaches size } a)$$
$$= [(\mu/\lambda)^{n_0} - 1]/[(\mu/\lambda)^a - 1]. \tag{2.62}$$

If $\lambda = \mu$ then (2.61) and (2.62) reduce to

$$q_{n_0} = (a - n_0)/a \quad \text{and} \quad p_{n_0} = n_0/a, \tag{2.63}$$

so the two probabilities depend linearly on the size of the initial population. Note that as $q_{n_0} + p_{n_0} = 1$ the chance of an unending walk is zero.

Figure 2.7. Representation of the birth–death process as a simple random walk.

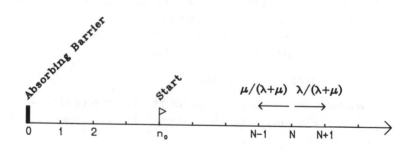

If we are just interested in extinction, then the absorbing barrier at a may be removed by letting $a \to \infty$ in expressions (2.61) and (2.63). This yields

$$q_{n_0} \to \begin{cases} 1 & \text{if } \lambda \leqslant \mu, \\ (\mu/\lambda)^{n_0} & \text{if } \lambda > \mu, \end{cases} \tag{2.64}$$

in direct agreement with (2.54).

The expected number of steps, d_{n_0}, before either state 0 or a is reached may be found from the general result

$$d_{n_0} = \left(\frac{n_0}{q-p}\right) - \left(\frac{a}{q-p}\right)\left\{\frac{1-(q/p)^{n_0}}{1-(q/p)^a}\right\}$$

since this immediately leads to

$$d_{n_0} = \left(\frac{\mu+\lambda}{\mu-\lambda}\right)\left\{n_0 - a\left(\frac{1-(\mu/\lambda)^{n_0}}{1-(\mu/\lambda)^a}\right)\right\} \qquad (\lambda \neq \mu) \tag{2.65}$$

and

$$d_{n_0} = n_0(a - n_0) \qquad (\lambda = \mu). \tag{2.66}$$

On letting $a \to \infty$, we see that for the unrestricted process with $\mu > \lambda$ the expected number of steps to extinction is $n_0(\mu + \lambda)/(\mu - \lambda)$.

2.4 The simple immigration–birth–death process

We have seen that if $\lambda \leqslant \mu$ then extinction is ultimately certain to occur; whilst if $\lambda > \mu$ then either the population becomes extinct, with probability $(\mu/\lambda)^{n_0}$, or else the population size $N(t)$ becomes infinite with probability $1 - (\mu/\lambda)^{n_0}$. In reality this last situation is not biologically reasonable, since $N(t)$ would reach a certain magnitude beyond which the rules of the simple birth–death process would no longer apply.

Introducing immigration into the system allows us to avoid these two extremes of population extinction and explosion, and also acknowledges the fact that few populations develop in true isolation.

2.4.1 *Deterministic model*

Suppose immigrants arrive randomly at rate α. Then the deterministic change in population size during a small time interval of length h is

$$N(t + h) - N(t) = (\lambda - \mu)hN(t) + \alpha h.$$

As h approaches zero this yields

$$dN(t)/dt = (\lambda - \mu)N(t) + \alpha \tag{2.67}$$

which integrates to give

$$N(t) = n_0 \exp\{(\lambda - \mu)t\} + [\alpha/(\lambda - \mu)][\exp\{(\lambda - \mu)t\} - 1]. \quad (2.68)$$

For $\lambda > \mu$, $N(t)$ becomes indefinitely large, as one would expect, whilst for $\lambda < \mu$, $N(t)$ approaches the constant value $\alpha/(\mu - \lambda)$ exponentially fast.

*2.4.2   Equilibrium probabilities

Extinction (though not a zero population size) is clearly impossible, since a new immigrant will always arrive to start the population off again. We therefore anticipate that when $\lambda < \mu$ a stochastic approach will lead to a limiting equilibrium distribution of population size.

On introducing the immigration component (α) the probability equations (2.45) become

$$dp_0(t)/dt = -\alpha p_0(t) + \mu p_1(t) \qquad (N = 0) \qquad (2.69)$$

$$dp_N(t)/dt = [\alpha + \lambda(N - 1)]p_{N-1}(t) - [\alpha + (\lambda + \mu)N]p_N(t)$$
$$+ \mu(N + 1)p_{N+1}(t) \qquad (N = 1, 2, \ldots). \qquad (2.70)$$

Although it is possible to integrate these equations (see Bailey, 1964), suppose that t is large enough for the probabilities $\{p_N(t)\}$ to have settled down to the constant equilibrium values $\{\pi_N\}$ ($N = 0, 1, 2, \ldots$). (A more rigorous analysis would first prove that this entirely reasonable assumption is indeed true.) Then

$$dp_N(t)/dt = d\pi_N/dt = 0,$$

whence from equations (2.69) and (2.70)

$$\alpha\pi_0 - \mu\pi_1 = 0 \qquad\qquad\qquad\qquad (N = 0) \qquad (2.71)$$

$$[\alpha + \lambda(N - 1)]\pi_{N-1} - \mu N\pi_N = (\alpha + \lambda N)\pi_N - \mu(N + 1)\pi_{N+1}$$
$$(N = 1, 2, \ldots). \qquad (2.72)$$

Now the two sides of equation (2.72) are identical, apart from $N - 1$ on the left being replaced by N on the right, and so

$$[\alpha + \lambda(N - 1)]\pi_{N-1} - \mu N\pi_N = \text{constant}$$

(independent of $N = 1, 2, \ldots$). But we know from (2.71) that

$$\alpha\pi_0 - \mu\pi_1 = \text{constant} = 0,$$

so

$$\mu N\pi_N = [\alpha + \lambda(N - 1)]\pi_{N-1} \quad (N = 0, 1, 2, \ldots). \qquad (2.73)$$

This equation may also be written down directly on intuitive grounds, since in equilibrium successive states are in 'balance'. The probability of (i) a death and the population previously being of size N, i.e. $(\mu N)\pi_N$, is the same as (ii) the probability of an immigration or birth and the population previously being of size $N - 1$, i.e. $[\alpha + \lambda(N - 1)]\pi_{N-1}$.

Solving equation (2.73) successively for $N = 1, 2, \ldots$ gives

$$\pi_1 = \alpha\pi_0/\mu$$
$$\pi_2 = (\alpha + \lambda)\pi_1/2\mu = (\alpha + \lambda)\alpha\pi_0/(\mu^2 2!)$$

and in general

$$\pi_N = \pi_0[\alpha + (N - 1)\lambda][\alpha + (N - 2)\lambda]\ldots[\alpha]/(\mu^N N!), \qquad (2.74)$$

i.e.

$$\pi_N = \pi_0(\lambda/\mu)^N[(N - 1) + \alpha/\lambda][(N - 2) + \alpha/\lambda]\ldots[\alpha/\lambda]/(N!). \quad (2.75)$$

On defining the binomial coefficients

$$\binom{n}{r} = \frac{n!}{(n - r)!r!} = n(n - 1)\ldots(n - r + 1)/r!,$$

expression (2.75) may be written as

$$\pi_N = \pi_0(\lambda/\mu)^N\binom{N - 1 + (\alpha/\lambda)}{N}. \qquad (2.76)$$

Now negative binomial probabilities have the standard form

$$\binom{r + N - 1}{N}p^r q^N \quad (N = 0, 1, 2, \ldots) \qquad (2.77)$$

where $p + q = 1$. Hence on writing $r = \alpha/\lambda, q = \lambda/\mu, p = 1 - \lambda/\mu$ and $\pi_0 = p^r$, comparison of (2.76) and (2.77) shows that the equilibrium distribution is negative binomial with

$$\pi_N = \binom{N - 1 + (\alpha/\lambda)}{N}\left(\frac{\lambda}{\mu}\right)^N\left(1 - \frac{\lambda}{\mu}\right)^{\alpha/\lambda} \quad (N = 0, 1, 2, \ldots). \qquad (2.78)$$

Use of standard results (for $\mu > \lambda$) immediately gives the equilibrium mean and variance of population size as $rq/p = \alpha/(\mu - \lambda)$ and $rq/p^2 = \alpha\mu/(\mu - \lambda)^2$, respectively. If $\mu < \lambda$, then the population grows indefinitely large and so an equilibrium situation cannot develop.

If $\lambda = 0$, so that we have a simple immigration–death process, then (2.74) reduces to

$$\pi_N = \pi_0\alpha^N/(\mu^N N!).$$

Since probabilities sum to 1

$$1 = \sum_{N=0}^{\infty} \pi_N = \pi_0 \sum_{N=0}^{\infty} (\alpha/\mu)^N/(N!) = \pi_0 \exp(\alpha/\mu),$$

and so $\pi_0 = \exp(-\alpha/\mu)$. Thus

$$\pi_N = \frac{(\alpha/\mu)^N \exp(-\alpha/\mu)}{N!} \quad (N = 0, 1, 2, \ldots) \tag{2.79}$$

which is the Poisson distribution with parameter α/μ. Suppose, for example, that particles move in and out of a given region. Then if we assume that immigration takes place at rate α, and departures (i.e. 'deaths') occur at rate μN, solution (2.79) applies. This result has been used by Chandrasekhar (1943) to model colloidal particles in suspension, and by Rothschild (1953) to determine the speed at which spermatozoa swim.

Appendix

The following FORTRAN program 'BIRTH' simulates the simple birth process using the NAG-routine G05CAF to generate the pseudo-random number sequence $\{Y\}$. If required, this routine may be replaced by any suitable uniform random number generator.

```
C     THIS PROGRAM IS USED TO SIMULATE A SIMPLE BIRTH
C     PROCESS WITH SEED S, INITIAL POPULATION SIZE N0,
C     MAXIMUM NUMBER OF EVENTS MAX, AND BIRTH RATE B;
C     NOTE THAT G05CBF(S) SETS THE SEED FOR G05CAF AS
C     2S + 1.
C
C

      REAL*8 G05CAF,Y
      REAL N
      INTEGER COUNT,S
C
C     READ IN PARAMETER VALUES
C

      READ(01,8000) S, N0,MAX,B
      CALL G05CBF(S)
C
C     SET INITIAL VALUES
C
```

```
        TIME = 0.0
        N = N0
        XN = ALOG(N)
        WRITE(02,8001) TIME,N,XN
C
C    SIMULATE INTER-EVENT TIMES AND OUTPUT WITH
C    POPULATION SIZE N AND XN = LOG(N)
C
        DO 10 COUNT = 1,MAX
        Y = G05CAF(Y)
        RATE = B*N
        REXP = -(DLOG(Y))/RATE
        TIME = TIME + REXP
        N = N + 1
        XN = ALOG(N)
  10    WRITE(02,8001) TIME,N,XN
C
 8000   FORMAT(3I8,F8.2)
 8001   FORMAT(F10.4,6X,F10.0,6X,F12.6)
        STOP
        END
```

3

General birth–death processes

The simple birth–death process is developed under the assumption that the probabilities that an organism will reproduce or die remain constant and are independent of population size. Obviously this can only be true if there is no interference amongst individual population members. However, in a restricted environment the growth of any expanding population must eventually be limited by a shortage of resources. A stage is then reached when the demands made on these resources preclude further growth and the population is then at its saturation level, a value determined by the 'carrying capacity' of the environment.

It might appear that this implies that a large number of separate theoretical models have to be analyzed, each one corresponding to a different type of interference between individual organisms competing for available resources. Fortunately this is not the case, as we can often take advantage of the fact that the total number of individuals (N) in a fixed region of space can change for only four reasons, namely:

- (a) birth   – rate depends on N
- (b) death ($N > 0$)   – rate depends on N
- (c) immigration   – rate independent of N
- (d) emigration ($N > 0$) – rate independent of N.

Indeed, if we combine (a) and (c) to form a general birth rate $B(N)$, and (b) and (d) to form a general death rate $D(N)$, then between them $B(N)$ and $D(N)$ encompass any modelling situation for which the population size N changes by one unit at each event. For example, in the simple immigration–birth–death process (Section 2.4) $B(N) = \alpha + \lambda N$ and $D(N) = \mu N$. In general, the wide variety of mathematical population equations to be found in the literature just reflects the plethora of special cases available for $B(N)$ and $D(N)$. Areas of application extend well beyond the boundaries of population biology.

3.1   General population growth
The general deterministic approach extends immediately from that

developed for the simple birth–death process (Section 2.3.1). Now

$$N(t + h) = N(t) + [B(N) - D(N)]h, \tag{3.1}$$

whence dividing both sides of (3.1) by h and letting h approach zero gives the general equation

$$dN(t)/dt = B(N) - D(N). \tag{3.2}$$

The feasibility of integrating (3.2) depends on the actual form of $B(N)$ and $D(N)$, though a crude numerical solution is always easy to compute by successive use of (3.1) with an appropriately small value of h.

To illustrate how equation (3.2) may be solved, consider a birth–death process with a severe density-dependent death rate, viz.

$$B(N) = \lambda N \quad \text{and} \quad D(N) = \mu N(N - 1). \tag{3.3}$$

Note that $D(1) = 0$ so the population can never become extinct. Then (3.2) becomes

$$dN(t)/dt = N[\lambda - \mu(N - 1)]. \tag{3.4}$$

On splitting this into partial fractions,

$$dN \left\{ \frac{1/(\lambda + \mu)}{N} + \frac{\mu/(\lambda + \mu)}{(\lambda + \mu) - \mu N} \right\} = dt$$

which integrates to give

$$\log_e(N) - \log_e[(\lambda + \mu) - \mu N] = (\lambda + \mu)t + K$$

for some constant K. Thus

$$N/[\lambda + \mu - \mu N] = K \exp\{(\lambda + \mu)t\} \tag{3.5}$$

whence

$$N(t) = [(\lambda + \mu)K \exp\{(\lambda + \mu)t\}]/[1 + \mu K \exp\{(\lambda + \mu)t\}].$$

Putting $t = 0$ in (3.5) gives $K = n_0/(\lambda + \mu - \mu n_0)$, so the deterministic population size at time t is

$$N(t) = \frac{(\lambda + \mu)n_0 \exp\{(\lambda + \mu)t\}}{(\lambda + \mu - \mu n_0) + \mu n_0 \exp\{(\lambda + \mu)t\}}. \tag{3.6}$$

3.1.1   *Probability equations*

The *deterministic* model (3.2) has the property that the population size $N(t)$ at any future time t may be calculated from knowledge of the population size $N(t_0)$ at the present or any previous time t_0. In contrast, if we know that a *stochastic* system is in a particular state at a given time then we cannot predict with any degree of certainty what its state will be at any future time. The best we can do is to calculate the probability distribution of population size by observing different realizations of the system starting from the same initial state and time. Consider a large number M of such realizations. Then we may define the probability

$$p_N(t) = \lim_{M \to \infty} \{(1/M) \times \text{number of realizations containing}$$

$$\text{exactly } N \text{ individuals at time } t\}. \tag{3.7}$$

Construction of the equations which determine the $\{p_N(t)\}$ is straightforward and parallels the argument already developed for the simple birth–death process in Section 2.3.2. In the small time increment h

    (i) $\Pr[N(t + h) = N(t) + 1] = B[N(t)]h$
    (ii) $\Pr[N(t + h) = N(t) - 1] = D[N(t)]h$
    (iii) $\Pr[N(t + h) = N(t)] = 1 - \{B[N(t)] + D[N(t)]\}h, \tag{3.8}$

whence it follows that

$$p_N(t + h) = p_{N+1}(t)D(N + 1)h + p_N(t)[1 - \{B(N) + D(N)\}h] + p_{N-1}(t)B(N - 1)h.$$

Dividing both sides of this equation by h and letting $h \to 0$ then gives the general stochastic equation

$$dp_N(t)/dt = D(N + 1)p_{N+1}(t) - \{B(N) + D(N)\}p_N(t) + B(N - 1)p_{N-1}(t) \tag{3.9}$$

for $N = 0, 1, 2, \ldots$ and $B(-1)$ defined to be zero.

Though this system of equations holds for any appropriate functions $B(N)$ and $D(N)$, its general solution depends on their particular form. Takashima (1957), for example, takes the birth and death rates to be linearly dependent on the inverse of $N(t)$, with $B(N) = \lambda[(a/N) - 1]$ and $D(N) = \mu[(b/N) - 1]$ where a, b, λ and μ are constants. Even in simple cases, however, solutions can be very difficult to obtain. Indeed, minor modifications to the simple birth–death rates $B(N) = N\lambda$ and $D(N) = N\mu$ can render a direct solution to (3.9) either impossible (in terms of present-day mathematics), or so messy that no meaningful behavioural understanding can be gained from it.

*3.1.2   General equilibrium solution

    If we are studying a biological process that has settled down into a state of equilibrium, i.e. neither extinction, population explosion, nor regular cycling can occur, then such problems of mathematical intractability no longer apply. The reason is that as $t \to \infty$ the $\{p_N(t)\}$ approach the equilibrium probabilities $\{\pi_N\}$, and so (3.9) reduces to the much simpler system of equations

$$D(N + 1)\pi_{N+1} - \{B(N) + D(N)\}\pi_N + B(N - 1)\pi_{N-1} = 0 \quad (3.10)$$

for $N = 0, 1, 2, \ldots$ since $d\pi_N/dt = 0$.

We have already analyzed a particular case of this general situation, namely the immigration–birth–death process (Section 2.4), and the solution of equation (3.10) proceeds on exactly the same lines. First rewrite (3.10) in the form

$$D(N + 1)\pi_{N+1} - B(N)\pi_N = D(N)\pi_N - B(N - 1)\pi_{N-1}. \quad (3.11)$$

Then since the two sides of this equation are identical, apart from N on the left being replaced by $N - 1$ on the right,

$$D(N)\pi_N - B(N - 1)\pi_{N-1} = \text{constant} \quad (\text{independent of } N).$$

Now $D(0) = 0$ (there are no individuals present so none can die), whilst $\pi_{-1} = 0$ (by definition, since N cannot be negative). Thus

$$D(0)\pi_0 - B(-1)\pi_{-1} = \text{constant} = 0,$$

and so

$$D(N)\pi_N = B(N - 1)\pi_{N-1}. \quad (3.12)$$

That is, in equilibrium the probability of a death and the population being of size N equals the probability of a birth and the population being of size $N - 1$.

Repeated application of relation (3.12) gives

$$\pi_1 = \frac{B(0)}{D(1)}\pi_0, \quad \pi_2 = \frac{B(0)B(1)}{D(1)D(2)}\pi_0, \quad \ldots$$

Hence on choosing π_0 to ensure that all the probabilities $\{\pi_N\}$ sum to 1, viz.

$$\pi_0 \left[1 + \frac{B(0)}{D(1)} + \frac{B(0)B(1)}{D(1)D(2)} + \cdots \right] = 1,$$

we have the general equilibrium solution

$$\pi_N = q_N / \sum_{i=0}^{\infty} q_i \quad (N = 0, 1, 2, \ldots) \quad (3.13)$$

where

$$q_0 = \frac{1}{B(0)}, \quad q_1 = \frac{1}{D(1)}$$

and

$$q_N = \frac{B(1)B(2)\dots B(N-1)}{D(1)D(2)\dots D(N)} \quad (N \geqslant 2). \tag{3.14}$$

To illustrate this approach let us consider the birth–death process with transition rates (3.3), namely $B(N) = \lambda N$ and $D(N) = \mu N(N-1)$. As $B(0) = D(1) = 0$ we return to the balance equation (3.12), i.e.

$$\mu N(N-1)\pi_N = \lambda(N-1)\pi_{N-1}$$

which yields the recurrence relation

$$\pi_N = \frac{(\lambda/\mu)}{N} \pi_{N-1} \quad (N = 2, 3, \dots). \tag{3.15}$$

Successive application of (3.15) then gives

$$\pi_N = \frac{(\lambda/\mu)^{N-1}}{N!} \pi_1.$$

We determine π_1 from

$$1 = \sum_{N=1}^{\infty} \pi_N = \pi_1 \sum_{N=1}^{\infty} \frac{(\lambda/\mu)^{N-1}}{N!} = \frac{\pi_1}{(\lambda/\mu)} (e^{\lambda/\mu} - 1),$$

yielding the solution

$$\pi_N = \frac{(\lambda/\mu)^N}{N!} (e^{\lambda/\mu} - 1)^{-1}. \tag{3.16}$$

This is a censored Poisson distribution over $N = 1, 2, \dots$.

Note that the different scenarios of Sections 2.4 and 3.1, namely $B(N) = \alpha + \lambda N$, $D(N) = \mu N$ and $B(N) = \lambda N$, $D(N) = \mu N(N-1)$, both lead to Poisson forms.

3.2   Logistic population growth

In many biological situations the finiteness of available resource ensures that an isolated population cannot grow without limit, and provided we wait long enough a sufficiently violent downward fluctuation in population size is bound to occur which will drive the population extinct. However, if a long time has to elapse before the probability of extinction becomes non-negligible, then although an equilibrium distribution $\{\pi_N\}$ of population size will not formally exist there will be a *quasi-equilibrium* distribution $\{\pi_N^{(Q)}\}$ defined in a similar manner to (3.7) by

$$\pi_N^{(Q)} = \lim_{M \to \infty} \{(1/M) \times \text{number of realizations containing}$$
$$\text{exactly } N > 0 \text{ individuals}\} \tag{3.17}$$

where M now refers to the number of *non-extinct* realizations at some large time t. Thus $\{\pi_N^{(Q)}\}$ is effectively a true equilibrium distribution over ecologically relevant periods of time.

3.2.1 *The Verhulst–Pearl equation*

To develop such a situation suppose that the net growth rate per individual, denoted by $f(N)$, is a function of the total population size $N(t)$. Then the deterministic rate of increase

$$dN/dt = Nf(N).\tag{3.18}$$

Now when N is large $df(N)/dN$ must be negative, since the larger the population becomes the greater must be its inhibitory effect on further growth. The simplest assumption to make is that $f(N)$ is linear, that is

$$f(N) = r - sN\tag{3.19}$$

for some positive constants r and s. Combining (3.18) and (3.19) then gives

$$dN/dt = N(r - sN),\tag{3.20}$$

which is the well-known Verhulst–Pearl logistic equation. The corresponding stochastic model has an interesting application in describing the herding behaviour of African elephants (Morgan, 1976; see also Holgate, 1967).

An alternative argument is to let r denote the intrinsic rate of natural increase for growth with unlimited resources, and to let K be the maximum attainable population size (the carrying capacity). Then when N is near 0 and K we require $f(N)$ to be near r and 0, respectively. So consider

$$f(N) = r[1 - (N/K)],\tag{3.21}$$

which drops linearly as N increases. Equation (3.18) now becomes

$$dN/dt = rN[1 - (N/K)] = N(r - sN)$$

for $s = r/K$, in agreement with (3.20).

Integrating (3.20) directly yields the solution

$$N(t) = K/[1 + \{(K - n_0)/n_0\} \exp(-rt)] \qquad (t \geqslant 0),\tag{3.22}$$

though a slightly neater representation is

$$N(t) = K/[1 + \exp\{-r(t - t_0)\}] \qquad (t \geqslant 0)\tag{3.23}$$

where t_0 is defined by

$$(K - n_0)/n_0 = \exp(rt_0),$$

i.e.

$$t_0 = (1/r) \log_e[(K/n_0) - 1].\tag{3.24}$$

At $t = t_0$ we have $N(t_0) = K/2$, so t_0 is the time taken for the population to reach half its maximum size.

Yet another representation may be obtained from (3.22) by replacing t by $t + 1$ and n_0 by $N(t)$, so forming

$$N(t + 1) = K/[1 + \{(K - N(t))/N(t)\} \exp(-r)]$$
$$= e^r N(t)/[1 + N(t)(e^r - 1)/K]. \qquad (3.25)$$

This expression is not only useful for sketching the shape of $N(t)$, but it may also be regarded as a possible model for a population growing at the discrete time points $t = 0, 1, 2, \ldots$.

On a historical note, although Verhulst suggested the use of the simple logistic curve (3.22) as early as 1838 to describe the growth of human populations, his work was virtually ignored until 1920 when Pearl and Reed derived it as an empirical curve which meets the realistic conditions:

   (i) asymptotic to the line $N(t) = K$ when $t = +\infty$,
  (ii) asymptotic to the line $N(t) = 0$ when $t = -\infty$, and
 (iii) has a point of inflexion at some time $t = t_0$.

Figure 3.1. (a) Growth of a yeast population ($\bigcirc$) and the superimposed logistic curve $N(t) = 665/[1 + \exp\{4.16 - 0.531t\}]$ (from data of Carlson, 1913). (b) Values ($\bigcirc$) of $Y(t) = \log_e\{[665 - N(t)]/N(t)\}$ and the superimposed regression line $Y(t) = 4.16 - 0.531t$.

(a)

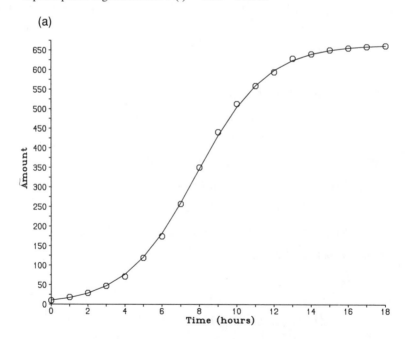

Amount

Time (hours)

It was not until shortly after this that Lotka (1925) provided a rational, as distinct from an empirical, derivation. Further early references are contained in Pearl (1930).

3.2.2 *Growth of a yeast population*

Many laboratory populations have been followed as they increase in size, and the success of the deterministic logistic curve (3.22) in summarizing the resulting data sets depends on the circumstances surrounding each particular experiment. One particularly good example (from a selection provided by Allee, Emerson, Park, Park and Schmidt, 1949) is the data of Carlson (1913) on yeast growth in laboratory cultures (see Table 3.1) and the subsequent analyses of Pearl (1927, 1930).

Figure 3.1a shows the amount of yeast $N(t)$ plotted against t, together with the superimposed logistic curve (3.22). Parameter values were estimated by first rearranging (3.22) in the form

$$Y(t) = \log_e[(K - N)/N] = \log_e[(K - n_0)/n_0] - rt. \tag{3.26}$$

If the logistic curve is appropriate then a plot of $Y(t)$ against t should therefore show a straight line of negative slope $(-r)$ and an intercept at $t = 0$ of $Y(0) = \log_e[(K - n_0)/n_0]$. Inspection of Figure 3.1a shows that K is likely

(b)

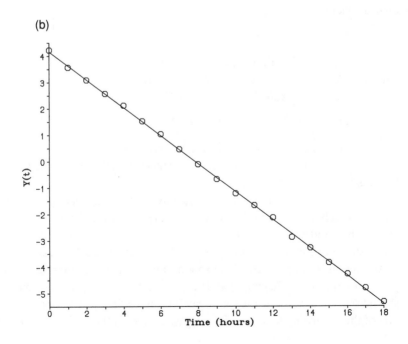

Table 3.1. Growth of a yeast population (from Carlson, 1913)

Hours (t)	Amount of yeast (N)	$\log_e[(665 - N)/N]$
0	9.6	4.223
1	18.3	3.565
2	29.0	3.088
3	47.2	2.572
4	71.1	2.123
5	119.1	1.522
6	174.6	1.033
7	257.3	0.460
8	350.7	-0.110
9	441.0	-0.677
10	513.3	-1.219
11	559.7	-1.671
12	594.8	-2.137
13	629.4	-2.872
14	640.8	-3.276
15	651.1	-3.847
16	655.9	-4.278
17	659.6	-4.805
18	661.8	-5.332

to be in the mid-660s, and with K chosen to be 665 a simple least squares analysis yields the regression model

$$Y(t) = 4.16 - 0.531t + e(t) \tag{3.27}$$

(under the assumption that the errors $e(t)$ are independent and have constant variance). Comparison of (3.26) and (3.27) provides the estimates $r = 0.531$ and $(K - n_0)/n_0 = \exp(4.16)$, whence (3.22) becomes

$$N(t) = 665/[1 + \exp\{4.16 - 0.531t\}]. \tag{3.28}$$

Figure 3.1b shows the regression line (3.27) together with the transformed data values (3.26), and the fit is clearly excellent with 99.96% of the variation being explained.

Choice of $K = 665$ is easily justified by first using a simple statistical package (such as MINITAB) to evaluate the regression for K equal to say 663, ..., 667 and then observing that $K = 665$ yields the line of best fit. An exact, though more formal, iterative method is described by Pearl (1930).

As yeast is a simple structure growing by cell division, there are good a priori reasons for believing that the (mathematical) representation (3.20) should successfully mimic the (biological) growth of the yeast population. Yet one of the surprising features of the logistic curve is that although for many situations the simple first-order arguments we have used *cannot* hold, it is a *fact* that population growth often closely follows the logistic curve even when it does not satisfy the assumptions underlying the Verhulst–Pearl model. Pearl (1927), for example, comments that the observed population growth of the United States from 1790 to 1920 follows the logistic curve with remarkable precision, even though immigration, which is a major factor in this instance, does not feature in equation (3.20).

As a further example, Sang (1950) points out that in Pearl's (1927) laboratory experiments on the growth of *Drosophila melanogaster* the yeast that was the source of food was itself a growing population, yet the logistic still fits the data reasonably well. The temptation to accept the logistic curve as a universal *law* of population growth must clearly be strongly resisted; its prime importance lies in the parameter estimates which provide useful *summary statistics*.

Care must be taken to view the logistic model's success in the light of developing populations only, since as Krebs (1985) remarks, most workers stopped their cultures as soon as N approached the upper asymptote K. A notable exception was Thomas Park who reared populations of *Tribolium* for several years. He discovered that the density did not stabilize around K after the initial sigmoid increase, but showed instead a long-term decline (see Park, Leslie and Mertz, 1964). Similar studies by Birch (1953) on *Calandra oryzae* showed initial logistic growth followed by large fluctuations in density with no indication of stabilization about an asymptote. Thus the logistic model may cease to be valid once the asymptotic value K has been attained.

3.2.3   *Growth of two sheep populations*

As the experiments on yeast growth were performed under ideal laboratory conditions, the assumptions underlying the derivation of the logistic model (3.20) can be expected to hold true. We shall now consider two classic field data sets which not only show good agreement with logistic growth during the initial sigmoid stage, but which also exhibit some degree of stabilization around the upper asymptote K. However, since the raising and killing of sheep by humans for food are not inherent parts of the logistic model, analysis must be made in purely descriptive terms.

The growth of a sheep population is subject to the controlling influence of the sheep farmer. The situations in South Australia and Tasmania are of particular interest, since the genial climate allows the sheep to lead a comparatively free life, and a century of annual records are available in each case. Initially both sheep populations grew dramatically. In Tasmania, for example, the initial population consisted of 300 merino lambs imported in 1820, yet by the middle 1850s, when all the areas suitable for sheep raising had been occupied, the population had reached a saturation density of 1.9 million. This value is determined primarily by the carrying (i.e. feeding) capacity of the pastures, though economic factors associated with supply and demand of the products of the sheep industry would also exert an influence.

Davidson (1938b) graphs the number (in thousands) of sheep recorded in Tasmania each year from 1818 to 1936 (note annual records are missing from 1820 to 1826) together with the superimposed logistic curve

$$N(t) = 1670/[1 + \exp\{240.81 - 0.131\,25t\}] \tag{3.29}$$

(see Figure 3.2). The parameters are obtained in a similar manner to those for the yeast data, the estimated upper asymptote $K = 1670$ (thousands) being the average population size for the period 1859–1924. This value is attained towards the end of the 1840s, and the population oscillates around it for the next 70 years. Davidson relates the change in sheep population to changing economic and environmental factors, so to regard such oscillations as being purely random is clearly naive; though we shall see later on that it

Figure 3.2. Growth of the Tasmanian sheep population (- - -) (thousands) from 1818 to 1936 and the superimposed logistic curve $N(t) = 1670/[1 + \exp\{240.81 - 0.131\,25t\}]$ (———) (redrawn from Davidson, 1938b).

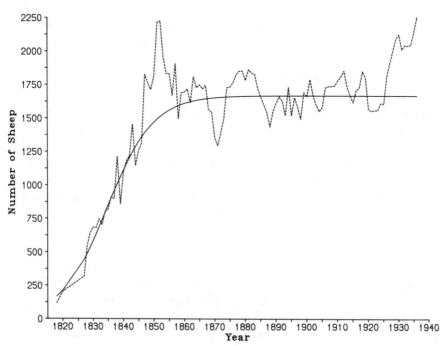

can be instructive to assume that they are random. He explains the factors controlling the reduction in the number of sheep carried on pastures after 1851 as being: an increased demand for fat sheep in local markets; a drop in imported sheep from 65 089 in 1851 to 11 786 in 1858; and the price of mutton in Hobart which appreciated from 3 pence per pound in 1851 to 6 pence in 1852 and $8\frac{1}{2}$ pence in 1853.

A similar picture unfolds from the annual number of sheep recorded in South Australia from 1838 to 1936 (Davidson, 1938a). Figure 3.3 shows the growth of this population together with Davidson's logistic curve

$$N(t) = 7115/[1 + \exp\{249.11 - 0.133\ 69t\}], \tag{3.30}$$

where the upper asymptote $K = 7115$ is determined not from the assumed equilibrium values (which here seem ill-determined) but from the initial sigmoid growth values (1848–52, 1863–67 and 1878–82).

Both sheep populations show good agreement between the data and fitted logistic curves over the initial sigmoid period. Moreover, the denominators in

Figure 3.3. Growth of the South Australian sheep population (- - -) (thousands) from 1838 to 1936 and the superimposed logistic curve $N(t) = 7115/[1 + \exp\{249.11 - 0.133\ 69t\}]$ (———) (redrawn from Davidson, 1938a).

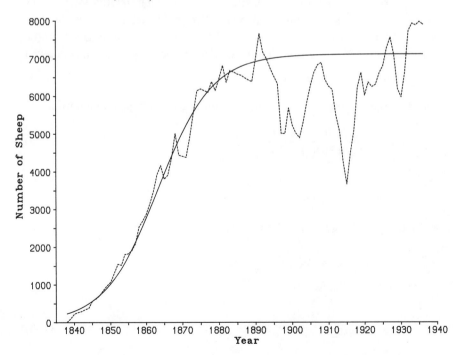

(3.29) and (3.30) are remarkably similar; only the carrying capacities (K) in the numerators differ, as one would anticipate due to the different sizes of the two geographic areas. However, the oscillations around the asymptote that are apparent in Figure 3.2 do not occur in Figure 3.3; instead there is a marked departure away from the asymptotic value $K = 7115$. The main reason for this behaviour is claimed to be the dependence of carrying capacity on the power of pasture to recover from overgrazing. In the arid climate of South Australia recovery is mainly dependent on rainfall, so the population size is affected by the prevalence of wet or dry spells. The good agreement between data and logistic model in the initial sigmoid region therefore provides little indication of likely behaviour thereafter.

It is not only these two sheep populations that show wide fluctuations in numbers around the upper asymptote K of the logistic curve. Krebs (1985) notes that laboratory populations of a single species living in a constant climate with a constant food supply can do likewise. Indeed, he claims that no cases have as yet been demonstrated where the population of any organism with a complex life history remains at a steady state K.

*3.3   Quasi-equilibrium probabilities

To determine the stochastic nature of such fluctuations let us investigate the quasi-equilibrium probabilities $\{\pi_N^{(Q)}\}$ (defined in Section 3.2). We shall use the balance equations (3.12) to develop a solution over the restricted range $N = 1, 2, \ldots$. Conditional on extinction not having occurred

$$D(N)\pi_N^{(Q)} = B(N-1)\pi_{N-1}^{(Q)} \qquad (N = 2, 3, \ldots), \tag{3.31}$$

whence repeated application of this equation gives the general solution

$$\pi_N^{(Q)} = \frac{B(1)B(2)\ldots B(N-1)}{D(2)D(3)\ldots D(N)}\pi_1^{(Q)} \qquad (N \geqslant 2). \tag{3.32}$$

Thus

$$\pi_N^{(Q)} = q_N D(1)\pi_1^{(Q)} \qquad (N \geqslant 2)$$

where the $\{q_i\}$ are given by (3.14). Since $D(1)\pi_1^{(Q)}$ may be determined from

$$1 = \sum_{N=1}^{\infty} \pi_N^{(Q)} = D(1)\pi_1^{(Q)} \sum_{N=1}^{\infty} q_N,$$

we have

$$\pi_N^{(Q)} = q_N / \sum_{i=1}^{\infty} q_i \qquad (N \geqslant 1) \tag{3.33}$$

which is identical to the equilibrium solution (3.13) except that q_0 is replaced by zero.

Under logistic growth $B(N)$ should decrease and $D(N)$ should increase as N increases, so let us put

$$B(N) = N(a_1 - b_1 N) \quad \text{and} \quad D(N) = N(a_2 + b_2 N) \qquad (3.34)$$

for some positive constants a_1, a_2, b_1 and b_2. Note that the associated deterministic equation

$$dN/dt = B(N) - D(N) = N[(a_1 - a_2) - (b_1 + b_2)N] \qquad (3.35)$$

is precisely the same as (3.20) with $r = a_1 - a_2$ and $s = b_1 + b_2$. Substituting for $B(N)$ and $D(N)$ from (3.34) into (3.14) then gives

$$q_1 = 1/D(1) = 1/(a_2 + b_2) \qquad (3.36)$$

and

$$
\begin{aligned}
q_N &= \frac{B(1)\dots B(N-1)}{D(1)\dots D(N)} \\
&= \frac{[a_1 - b_1]\dots[a_1 - (N-1)b_1]}{[a_2 + b_2]\dots[a_2 + Nb_2]N} \qquad (N > 1)
\end{aligned}
\qquad (3.37)
$$

as the component terms in solution (3.33). Since $B(N) = N(a_1 - b_1 N)$ is a birth rate, and hence cannot be negative, N is clearly restricted to the range $1 \leqslant N \leqslant a_1/b_1$.

To illustrate this procedure consider the parameter values $a_1 = 1.0$, $a_2 = 0.2$ and $b_1 = b_2 = 0.1$. The corresponding carrying capacity

$$K = r/s = (a_1 - a_2)/(b_1 + b_2) = 4,$$

whilst N can be at most $a_1/b_1 = 10$. Now (3.37) provides the recursive relation

$$q_{N+1} = \frac{B(N)}{D(N+1)} q_N = \frac{N(1 - 0.1N)}{(N+1)(0.2 + 0.1N)} q_N,$$

and (3.36) the value $q_1 = 1/(a_2 + b_2) = 1/0.3$. Whence $\{q_N\}$, and hence $\{\pi_N^{(Q)}\}$, may be derived sequentially as shown in Table 3.2.

3.4   Simulation of the general population process

Simulation of the general process with birth and death rates $B(N)$ and $D(N)$ exactly parallels that of the simple birth–death process with rates λ and μ (Section 2.3.4). If the population is of size N at time t then the next event will either be a birth with probability $B(N)/R(N)$, or a death with probability $D(N)/R(N)$, where $R(N)$ denotes $B(N) + D(N)$. Moreover, we

Table 3.2. Construction of the quasi-equilibrium probabilities $\{\pi_N^{(Q)}\}$ for $a_1 = 1.0$, $a_2 = 0.2$ and $b_1 = b_2 = 0.1$.

N	q_N		$\pi_N^{(Q)} = q_N / \sum_{i=1}^{10} q_i$
1	$(1/0.3)$	$= 3.3333$	0.1768
2	$(1/2)(0.9/0.4)q_1$	$= 3.7500$	0.1989
3	$(2/3)(0.8/0.5)q_2$	$= 4.0000$	0.2121
4	$(3/4)(0.7/0.6)q_3$	$= 3.5000$	0.1856
5	$(4/5)(0.6/0.7)q_4$	$= 2.4000$	0.1273
6	$(5/6)(0.5/0.8)q_5$	$= 1.2500$	0.0663
7	$(6/7)(0.4/0.9)q_6$	$= 0.4762$	0.0253
8	$(7/8)(0.3/1.0)q_7$	$= 0.1250$	0.0066
9	$(8/9)(0.2/1.1)q_8$	$= 0.0202$	0.0011
10	$(9/10)(0.1/1.2)q_9$	$= 0.0015$	0.0001
$\geqslant 11$	0	$= 0.0000$	0.0000
	$\sum q_i = 18.8562$		1.0001

shall show that the time S to the next event is an exponentially distributed random variable with

$$\Pr(S \geqslant s) = \exp\{-R(N)s\}, \qquad (3.38)$$

and that to simulate a value $S = s$ we may put

$$\exp\{-R(N)s\} = Y \qquad (3.39)$$

where Y is a uniform random variable in the range 0 to 1. Taking logarithms of both sides of (3.39) then gives the simulated inter-event times

$$s = -[\log_e(Y)]/R(N). \qquad (3.40)$$

3.4.1   *Simulation results*

These generalized procedures are incorporated into the program GENONE (see the Appendix to this chapter), and a simulation run is terminated when either a given time (MAXTIME) has elapsed or, if $B(0) = 0$, the population has become extinct. The procedure involves the following steps:

(i) enter the functions $B(N)$ and $D(N)$;
(ii) choose two new random numbers Y_1 and Y_2;
(iii) if $0 \leqslant Y_1 \leqslant B(N)/R(N)$ take the next event to be a birth, otherwise a death;
(iv) evaluate the inter-event time $s = -[\log_e(Y_2)]/R(N)$;

(v) change N to $N + 1$ (birth) or $N - 1$ (death);
(vi) return to step (ii). (3.41)

Since comparisons will constantly be made between deterministic and stochastic behaviour, to avoid possible confusion we shall now adopt the convention that $N(t)$ and $X(t)$ denote deterministic and stochastic population size, respectively, at time t.

A simulated realization $\{X(t)\}$ is shown in Figure 3.4 for the stochastic logistic process with parameter values $n_0 = 1$, $a_1 = 2.2$, $a_2 = 0.2$ and $b_1 = b_2 = 0.1$. $X(t)$ must lie between 0 and $a_1/b_1 = 22$, whilst the carrying capacity $K = (a_1 - a_2)/(b_1 + b_2) = 10$. Also shown is the deterministic solution $N(t)$, and for $0 \leqslant t \leqslant 3$ we see that $X(t)$ follows $N(t)$ quite closely. From $t = 3$ onwards $N(t)$ is virtually equal to the carrying capacity 10, whilst $X(t)$ swings quite considerably on either side of it. Indeed, for this particular run $X(t)$ reaches 19, almost twice K, before $t = 5.5$.

This example neatly exposes the trap facing devotees of deterministic solutions. Whilst an *average* of many realizations will indeed lie close to the

Figure 3.4. A simulated stochastic realization $\{X(t)\}$ (———), and the associated deterministic curve $\{N(t)\}$ (−−−) for a logistic growth process with $n_0 = 1$, $a_1 = 2.2$, $a_2 = 0.2$ and $b_1 = b_2 = 0.1$.

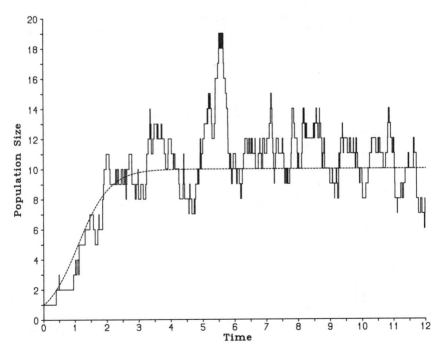

associated $N(t)$-curve, a *single* realization may well differ quite markedly from it. In practice inferences are often drawn from a single set of observations, and blind obedience to deterministic results is clearly courting trouble. A deterministic analysis should always be conducted in parallel with a simulation study of the corresponding stochastic model.

However, once the population size $X(t)$ reaches the vicinity of the carrying capacity $N(\infty) = K$, then a single realization can provide information on the quasi-equilibrium probability structure $\{\pi_N^{(Q)}\}$ since (3.17) is equivalent to

$$\pi_N^{(Q)} = \lim_{t \to \infty} \{\text{proportion of time spent in state } N$$

given that extinction has not occurred by time $t\}$. 　　(3.42)

For example, Figure 3.5 shows the result of computing $\{\pi_N^{(Q)}\}$ from the inter-event times of a single realization, over a time length of $t = 1000$ for the logistic model of Figure 3.4. The starting value $X(0)$ was taken to be the carrying capacity 10, and the realization comprised 22 863 events. The simulated values are clearly in good agreement with those computed from the theoretical result (3.33). Though even better agreement would be obtained from a longer simulation run, say over a time length of $t = 10\,000$, subsequent simulation attempts showed that extinction was quite likely to have occurred before this time was reached.

Figure 3.5. A comparison of simulated (– – –) and theoretical (———) quasi-equilibrium probabilities $\{\pi_N^{(Q)}\}$ computed for the logistic process of Figure 3.4.

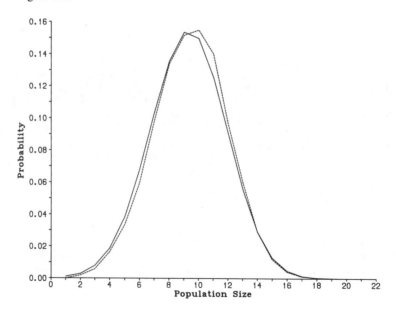

   Let us now justify the inter-event time result (3.40). Suppose that the population reaches size N at time t, and let $G_N(s)$ denote the probability that no event occurs in the subsequent time interval of length s. Then

$$G_N(s + h) = \Pr\{\text{no event occurs in the time interval } (t, t + s + h)\}$$

which, as the two time intervals $(t, t + s)$ and $(t + s, t + s + h)$ are non-overlapping, equals

$$\Pr\{\text{no event occurs in } (t, t + s)\} \times \Pr\{\text{no event occurs}$$
$$\text{in } (t + s, t + s + h)\}.$$

Thus if h is small and $R(N) = B(N) + D(N)$ we have

$$G_N(s + h) \simeq G_N(s) \times [1 - R(N)h],$$

i.e.

$$(1/h)[G_N(s + h) - G_N(s)] \simeq -G_N(s)R(N).$$

As h approaches zero this becomes

$$dG_N(s)/ds = -G_N(s)R(N),$$

which integrates to give

$$G_N(s) = \exp\{-R(N)s\}. \tag{3.43}$$

The probability that the next event does occur in the time interval $(t, t + s)$ is therefore given by

$$F_N(s) = 1 - G_N(s) = 1 - \exp\{-R(N)s\}. \tag{3.44}$$

Note that this is independent of the starting time t.
   To simulate a random inter-event time $S = s$, first choose a uniform random number $0 \leqslant Y \leqslant 1$ and put

$$S = F_N^{-1}(Y) \tag{3.45}$$

where F_N^{-1} denotes the inverse of F_N. Then

$$\Pr(S \leqslant s) = \Pr(F_N^{-1}(Y) \leqslant s) = \Pr\{F_N(F_N^{-1}(Y)) \leqslant F_N(s)\}$$

(since $F(s)$ never decreases as s increases),

i.e.

$$\Pr(S \leqslant s) = \Pr(Y \leqslant F_N(s)).$$

As $0 \leqslant Y \leqslant 1$ is a uniform random variable

$$\Pr(Y \leqslant y) = y.$$

So on putting $y = F_N(s)$ we have

$$\Pr(Y \leqslant F_N(s)) = F_N(s),$$

i.e.

$$\Pr(S \leqslant s) = F_N(s). \tag{3.46}$$

This proves that the random variable $S = F_N^{-1}(Y)$ has the required distribution.

This result is completely general in its applicability. For the case of the exponential form (3.44) we set

$$Y = F_N(s) = 1 - \exp\{-R(N)s\}$$

and solve for s, giving

$$s = -[\log_e(1 - Y)]/R(N). \tag{3.47}$$

On noting that if $0 \leqslant Y \leqslant 1$ is a uniform random variable then so is $(1 - Y)$, we may replace $(1 - Y)$ in (3.47) by Y whence

$$s = -[\log_e(Y)]/R(N)$$

as required.

3.5   The Normal approximation to the quasi-equilibrium probabilities

The symmetric unimodal shape of the distribution $\{\pi_N^{(Q)}\}$ shown in Figure 3.5 enables us to exploit the Normal distribution

$$\Phi(x) = \Pr(X \leqslant x) = \int_{-\infty}^{x} (2\pi\sigma^2)^{-\frac{1}{2}} \exp\{-(y - \mu)^2/2\sigma^2\} \, dy. \tag{3.48}$$

For example, here the first two moments are given by

$$\text{mean}(\mu) = \sum N\pi_N^{(Q)} = 9.2963$$
$$\text{variance}(\sigma^2) = \sum N^2\pi_N^{(Q)} - \mu^2 = 6.5511.$$

Thus in quasi-equilibrium the Normal approximation to (say)

$$\Pr\{5 \leqslant X(t) \leqslant 15\} \simeq \Phi[(15 + \tfrac{1}{2} - \mu)/\sigma] - \Phi[(5 - \tfrac{1}{2} - \mu)/\sigma]$$
$$= \Phi(2.4238) - \Phi(-1.8739) = 0.9619$$

(from tables of the Normal distribution), in good agreement with the exact theoretical probability of 0.9630 determined from (3.33).

We may determine the likely range over which the population size may fluctuate by noting that over a long realization $\{X(t)\}$ will lie in the interval $\mu \pm 2\sigma$ roughly 95% of the time, and in $\mu \pm 3\sigma$ virtually all of the time. So in the above example

$$\Pr\{4.2 \leqslant X(t) \leqslant 14.4\} \simeq 0.95,$$

though as $X(t)$ is integer we should really calculate

$$\Pr\{4 \leqslant X(t) \leqslant 14\} \simeq 0.97.$$

This shows that $\{X(t)\}$ wanders below 4 and above 14 for only 3% of the elapsed time.

*3.6   Probability of ultimate extinction

If the population process does not involve immigration then we are dealing with a closed system of developing individuals which may become extinct. Let us therefore now calculate the probability of ultimate extinction, $p_0(\infty)$, in terms of the general birth and death rates $B(N)$ and $D(N)$.

Denote E_i to be the value of $p_0(\infty)$ for an initial population of size i. Then since the first event is either a birth (i increases to $i + 1$), or a death (i decreases to $i - 1$), we have for $i = 1, 2, \ldots$

$$E_i = \Pr\{\text{first event is a birth}\} \times E_{i+1}$$
$$\qquad + \Pr\{\text{first event is a death}\} \times E_{i-1},$$

i.e.

$$E_i = [B(i)E_{i+1} + D(i)E_{i-1}]/[B(i) + D(i)]. \qquad (3.49)$$

Since there is no immigration, $E_0 = 1$, whilst E_i can be calculated for $i = 2, 3, \ldots$ in terms of E_1 by repeated application of equation (3.49), viz.

$$E_n = E_1 + (E_1 - 1)(S_1 + S_2 + \cdots + S_{n-1}) \qquad (3.50)$$

where

$$S_i = \frac{D(1)D(2)\ldots D(i)}{B(1)B(2)\ldots B(i)}. \qquad (3.51)$$

Although result (3.50) is not a complete solution, since it is expressed in terms of E_1 which is still unknown, it does lead to two useful results.

First, since E_n is a probability and is therefore constrained to take values between 0 and 1 regardless of how large n is, we see from (3.50) that if the sum $S_1 + S_2 + \cdots + S_{n-1}$ does not converge to a finite limit as $n \to \infty$ then we must have $E_1 - 1 = 0$. Thus $E_1 = 1$ and hence $E_n = 1$ for all $n = 2, 3, \ldots$. So in a closed population ultimate extinction is certain if

$$\sum_{i=1}^{\infty} S_i = \sum_{i=1}^{\infty} \frac{D(1)D(2)\ldots D(i)}{B(1)B(2)\ldots B(i)} = \infty. \qquad (3.52)$$

Second, we can evaluate $p_0(\infty)$ for those models in which condition (3.52) is not satisfied. For example, if the population can grow indefinitely large we may argue that as n increases E_n must decay to zero (i.e. $E_\infty = 0$). Expression (3.50) then yields

$$0 = E_1 + (E_1 - 1) \sum_{i=1}^{\infty} S_i,$$

so

$$E_1 = \sum_{i=1}^{\infty} S_i \bigg/ \left[1 + \sum_{i=1}^{\infty} S_i \right]$$

whence (3.50) becomes

$$E_n = \sum_{i=n}^{\infty} S_i \bigg/ \left[1 + \sum_{i=1}^{\infty} S_i \right]. \tag{3.53}$$

To illustrate how these two results may be applied, let us return to the simple birth–death process for which we have already shown in Section 2.3.3 that ultimate extinction is certain if and only if $\lambda \leqslant \mu$. On substituting $B(N) = \lambda N$ and $D(N) = \mu N$ into (3.51) we see that

$$S_i = (\mu/\lambda)^i,$$

whence (3.52) becomes

$$\sum_{i=1}^{\infty} S_i = \sum_{i=1}^{\infty} (\mu/\lambda)^i.$$

This summation clearly diverges if and only if $\lambda \leqslant \mu$, as required. If $\lambda > \mu$, so that indefinitely prolonged growth is possible, then (3.53) gives

$$E_n = \sum_{i=n}^{\infty} (\mu/\lambda)^i \bigg/ \left[1 + \sum_{i=1}^{\infty} (\mu/\lambda)^i \right] = (\mu/\lambda)^n, \tag{3.54}$$

in exact agreement with result (2.54).

Finally, we note that in most closed populations with a single deterministically stable steady state $N(\infty) = K$, the death rate $D(N)$ will exceed the birth rate $B(N)$ for all N greater than K. Result (3.52) shows that in such situations ultimate extinction is certain.

3.7   Mean time to extinction

Of major interest is the mean time to extinction, $T_E(n)$, of a population that initially consists of n individuals and for which eventual extinction is certain. Specifically, T_E can be used to determine an index of stability. We may define a population as being *ecologically stable* if it persists for a large number of generations, and *ecologically unstable* if it persists for only a few. Large values of T_E relate to stability, and small values to instability, and since the order of magnitude of T_E represents the degree of stability we define the *stability index*

$$\xi = \log_e(T_E). \tag{3.55}$$

Consider the first event in a population of initial size n. Then

$$
\begin{aligned}
T_E(n) = {} & \text{mean time to the first event} \\
& + \Pr(\text{first event is a birth}) \times T_E(n + 1) \\
& + \Pr(\text{first event is a death}) \times T_E(n - 1).
\end{aligned} \tag{3.56}
$$

Since the time to the first event is distributed exponentially with parameter $B(n) + D(n)$, the mean time to the first change in population size is $1/[B(n)+D(n)]$. Thus (3.56) becomes

$$T_E(n) = \{1 + B(n)T_E(n + 1) + D(n)T_E(n - 1)\}/[B(n) + D(n)] \tag{3.57}$$

for $n = 1, 2, \ldots$. A considerable amount of algebraic juggling then leads to the solution

$$
T_E(n) =
\begin{cases}
\displaystyle\sum_{i=1}^{\infty} q_i & \text{if } n = 1 \\[3mm]
\displaystyle\sum_{i=1}^{\infty} q_i + \sum_{m=1}^{n-1} S_m \sum_{i=m+1}^{\infty} q_i & \text{if } n \geqslant 2
\end{cases}
\tag{3.58}
$$

where the $\{q_i\}$ are given by (3.14) and the S_i by (3.51) (see Nisbet and Gurney, 1982, pp. 194–6 for full details).

We can now see why the simulated logistic example in Section 3.4.1 was still running at time $t = 1000$, yet several attempts to obtain a simulation run lasting until time $t = 10\,000$ were unsuccessful. Using the simplest version of (3.58) (i.e. $n = 1$) gives $T_E(1) = 2972$. Thus whilst $t = 1000$ is only one-third of the average survival time for a simulation run, $t = 10\,000$ is over three times greater and we might expect that by this time a fair proportion of such runs would have become extinct.

The cumbersome nature of result (3.58) not only poses problems of numerical evaluation for large n, but also prevents any intuitive feeling being gained about the mean time to extinction. Consider, therefore, logistic growth but with the birth and death rates (3.34) simplified to

$$B(N) = rN \quad \text{and} \quad D(N) = rN^2/K. \tag{3.59}$$

Then (3.14) gives

$$q_i = K^i/[ri(i!)] \qquad (i \geqslant 1), \tag{3.60}$$

whilst from (3.51)

$$S_i = (i!)/K^i \qquad (i \geq 1). \tag{3.61}$$

For $i > K$ we see that $S_i > 1$, and so $\Sigma\, S_i = \infty$ which implies that extinction is certain (result (3.52)). Expression (3.58) with $n = 1$ then gives

$$T_E(1) = \sum_{i=1}^{\infty} q_i = (1/r) \sum_{i=1}^{\infty} K^i/[i(i!)]. \tag{3.62}$$

This summation is numerically troublesome unless K is small, but an approximation to it, namely

$$T_E(1) \simeq (rK)^{-1} \exp(K) \tag{3.63}$$

(see Murray, 1974, p. 6), is particularly useful. From (3.55)

$$\xi = \log_e(T_E) \simeq K - \log_e(rK) \simeq K \qquad \text{(for large } K\text{)}, \tag{3.64}$$

and so here the carrying capacity K is itself an index of stability.

3.8　Probability of extinction by time t

Though we have derived results for the probability of, and mean time to, *ultimate* extinction, exact results for the probability of extinction $p_0(t)$ in a *finite* time t are much harder to determine. Indeed, solution (2.53) for the simple birth–death process is one of the few success stories. Fortunately we can obtain a useful simplification for processes which, prior to extinction, have settled down to a quasi-equilibrium state, namely that

$$p_0(t) \simeq 1 - \exp\{-D(1)\pi_1^{(Q)}t\} \tag{3.65}$$

(result (3.69) of Section 3.8.1). If we consider normalized time $\tau = t/T_E$, then (3.65) simplifies still further to give

$$p_0(\tau) \simeq 1 - \exp(-\tau). \tag{3.66}$$

Inverting (3.66) to obtain

$$\tau \simeq -\log_e[1 - p_0(\tau)]$$

yields the following τ-values for given probabilities of extinction:

$p_0(\tau)$	0.001	0.01	0.1	0.5	0.9	0.99	0.999
τ	0.001	0.01	0.11	0.69	2.30	4.61	6.91

We see that extinction is unlikely to occur before $\tau = 0.1$ and is likely to have occurred by $\tau = 2.3$. Note that the skewness of the exponential distribution (3.66) results in the median time to extinction ($\tau = 0.69$) being considerably less than the mean time to extinction ($\tau = 1$).

Continuing our numerical illustration of logistic growth with $B(N) = N(2.2 - 0.1N)$ and $D(N) = N(0.2 + 0.1N)$, combining (3.14) and (3.33) gives

$$D(1)\pi_1^{(Q)} = \pi_1^{(Q)}/q_1 = 1/\sum_{i=1}^{22} q_i = 1/2972.36 = 0.000\ 336\ 4.$$

Thus (3.65) becomes

$$p_0(t) \simeq 1 - \exp\{-0.000\ 336\ 4t\},$$

whence

$$p_0(1000) \simeq 1 - \exp\{-0.3364\} = 0.2857$$

and

$$p_0(10\ 000) \simeq 1 - \exp\{-3.364\} = 0.9654.$$

So whilst the majority of simulations will still be running at time $t = 1000$, more than 96% will have become extinct by time $t = 10\ 000$. This quantifies our simulation experience of Section 3.4.1.

*3.8.1 *Derivation of an approximation for $p_0(t)$*

Following Nisbet and Gurney (1982, pp. 197–8), suppose that the process has settled down to a quasi-equilibrium state. Then the probability that $N(t) = 1$ given that extinction has not occurred is

$$p_1(t)/[1 - p_0(t)] \simeq \pi_1^{(Q)} \qquad \text{(for large } t\text{).} \tag{3.67}$$

Now we know from the first of the general equations (3.9) that provided $B(0) = D(0) = 0$

$$dp_0(t)/dt = D(1)p_1(t). \tag{3.68}$$

Hence on combining (3.67) and (3.68)

$$dp_0(t)/dt \simeq D(1)\pi_1^{(Q)}[1 - p_0(t)],$$

and since $D(1)\pi_1^{(Q)}$ is constant this integrates to give

$$p_0(t) \simeq 1 - \exp\{-D(1)\pi_1^{(Q)}t\}. \tag{3.69}$$

As $p_0(t)$ is the probability of extinction up to and including time t, $dp_0(t)/dt$

is the probability density of extinction at time t. Thus the mean time to extinction for a population of initial size n is

$$T_E(n) = \int_0^\infty t\left[\frac{dp_0(t)}{dt}\right] dt = -\int_0^\infty t\left[\frac{d}{dt}\{p_0(\infty) - p_0(t)\}\right] dt,$$

which integrates by parts to give

$$T_E(n) = \int_0^\infty \{p_0(\infty) - p_0(t)\} \, dt = \int_0^\infty \{1 - p_0(t)\} \, dt \qquad (3.70)$$

provided that ultimate extinction is certain. Inserting the approximation (3.69) then gives

$$T_E(n) \simeq \int_0^\infty \exp\{-D(1)\pi_1^{(Q)}t\} \, dt = 1/[D(1)\pi_1^{(Q)}]. \qquad (3.71)$$

Note that this expression is independent of n.

Now (3.58) states that

$$T_E(1) = \sum_{i=1}^\infty q_i.$$

Hence as expression (3.33) with $N = 1$ gives

$$\sum_{i=1}^\infty q_i = q_1/\pi_1^{(Q)},$$

whilst from (3.14)

$$q_1 = 1/D(1),$$

we have

$$T_E(1) = 1/[D(1)\pi_1^{(Q)}]. \qquad (3.72)$$

The approximate mean time to extinction (3.71) for a population of initial size n is therefore identical to the exact mean time to extinction (3.72) for a population of initial size 1.

This result is intuitively reasonable, since if $D(1)$ is a lot smaller than $B(1)$ then the single initial population member (i.e. $N = 1$) is unlikely to die in the opening stage of the process. During this time the population size will soon rise to say $N = n$, and so $T_E(n)$ is effectively the same as $T_E(1)$.

3.9   Comparison of approximate quasi-equilibrium probability distributions

The close agreement between simulated and theoretical values for the logistic probabilities $\{\pi_N^{(Q)}\}$ illustrated in Figure 3.5 suggests that if numerical estimates are required then simulation is an ideal way of obtaining them, especially if T_E is large. However, should theoretical results be needed, then since the general solution (3.33) is too opaque to be of much practical use approximate representations would be of considerable benefit.

The Normal approximation has already been introduced in Section 3.5, but as this implies symmetry we shall also study two simple non-symmetric forms. Denote by m and V the mean and variance, respectively, of the quasi-equilibrium distribution $\{\pi_N^{(Q)}\}$. Then

(1) *the binomial distribution* ($V < m$) has probabilities

$$p_r = \binom{n}{r} p^r q^{n-r} \qquad (r = 0, \ldots, n)$$

with mean np and variance npq. On putting $m = np$ and $V = npq$ (which is less than m), we have $q = V/m$, $p = 1 - (V/m)$ and n is taken to be the nearest integer to $m/p = m^2/(m - V)$. Thus provided n is sufficiently large for $\pi_n^{(Q)} \simeq 0$,

$$\pi_N^{(Q)} \simeq \binom{n}{N} [1 - (V/m)]^N [V/m]^{n-N} \tag{3.73}$$

over the effectively unrestricted range $N = 1, 2, \ldots, n$.

(2) *the negative binomial distribution* ($V > m$) has probabilities

$$p_r = \binom{n + r - 1}{r} p^n q^r \qquad (r = 0, 1, 2, \ldots)$$

with mean $m = nq/p$ and variance $V = nq/p^2$ (which is greater than m). This time we put $p = m/V$, $q = 1 - (m/V)$, and take n to be the nearest integer to $m^2/(V - m)$, whence

$$\pi_N^{(Q)} \simeq \binom{n + N - 1}{N} [m/V]^n [1 - (m/V)]^N \qquad (N = 1, 2, \ldots). \tag{3.74}$$

(3) *the Normal approximation* (3.48), written in the form

$$\pi_N^{(Q)} \simeq \Pr(N - \tfrac{1}{2} \leqslant X \leqslant N + \tfrac{1}{2}),$$

gives

$$\pi_N^{(Q)} \simeq (2\pi V)^{-\frac{1}{2}} \exp\{-(N - m)^2/2V\} \qquad (N = 1, 2, \ldots). \tag{3.75}$$

3.9.1 *Logistic mean and variance values*
To determine logistic values for m and V, consider the Verhulst–

Pearl equation

$$dN(t)/dt = N(t)[r - sN(t)] \tag{3.20}$$

i.e.

$$N(t + \delta t) - N(t) \simeq N(t)[r - sN(t)]\delta t \tag{3.76}$$

where δt is a small time increment. To turn this into a stochastic equation we write the random population change in time δt as

$$X(t + \delta t) - X(t) = X(t)[r - sX(t)]\delta t + \delta Z. \tag{3.77}$$

Here δZ represents chance fluctuation, called the noise component, with

$$E(\delta Z) = 0. \tag{3.78}$$

If the population lies in a state of quasi-equilibrium around the mean value m, i.e. $\{X(t)\}$ is centred around m, then

$$E[X(t)] = m \quad \text{and} \quad E[X(t + \delta t)] = m. \tag{3.79}$$

So taking expectations on both sides of equation (3.77) yields

$$0 = E[rX(t) - sX^2(t)]\delta t + E(\delta Z),$$

i.e.

$$rE[X(t)] - sE[X^2(t)] = 0.$$

Thus as $E[X(t)] = m$ and $E[X^2(t)] = m^2 + \text{Var}[X(t)]$, we have

$$rm - s\{m^2 + \text{Var}[X(t)]\} = 0. \tag{3.80}$$

Hence as $m \simeq r/s$, an improved approximation to m is given by

$$\hat{m} = (r/s) - (1/m)\,\text{Var}[X(t)],$$

i.e.

$$\hat{m} = (r/s) - (s/r)\,\text{Var}[X(t)]. \tag{3.81}$$

An extension of this argument (proved in Section 3.9.2) shows that

$$\text{Var}[X(t)] \simeq \hat{V} = r\gamma/2s^2 \tag{3.82}$$

where

$$\gamma \simeq (s^2/r^2)[(a_1 + a_2)m - (b_1 - b_2)m^2], \tag{3.83}$$

i.e. since $m \simeq r/s$

$$\gamma \simeq (s/r)(a_1 + a_2) - (b_1 - b_2). \tag{3.84}$$

Substitution of (3.82) into (3.81) then gives

$$\hat{m} = (2r - \gamma)/(2s). \tag{3.85}$$

To illustrate these results let us return to our logistic example with a_1 = 2.2, a_2 = 0.2 and $b_1 = b_2 = 0.1$, with its exact mean $m = 9.2963$ and exact variance $V = 6.5511$ (Section 3.5). As $r = a_1 - a_2 = 2$ and $s = b_1 + b_2 = 0.2$, from (3.84)

$$\gamma \simeq (0.2/2)(2.4) = 0.24.$$

Thus (3.85) yields

$$\hat{m} = (4 - 0.24)/(0.4) = 9.40 \quad (1.1\% \text{ too high}),$$

whilst (3.82) becomes

$$\hat{V} = (2 \times 0.24)/(2 \times 0.04) = 6.00 \quad (8.4\% \text{ too low}).$$

So although $\hat{m}$ estimates m quite accurately, the variance estimate $\hat{V}$ is relatively poor.

Incidentally, $\hat{m}$ is more precise than the simulated mean $\hat{m}_{sim} = 9.4485$ (1.6% too high) obtained by averaging the equilibrium portion of Figure 3.4, whilst the simulated variance $\hat{V}_{sim} = 6.1349$ (6.4% too low) is comparable with $\hat{V}$.

*3.9.2   *Derivation of the variance approximation*
We shall determine a local approximation for the variance of $\{X(t)\}$ by assuming that $X(t)$ deviates only slightly from its mean value m. That is

$$X(t) = m[1 + u(t)] \tag{3.86}$$

where the random variable $u(t)$ is presumed sufficiently small for u^2 to be negligible in comparison with u. This linearization technique is popular in mathematical ecology, though the crude nature of the underlying assumption is all too often forgotten. We have already seen in our numerical logistic example that a 95% confidence interval for $X(t)$ is $4.2 \leqslant X(t) \leqslant 14.4$, i.e. $-0.55 \leqslant u \leqslant 0.55$, and to claim that 'when $u \simeq \frac{1}{2}$, u^2 is negligible' is clearly ridiculous. Fortunately, the approach generally performs far better than one has a right to expect!

On writing the stochastic equation (3.77) in the form

$$\delta X = X(r - sX)\delta t + \delta Z,$$

and differentiating (3.86) to obtain

$$\delta X = m\delta u,$$

we have

$$m\delta u = m(1 + u)(r - sm - smu)\delta t + \delta Z.$$

As $m \simeq r/s$,

$$m\delta u \simeq -sm^2 u(1 + u)\delta t + \delta Z \tag{3.87}$$

which, on ignoring u^2, reduces to

$$\delta u = -smu\delta t + (\delta Z/m).$$

Thus

$$(u + \delta u) = u(1 - r\delta t) + (s/r)\delta Z,$$

whence squaring gives

$$\begin{aligned}(u + \delta u)^2 &= u^2(1 - r\delta t)^2 + 2u(s/r)(1 - r\delta t)\delta Z \\ &\quad + (s/r)^2\delta Z^2.\end{aligned} \tag{3.88}$$

We now take expectations on both sides of (3.88), bearing in mind that:

(i) $E(\delta Z) = (+1)B(X)\delta t + (-1)D(X)\delta t$
$\simeq [B(m) - D(m)]\delta t = 0$;

(ii) $E(\delta Z^2) \simeq \mathrm{Var}(\delta Z)$
$= (+1)^2 B(X)\delta t + (-1)^2 D(X)\delta t$
$\simeq [(a_1 + a_2)m - (b_1 - b_2)m^2]\delta t$; (3.89)

(iii) $E(u) = 0$ and $E[(u + \delta u)^2] = E(u^2) = \mathrm{Var}(u)$,
since the process is in quasi-equilibrium; and

(iv) terms in δt^2 may be neglected.

This procedure gives (approximately)

$$\mathrm{Var}(u) = (1 - 2r\delta t)\,\mathrm{Var}(u) + 0 + (s/r)^2\,\mathrm{Var}(\delta Z),$$

i.e.

$$(2r\delta t)\,\mathrm{Var}(u) = \gamma\delta t \qquad \text{(from (3.83) and (3.89))},$$

whence

$$\mathrm{Var}(u) = \gamma/2r.$$

Consequently

$$\mathrm{Var}(X) = m^2\,\mathrm{Var}(u) \simeq (r/s)^2(\gamma/2r) = r\gamma/2s^2.$$

3.9.3   *An approximate skewness result*
As well as measuring the location (mean) and spread (variance) of

the quasi-equilibrium distribution, we can also determine how lop-sided (skew) it is. An approximate measure of skewness for large m is

$$\mu_3 \simeq V(b_2 - b_1)/(b_1 + b_2) \qquad (3.90)$$

(Bartlett, Gower and Leslie, 1960). This implies that μ_3 can range from $-V$ when $b_2 = 0$ (skew left) to $+V$ when $b_1 = 0$ (skew right). With $B(N) = N(a_1 - b_1 N)$ and $D(N) = N(a_2 + b_2 N)$, it follows from (3.90) that positive skewness is associated with competition affecting the death rate more than the birth rate (i.e. $b_2 > b_1$), whilst with negative skewness the birth rate is affected more than the death rate (i.e. $b_1 > b_2$).

3.9.4 *A numerical comparison*

To illustrate how our approximations fare when compared with exact probabilities $\{\pi_N^{(Q)}\}$, let us now consider two particular sets of parameter values a_i, b_i $(i = 1, 2)$. To make life difficult for the estimation procedures we accentuate skewness by putting $b_1 = 0$, whence from (3.90)

$$\mu_3 \simeq V > 0. \qquad (3.91)$$

For case (i) we take $a_1 = 1.1$ and $a_2 = b_2 = 0.1$, so the carrying capacity $K = 10$; for case (ii) $a_1 = 10.1$ and $a_2 = b_2 = 0.1$, so $K = 100$. In (i) $m = 8.688$ and $V = 11.420 > m$, whilst in (ii) $m = 98.979$ and $V = 101.016 > m$, so the negative binomial distribution is appropriate for both.

Repetitive evaluation of expressions (3.73) (binomial) and (3.74) (negative binomial) for the approximating probabilities $\{\pi_N^{(a)}\}$ can be avoided by writing them in the sequential form

binomial: $\pi_1^{(a)} = npq^{n-1}$ then $\pi_N^{(a)} = \pi_{N-1}^{(a)}(n - N + 1)p/Nq$

negative binomial: $\pi_1^{(a)} = np^n q$ then $\pi_N^{(a)} = \pi_{N-1}^{(a)}(n + N - 1)q/N$.

Note that it is just as easy to generate the exact values $\{\pi_N^{(Q)}\}$ as it is the approximate values $\{\pi_N^{(a)}\}$. The benefit of constructing an approximate distribution is that it provides a concise summary of the quasi-equilibrium probabilities through the parameter values.

Figure 3.6 shows the exact, negative binomial and Normal probabilities for model (i), i.e. $a_1 = 1.1$, $a_2 = b_2 = 0.1$ and $b_1 = 0$. For comparison, simulated probabilities are also shown, these being calculated from a single run over an elapsed time $t = 600$. This value was chosen since it lies near to the mean time to extinction (3.71), namely

$$T_E \simeq [D(1)\pi_1^{(Q)}]^{-1} = [0.2 \times 0.008\ 780]^{-1} = 570.$$

Except at $N = 1$ and 2 the negative binomial form is fairly accurate, and even the Normal probabilities are reasonable bearing in mind that the $\{\pi_N^{(Q)}\}$ are skew. Indeed, Whittle (1957) claims that use of the Normal approximation is justified for most 'linearizable' stochastic processes. The simulation run is too short to yield accurate estimates; for example, $\pi_9^{(\text{sim})} = 0.1081$ is 6.3% lower than $\pi_9^{(Q)} = 0.1153$.

Note that interest in biological populations often focusses around extreme behaviour, that is in the tails of the probability distribution, and so here the scope for wrong interpretation is considerable. For example, whilst the negative binomial form does indeed give a good general description of the quasi-equilibrium distribution, it clearly behaves poorly at $N = 1$ and 2. The estimate $\pi_1^{(a)} = 0.003\,167$ is 64% lower than $\pi_1^{(Q)} = 0.008\,780$, and yields a mean extinction time of $[0.2 \times 0.003\,167]^{-1} = 1579$ which is 2.8 times larger than it should be.

On increasing the carrying capacity K from 10 to 100 by changing a_1 from 1.1 to 10.1 (case (ii)), we see from Figure 3.7 that exact and negative binomial probabilities are almost indistinguishable. Even the Normal form is a good approximation, the exact probability distribution $\{\pi_N^{(Q)}\}$ being nearly sym-

Figure 3.6. Comparison of exact (◯), negative binomial (———), Normal (– – –) and simulated (- - -) probabilities for logistic parameters $a_1 = 1.1$, $a_2 = b_2 = 0.1$ and $b_1 = 0$.

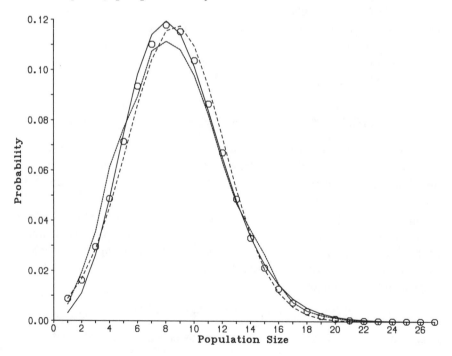

metric. Indeed, on defining the coefficient of skewness (β) by the dimension-less form

$$\beta = \mu_3 V^{-1.5}, \tag{3.92}$$

we see from (3.90) that

$$\beta \simeq (b_2 - b_1)/[(b_1 + b_2)\sqrt{V}] \tag{3.93}$$

and so $\beta \simeq 0$ for large variance V.

On evaluating $\pi_1^{(2)}$ for case (ii), result (3.71) gives

$$T_E \simeq (0.2 \times 0.6909 \times 10^{-38})^{-1} = 7.24 \times 10^{38},$$

which for all practical purposes is effectively infinite! Thus extinction within any 'reasonable' time is virtually impossible, and simulated probabilities are limited in their accuracy only by the length of run that can be afforded.

The following probability values illustrate the relative numerical accuracy of the two approximating distributions under case (ii) as we move towards their tails.

Figure 3.7. Comparison of exact ($\bigcirc$), negative binomial (———) and Normal (– – –) probabilities for logistic parameters $a_1 = 10.1$, $a_2 = b_2 = 0.1$ and $b_1 = 0$.

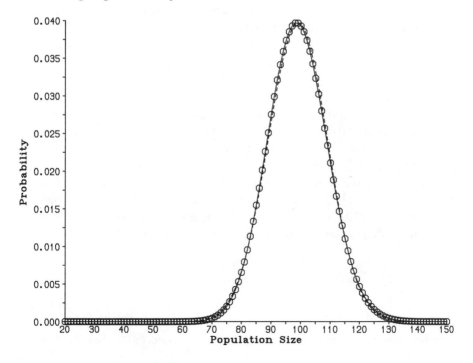

	Exact	Negative binomial	Normal
$N = 70$	0.000 461	0.000 454 (-1.52%)	0.000 622 $(+34.9\%)$
$N = 85$	0.015 478	0.015 479 $(+0.01\%)$	0.015 089 (-2.51%)
$N = 100$	0.039 259	0.039 254 (-0.01%)	0.039 489 $(+0.59\%)$

Since $m \simeq 100$ and $\sqrt{V} \simeq 10$, the values $N = 70$, 85 and 100 roughly correspond to 0, $1\frac{1}{2}$ and 3 standard deviations from the mean. The negative binomial form is exceedingly accurate for $N = 85$ and 100, and is good even when N is as low as 70. In contrast, the Normal form is clearly inferior at $N = 85$, and is hopeless by $N = 70$.

3.10   The diffusion approximation

The binomial, negative binomial and Normal approximations were chosen purely on the grounds that they have simple structures which can be easily manipulated to give the correct mean and variance. If we wish to generate more complex mathematical forms to describe the quasi-equilibrium probabilities $\{\pi_N^{(Q)}\}$, then since mathematical analysis of models with a discrete population size N (i.e. $N = 0, 1, 2, \ldots$) is intrinsically more demanding than that required for continuous N (i.e. any $N \geqslant 0$) it is sensible to envisage N as a continuous variable.

One approach is to replace the set of differential equations (3.9), viz.

$$dp_N(t)/dt = D(N + 1)p_{N+1}(t) - [B(N) + D(N)]p_N(t) + B(N - 1)p_{N-1}(t)$$

for $N = 0, 1, 2, \ldots$ and $t \geqslant 0$, by a *diffusion equation* in which N is allowed to take any non-negative value. Unfortunately, the diffusion technique, which involves expanding (3.9) in terms of Taylor series expansions and then taking mathematical limits in rather subtle ways, is well beyond the scope of this book. We shall therefore just quote the main result, and recommend those readers who are interested in the details of the mathematical derivation to consult the excellent account given by Nisbet and Gurney (1982, chapter 6). They not only provide a clear presentation of the related diffusion equation

$$\frac{\partial p_N(t)}{\partial t} = -\frac{\partial}{\partial N}\{[B(N) - D(N)]p_N(t)\}$$
$$+ \frac{1}{2}\frac{\partial^2}{\partial N^2}\{[B(N) + D(N)]p_N(t)\} \tag{3.94}$$

for $N \geqslant 0$ and $t \geqslant 0$, but they also highlight the various pitfalls that can beset the unwary. Here $\partial/\partial t$, $\partial/\partial N$ means differentiate with respect to t, N whilst regarding N, t as constant.

The equivalent equation for the probabilities

$$p_N^{(Q)}(t) = p_N(t)/[1 - p_0(t)] \qquad (N > 0, t \geq 0), \tag{3.95}$$

i.e. conditional on extinction not having occurred by time t, is also stated. However, they warn that '[it is] a rather nasty non-linear equation which is intractable analytically and more difficult to solve numerically than the discrete equations which it purports to approximate'.

*3.10.1 *Equilibrium probabilities*

Before seeing how the diffusion equation (3.94) can be used to construct an approximation to the quasi-equilibrium probabilities $\{\pi_N^{(Q)}\}$, let us first consider processes which possess a true equilibrium structure, that is they are subject to neither extinction nor population explosion. As t becomes large the probabilities $p_N(t)$ settle down to constant non-zero values, and so we may set $\partial p_N(t)/\partial t = 0$. Hence on writing

$$f(N) = B(N) - D(N) \quad \text{and} \quad g(N) = B(N) + D(N), \tag{3.96}$$

for large t (3.94) becomes

$$\frac{1}{2} \frac{\partial^2}{\partial N^2} [g(N)\pi(N)] - \frac{\partial}{\partial N} [f(N)\pi(N)] = 0 \tag{3.97}$$

where

$$\pi(N) = \lim_{t \to \infty} \{p_N(t)\}.$$

Equation (3.97) integrates directly to give

$$\frac{1}{2} \frac{\partial}{\partial N} [g(N)\pi(N)] - f(N)\pi(N) = C \tag{3.98}$$

for some constant C. Now we may argue that when N is very large, i.e. well in excess of the carrying capacity, both $\pi(N)$ and $\partial\pi(N)/\partial N$ must be very small. Thus $C = 0$ and (3.98) becomes

$$\frac{\partial}{\partial N} [g(N)\pi(N)] - 2f(N)\pi(N) = 0.$$

Writing this equation in the standard form

$$\frac{\partial}{\partial N} [g(N)\pi(N)] - 2\left[\frac{f(N)}{g(N)}\right][g(N)\pi(N)] = 0$$

then yields the solution

$$\pi(N) = A[g(N)]^{-1} \exp\left\{ 2 \int_0^N \frac{f(x)}{g(x)} \, dx \right\} \qquad (N \geqslant 0), \qquad (3.99)$$

where the normalizing constant A is chosen to ensure that

$$\int_0^\infty \pi(x) \, dx = 1. \qquad (3.100)$$

Result (3.99) provides a means of obtaining an approximating form for the quasi-equilibrium probabilities $\{\pi_N^{(Q)}\}$. If t is a lot less than T_E, then $p_0(t)$ in expression (3.95) will be almost zero and so $p_N^{(Q)}(t)$ and $p_N(t)$ will be virtually identical. Thus for large $t \ll T_E$

$$\pi_N^{(Q)} \simeq p_N^{(Q)}(t) \simeq p_N(t) \simeq \pi(N),$$

with a marginal change in the normalizing condition (3.100) to

$$\int_1^\infty \pi(x) \, dx = 0. \qquad (3.101)$$

A possible drawback to solution (3.99) is that it necessitates evaluating the integral

$$\int_0^N [f(x)/g(x)] \, dx,$$

though this presents no problem when $B(N) = N(a_1 - b_1 N)$ and $D(N) = N(a_2 + b_2 N)$. In particular, when $b_1 = b_2 = b$ (say)

$$\pi(N) = \frac{A_1}{N} \exp\left\{ -\left(\frac{2b}{a_1 + a_2} \right) \left[N - \frac{(a_1 - a_2)}{2b} \right]^2 \right\} \qquad (3.102)$$

where A_1 is an appropriate normalizing constant.

Now if X is a Normally distributed random variable with mean μ and variance σ^2, then its probability density function is given by

$$(2\pi\sigma^2)^{-\frac{1}{2}} \exp\{ -(x - \mu)^2/2\sigma^2 \}. \qquad (3.103)$$

Comparison of (3.102) and (3.103) shows that they are in reasonable agreement provided that N remains near to the equilibrium value and

$$\mu = (a_1 - a_2)/(2b) \quad \text{with} \quad \sigma^2 = (a_1 + a_2)/(4b). \qquad (3.104)$$

Note that these moments agree with our approximating values determined in Section 3.9.1, namely $\mu = r/s \simeq m$ and $\sigma^2 = r\gamma/2s^2 = \hat{V}$.

*3.10.2   **Justifying the Normal approximation***
   When $b_1 \neq b_2$ we may justify the use of the Normal approximation by linearizing about the deterministic steady state N^*, defined by $B(N^*) = D(N^*)$. On setting

$$N(t) = N^*[1 + n(t)] \qquad (3.105)$$

and regarding n small, we have from (3.96) that

$$
\begin{aligned}
f[N^*(1 + n)] &= B[N^*(1 + n)] - D[N^*(1 + n)] \\
&= N^*(1 + n)[(a_1 - a_2) - (b_1 + b_2)N^*(1 + n)] \\
&\simeq -(N^*)^2 n(b_1 + b_2)
\end{aligned}
$$

together with

$$g[N^*(1 + n)] \simeq g(N^*).$$

Hence as $dN = N^* \, dn$ and $n = (N - N^*)/N^*$, the general solution (3.99) becomes

$$\pi(N) \simeq \text{constant} \times \exp\left\{2 \int \frac{[-(N^*)^2 n(b_1 + b_2)]}{g(N^*)} N^* \, dn\right\},$$

i.e.

$$\pi(N) \simeq \text{constant} \times \exp\{-(N - N^*)^2/2R\} \qquad (3.106)$$

where

$$R = g(N^*)/[2N^*(b_1 + b_2)].$$

This is a Normal distribution with mean N^* and variance R; in the logistic case $N^* = r/s$ and $R = r\gamma/2s^2$ (in agreement with (3.82)).

3.11   Application to the yeast and sheep data

   In Section 3.2 three data sets are described in terms of logistic growth: whilst the yeast population lies remarkably close to its associated deterministic growth curve, the two sheep populations exhibit considerable stochastic variation. These three data sets are therefore ideal for illustrating our results.

3.11.1   *Yeast data*

   We have already seen (Figure 3.1a) that the observed amounts of yeast $X(t)$ are in very close agreement with the deterministic values

$$N(t) = 665/[1 + \exp\{4.16 - 0.531t\}]. \qquad (3.28)$$

Now since the difference between $X(t)$ and $N(t)$ is less than 1 over the later times $t = 14, \ldots, 18$ (i.e. near the asymptote), and since $X(t)$ increases for all values of $t = 0, \ldots, 18$, deaths may be ignored. When $D(N) = 0$ the equilibrium variance is zero: any change in population size must be the result of a birth, yet at $N = K$ the birth rate $B(K) = 0$ and so the population size must remain fixed at K thereafter. Moreover, we have already determined $K = r/s = 665$ and $r = 0.531$, hence $s = 0.000\,799$. Thus an appropriate descriptive stochastic model has transition rates

$$B(N) = N(0.531 - 0.000\,799N) \quad \text{and} \quad D(N) = 0. \tag{3.107}$$

However, since the amount of variability is so low, the deterministic representation (3.28) is clearly better in view of its simplicity.

3.11.2   *Tasmanian sheep data*

Though extinction is clearly impossible under model (3.107), the same cannot hold true for the Tasmanian and South Australian sheep models since considerable natural variation is present in both populations. Inspection of Figure 3.2 shows that the observed Tasmanian values $X(t)$ appear to be well described by

$$X(t) = N(t) + \text{natural variation},$$

where $N(t)$ is the deterministic expression (3.29).

Davidson based his estimate of the carrying capacity $K = 1670$ on the period 1859–1924, and we shall use the same period to illustrate how the parameters a_1, a_2, b_1 and b_2 of the stochastic logistic model may be estimated. Direct comparison of (3.23) and (3.29) gives $r = 0.131\,25$, so

$$a_1 - a_2 = r = 0.131\,25$$
$$b_1 + b_2 = s = r/K = 0.000\,078\,593 \tag{3.108}$$

(from representation (3.35)). We can now use the sample variance and skewness to develop two more equations.

From (3.82)

$$\text{Var}(X) \simeq (1/65) \sum_{t=1859}^{1924} [X(t) - 1670]^2 = 16\,254 \simeq r\gamma/2s^2,$$

and so $\gamma \simeq 0.001\,529\,9$. Use of (3.84) then gives

$$0.001\,529\,9 = 0.000\,598\,8(a_1 + a_2) - (b_1 - b_2). \tag{3.109}$$

Now whilst reasonable ball-park estimates of variance may be obtained from time-series containing only 66 observations, estimating skewness is another matter. The estimate

$$\text{skewness}(X) \simeq (1/65) \sum_{t=1859}^{1924} [X(t) - 1670]^3 = -1\,357\,992$$

is severely affected by any substantial departure of $X(t)$ from K. If we do attempt to use it in combination with approximation (3.90), i.e.

$$\text{skewness}(X) \simeq \text{Var}(X)(b_2 - b_1)/s,$$

then we have

$$b_2 - b_1 \simeq -0.006\ 566. \tag{3.110}$$

But solving for b_2 from (3.108) and (3.110) gives $b_2 = -0.003\ 24$, which is in direct conflict with the requirement that $b_2 \geqslant 0$. The most appropriate choice here is to put $b_2 = 0$; this simplifies logistic death to simple death yet still provides for maximum negative skewness. Then (3.108) gives $b_1 = s$ $= 0.000\ 078\ 59$, whence expression (3.109) becomes

$$a_1 + a_2 = 2.6862. \tag{3.111}$$

Combining (3.108) and (3.111) gives $a_1 = 1.4087$ and $a_2 = 1.2775$, and so the estimated transition rates are

$$B(N) = N(1.4087 - 0.000\ 078\ 59N) \quad \text{and} \quad D(N) = 1.2775N. \tag{3.112}$$

Note that this fitted model provides a good description of the data in terms of the initial sigmoid shape, the level of the attained asymptote (i.e. the carrying capacity), and the variance of the fluctuations about this level. It does not, and cannot, provide information on the process generating mechanism unless we also take account of the biological, social and economic factors which influenced the actual development of the sheep population.

3.11.3 *South Australian sheep data*

The question of model validity clearly arises with the South Australian sheep data (Figure 3.3). Although these data follow Davidson's deterministic curve (3.30) quite closely until about 1895, thereafter the apparent large stochastic variation could either be natural or else it could be due to changed circumstances. We can never be certain which, but for climatic reasons already mentioned in Section 3.2.3 it does seem quite likely that the fitted logistic model does not hold in the second half of the series.

Nevertheless, for the purpose of illustration let us suppose that the deterministic model

$$N(t) = 7115/[1 + \exp\{249.11 - 0.133\ 69t\}] \tag{3.30}$$

does hold from 1838 onwards, and that an approximate steady state is achieved by 1890. Then with the mean m taken to be $K = 7115$, the variance

V calculated from the last 47 data values is $1\,774\,934$. Taking $b_2 = 0$ as above, so that $B(N) = N(a_1 - b_1 N)$ and $D(N) = a_2 N$, we have

$$a_1 - a_2 = r = 0.133\,69$$

and

$$b_1 = s = r/K = 0.000\,018\,790.$$

Since

$$V \simeq r\gamma/2s^2 = 1\,774\,934,$$

we therefore have

$$\gamma \simeq 0.009\,374\,8 = (s/r)(a_1 + a_2) - b_1$$

giving

$$a_1 + a_2 = 66.8353.$$

So $a_1 = 33.4845$ and $a_2 = 33.3508$, whence

$$B(N) = N(33.4845 - 0.000\,018\,79N)$$

and

$$D(N) = 33.3508N. \tag{3.113}$$

Although commonsense consideration of this sheep population dictates that extinction is extremely unlikely to occur within any sensible time-scale (we hope!), it is instructive to follow the consequence of assuming that the underlying logistic description remains true over the foreseeable future. Computing the quasi-equilibrium distribution $\{\pi_N^{(Q)}\}$ from (3.33), (3.36) and (3.37) for the parameter values (3.113) gives $\pi_1^{(Q)} = 0.134 \times 10^{-5}$. So on using (3.71), the mean time to extinction $T_E = 1/[D(1)\pi_1^{(Q)}] = 22\,400$ years, which on a human time-scale is enormous. To evaluate the probability of extinction in say the period 1890 to 1990, we use results (3.69) and (3.72) to obtain

$$p_0(100) \simeq 1 - \exp\{-100/T_E\} = 0.0045,$$

i.e. a chance of less than 1 in 200.

Appendix

The following FORTRAN program 'GENONE' simulates the general birth and death process with rates $B(N)$ and $D(N)$, respectively; these should be inserted in place of the logistic example rates $B(N) = N(2.2 - 0.1N)$ and $D(N) = N(0.2 + 0.1N)$ below. As with the program 'BIRTH', the NAG-routine G05CAF may be replaced by any suitable random number generator.

```
C       THIS PROGRAM IS USED TO SIMULATE A GENERAL BIRTH
C       AND DEATH PROCESS WITH RATES B(N) AND D(N),
C       SEED S, INITIAL POPULATION SIZE N0, MAXIMUM
C       DURATION TIME MAXTIME, AND MAXIMUM NUMBER
C       OF EVENTS MAXEVENTS.
C
C
        REAL*8 G05CAF,Y
        REAL N,NX,MAXTIME
        INTEGER COUNT,S
C
C       READ IN PARAMETER VALUES
C
        READ(01,8000) S,N0,MAXTIME
        CALL G05CBF(S)
C
C       SET INITIAL VALUES
C
        MAXEVENT = 100000
        TIME = 0.0
        N = N0
C
C       SIMULATE INTER-EVENT TIMES AND OUTPUT WITH
C       POPULATION SIZE NX
C
        DO 10 COUNT=1,MAXEVENT
C
C       INSERT BIRTH AND DEATH RATES B(N) AND D(N)
C
        B = N*(2.2 − (0.1*N))
        D = N*(0.2 + (0.1*N))
C
        WRITE(02,8001) TIME,N
        Y = G05CAF(Y)
        RATE = B+D
        TEST = B/RATE
        REXP = −(DLOG(Y))/RATE
        TIMEX = TIME
        TIME = TIME + REXP
        NX = N
```

```
        IF(TIME.GT.MAXTIME) STOP
C
        Y = G05CAF(Y)
        N = N-1
        IF(Y.LE.TEST) N = N+2
        WRITE(02,8001) TIME,NX
10      CONTINUE
C
 8000   FORMAT(2I8,F8.2)
 8001   FORMAT(F10.4,6X,F8.0)
        STOP
        END
```

4

Time-lag models of population growth

The variety of dynamic behaviour exhibited by many species of plants, insects and animals is far wider than can possibly be described by the simple birth–death and logistic models. For example, fluctuations in population density of the large pine moth *Dendrolimus pini* from 1880 to 1940 (Figure 4.1) show long intervals at very low values, interspersed with erratic episodes of outbreak and subsequent crash; whilst the larch bud-moth *Zeiraphera diniana* shows regular cyclic oscillations from 1949 to 1966 (Figure 4.2) with a roughly ten-year period that has been documented back into the middle of the nineteenth century (May, 1986). May's paper provides an entertaining and thought-provoking overview of how simple deterministic non-linear models can generate an astonishing array of population behaviour. In this chapter we shall see how such models can help us to understand real-life diversity, and we shall explore the effect of including natural variation within them.

Figure 4.1. Fluctuations in the population density of the large pine moth, *Dendrolimus pini*, in Central Germany, from 1880 to 1940. (Reproduced from May, 1986, by permission of The Royal Society; data from Schwerdtfeger, 1935.)

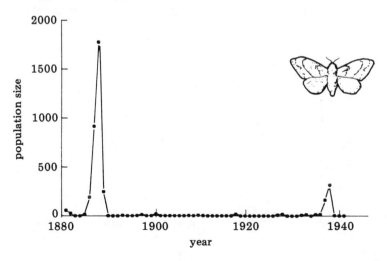

4.1   Introduction

In the deterministic logistic model (3.20), namely

$$dN(t)/dt = rN(t)[1 - N(t)/K], \qquad (4.1)$$

population change depends solely on the population size at the current time t. In practice, however, there is often a time-lag between the inception of an action and the resulting change: seeds take time to change into plants, whilst newly born animals take time to develop into adults capable of reproducing. Such delays are likely to cause oscillations in population size, in exactly the same way as the delay between controlling a shower tap and receiving the corresponding stream of water means that it is possible to alternately freeze and scald oneself. In general, if the duration of the delay in this feedback loop is longer than the 'natural period' of the system, then large amplitude oscillations will result. If in the absence of regulation population growth is determined by the equation

$$dN(t)/dt = rN(t), \qquad (4.2)$$

then this natural period is defined as $1/r$.

Delayed regulation may occur in various ways, and two of the simplest are outlined below.

(i)   *Reaction time-lag*

Suppose the individual growth rates are subject to a time-delay t_D, so that

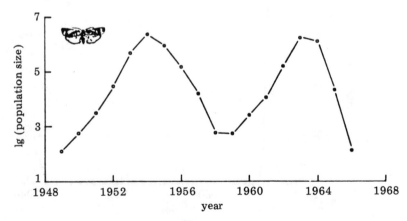

Figure 4.2. Regular oscillations in the population density (plotted on a logarithmic scale) of the larch bud-moth, *Zeiraphera diniana*, in the Upper Engadine, Switzerland, from 1949 to 1966. (Reproduced from May, 1986, by permission of The Royal Society; data from Baltensweiler, 1971.)

equation (4.1) is replaced by

$$dN(t)/dt = rN(t)[1 - N(t - t_D)/K].$$ (4.3)

This prototype model was introduced by Hutchinson (1948) to describe the biocoenosis of lakes, though it appeared first in economic studies of the stability of business cycles. May (1974b) suggests thinking of this system as herbivores grazing upon vegetation which takes a time t_D to recover, although apart from a few microbiological systems equation (4.3) is not generally biologically meaningful. However, not only is it fairly simple to solve, but also the solution is easy to interpret. This basic model therefore plays a useful role in providing general insight into the behavioural characteristics of time-delay processes.

(ii) *Delayed birth rates*
Far more plausible is a model in which developmental delays affect the birth rate, but not the death rate. We may therefore replace the general birth–death equation

$$dN(t)/dt = B[N(t)] - D[N(t)]$$ (3.2)

by

$$dN(t)/dt = B[N(t - t_D)] - D[N(t)].$$ (4.4)

Models (4.3) and (4.4) are both special cases of the equation

$$dN(t)/dt = H[N(t), N(t - t_D)]$$ (4.5)

for some general function H, and this form clearly generates a whole host of variations. Moreover, there is no reason why just one time-lag (t_D) should be used. For example, the logistic type equation

$$dN(t)/dt = rN(t - t_G)[1 - N(t - t_D)/K]$$ (4.6)

also allows for a reproductive time-lag (t_G) which may be measured by the gestation time or its equivalent; in the early stages of population growth this reproductive time-lag may be important in slowing down the rate of population increase.

In many situations the birth rate may depend not just on one particular time t_D, but on a weighted average of previous times s, whence (4.4) becomes

$$dN(t)/dt = r \int_0^\infty z(s)N(t - s)\, ds - D[N(t)]$$ (4.7)

(Nisbet and Gurney, 1982). The weighting function $z(s)$ is normalized to ensure that

$$\int_0^\infty z(u)\,du = 1.$$

The field of all possible time-lag models is obviously vast, and all we can do here is to introduce basic principles and ideas. For a more detailed mathematical study of the effect of time-delays on the stability of equilibria and the nature of oscillations of species density, in both single- and multi-species situations, see Cushing (1977).

4.2   Reaction time-lag – deterministic analysis

First observe that equation (4.3) has a steady state population size K. Then linearizing this equation about K, by writing

$$N(t) = K[1 + n(t)] \tag{4.8}$$

and ignoring terms in n^2, reduces (4.3) to the locally linear form

$$dn(t)/dt = -rn(t - t_D). \tag{4.9}$$

*4.2.1   *Stability conditions*

To investigate the stability of this local equation let us assume that it has a solution of the form

$$n(t) = \exp\{-ct\}\exp\{iwt\} \qquad \text{(here } i = \sqrt{(-1))}. \tag{4.10}$$

Since there are an infinite number of such solutions the following argument is not mathematically rigorous. However, the conclusions are correct! On using the standard result that

$$\exp\{iwt\} = \cos(wt) + i\sin(wt),$$

substituting the proposed solution (4.10) into (4.9) leads to

$$(c - iw)[\cos(wt) + i\sin(wt)] = r\exp\{ct_D\}[\cos\{w(t - t_D)\} + i\sin\{w(t - t_D)\}].$$

Hence on equating real and imaginary parts we have two simultaneous equations for the unknown c and w, namely

$$c\cos(wt) + w\sin(wt) = r\exp\{ct_D\}\cos[w(t - t_D)]$$
$$-w\cos(wt) + c\sin(wt) = r\exp\{ct_D\}\sin[w(t - t_D)].$$

Expanding $\cos[w(t - t_D)]$ and $\sin[w(t - t_D)]$ then shows that these equations are compatible if and only if

$$c = r\exp\{ct_D\}\cos(wt_D) \tag{4.11}$$

and

$$w = r \exp\{ct_D\} \sin(wt_D).\tag{4.12}$$

For the system to be stable and overdamped (i.e. exponential damping with no oscillations) we see from (4.10) that we require $c > 0$ and $w = 0$. Now (4.11) with $w = 0$ gives

$$\log_e(c/r) = ct_D = (c/r)(rt_D).$$

Hence on writing $x = c/r$,

$$rt_D = [\log_e(x)]/x,\tag{4.13}$$

and for a solution of this equation to exist we must clearly have

$$rt_D \leqslant \max\{[\log_e(x)]/x\}.$$

To find this maximum value put $y = [\log_e(x)]/x$. Then $dy/dx = 0$ at $\log_e(x) = 1$, i.e. at $x = e$ which corresponds to a maximum y-value of e^{-1}. Thus for the system to be stable and overdamped we need

$$rt_D \leqslant e^{-1} = 0.3679.\tag{4.14}$$

Similarly, for the system to be stable and underdamped (i.e. exponential damping with oscillations) we require $c > 0$ and $w \neq 0$. Now transition from stability ($c > 0$) to instability ($c < 0$) occurs if equation (4.11) has a solution with $c = 0$. This implies that $\cos(wt_D) = 0$, i.e. $wt_D = (2n + 1)\pi/2$ for some integer n. Equation (4.12) then becomes

$$w = r \sin[(2n + 1)\pi/2] = r(-1)^n.$$

Taking $n = 0$, so that $w = +r$, we see that the point of instability is at $rt_D = \pi/2$. Thus for the system to be stable and underdamped we need

$$e^{-1} < rt_D < \pi/2 = 1.5708,\tag{4.15}$$

whilst for instability we need

$$rt_D > \pi/2.\tag{4.16}$$

When rt_D is greater than $\pi/2$ the steady state K is unstable, and there is a limit cycle solution in which the oscillations become increasingly severe as rt_D increases. If we assume that the period of oscillation does not change substantially as rt_D increases, then we may estimate it by taking rt_D to be just slightly greater than $\pi/2$. Thus $w \simeq r$, and so

$$\text{period} = 2\pi/w \simeq 2\pi/r \simeq 4t_D.$$

Table 4.1. Limit cycle periods and ratios of maximum to minimum population values for the time-delayed logistic equation (4.3) (from May, 1974b)

rt_D	$N(\max)/N(\min)$	(Cycle period)/t_D
1.57	1.00	—
1.6	2.56	4.03
1.7	5.76	4.09
1.8	11.6	4.18
1.9	22.2	4.29
2.0	42.3	4.40
2.1	84.1	4.54
2.2	178	4.71
2.3	408	4.90
2.4	1040	5.11
2.5	2930	5.36

4.2.2   *Numerical solutions*

Though non-linear equations such as (4.3) are receiving extensive coverage in the mathematical literature, we shall not pursue mathematical analysis here. Instead we shall rely on numerical solutions since they are usually easy to obtain and provide at least a partial insight into behavioural characteristics. For example, May (1974b) computed both the limit cycle period and the ratio of maximum to minimum population values from (4.3). These values are shown in Table 4.1, and the limit cycle period clearly lies between $4.0t_D$ and $5.4t_D$ even for values of rt_D which give rise to large amplitude limit cycles.

Unless very accurate results are required, such first-order differential equations are simple to integrate numerically by replacing t by ih and $dN(t)/dt$ by $\{N[(i + 1)h] - N(ih)\}/h$ where $i = 0, 1, 2, \ldots$ and h is small (e.g. 0.01). The logistic time-delay equation (4.3) then becomes

$$N[(i + 1)h] = N[ih] + rhN[ih]\{1 - N[(i - k)h]/K\} \qquad (4.17)$$

where the time-delay t_D is written as kh, i.e. $k = t_D/h$. The only slight problem is in deciding how to start the process, since the right-hand side of (4.17) depends not only on i but also on $i - k$. The easiest way is to hold the population size constant for a time t_D, and then release it. Equation (4.17) can then be computed over $i = k + 1, k + 2, \ldots$.

Figure 4.3 shows such a numerical solution to the time-delay logistic equation

$$dN(t)/dt = N(t)[1 - 0.02N(t - 2)]. \qquad (4.18)$$

Here $r = 1$, $t_D = 2$ and there is an unstable equilibrium point at $K = 50$. The process was held at $N(t) = 49$ for $0 \leqslant t \leqslant 2$, and then released. The oscillations grow quickly in magnitude, effectively reaching the limit cycle stage by the ninth cycle, after which $N(t)$ oscillates between 3.4 and 144.8. The 14 full limit cycles shown took 123.9 time units to complete, giving a period $(T_{\lim})$ of 8.85 time units in agreement with May's more refined value of $4.40t_D$ $= 8.80$ given in Table 4.1.

Taking $T_{\lim} = 8.80 = 2\pi/w_{\lim}$ gives the limit cycle frequency $w_{\lim} = 0.7140$. This is rather less than the approximate value $w = 0.8368$ obtained by solving the local equations (4.11) and (4.12) numerically, though the values are not too dissimilar. The corresponding (local) value for c is -0.0864, which combines with the (local) period $T = 2\pi/w = 7.51$ to give an amplitude increase of $\exp\{Tc\} = 1.91$ between successive cycles. This value is in excellent agreement with the opening stages of Figure 4.3, for which the first six cycles have successive amplitude ratios in the range 1.93–1.98. So yet again the local linear approximation has successfully predicted the main features of the process.

Figure 4.3. Approach to the limit cycle for the deterministic time-delayed logistic equation $dN(t)/dt = N(t)[1 - 0.02N(t - 2)]$.

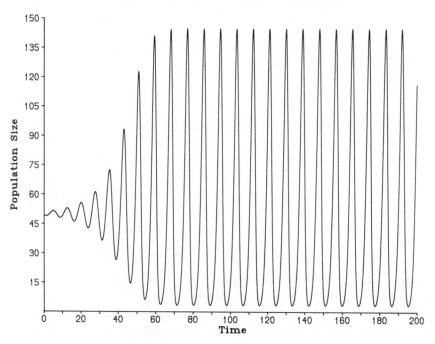

4.3   Reaction time-lag – stochastic analysis
     Every time the deterministic limit cycle in the above example drops down near to $N(\text{min}) = 3.4$, the corresponding stochastic process $\{X(t)\}$ may wander onto the $N = 0$ axis (extinction). Let us therefore simulate $\{X(t)\}$ to discover the difference between stochastic and deterministic behaviour.
     Over a small time interval $(t, t + h)$ the transition probabilities are given by

$$\Pr\{X(t + h) = X(t) + 1\} = rX(t)h = B[X(t)]h$$
$$\Pr\{X(t + h) = X(t) - 1\} = (r/K)X(t)X(t - t_D)h = D[X(t)]h,$$
$$(4.19)$$

and so the process may be simulated by inserting these rates B and D into the general simulation program GENONE (Appendix to Chapter 3). One way of incorporating the time-delay t_D is to store $\{X(t)\}$ in an array $\{A(j)\}$, where $j = 0, 1, 2, \ldots$ corresponds to (say) the discrete time points $t = 0, 0.01, 0.02, \ldots$. Then

$$D[X(t)] \simeq (r/K)X(t)A(j_D)$$

where j_D is the integer part of $100(t - t_D)$. Alternatively, we can avoid this minor complication by considering a different simulation approach.
     Suppose we examine each of the small time intervals $0 \leqslant t \leqslant h$, $h < t \leqslant 2h, \ldots$ in turn. Then we see from (4.19) that for a uniform random number $0 \leqslant Y \leqslant 1$, and $t = ih$,

(i) $X[(i + 1)h] = X(ih) + 1$:   if $0 \leqslant Y \leqslant B[X(ih)]h$
(ii) $X[(i + 1)h] = X(ih) - 1$:   if $B[X(ih)]h < Y \leqslant B[X(ih)]h$
                                                                    $+ D[X(ih)]h$
(iii) $X[(i + 1)h] = X(ih)$:   otherwise. (4.20)

These three conditional statements are easily transformed into a simulation program cycling over $i = 0, 1, 2, \ldots$. The only computational problem is to find a suitable value for h. This value must be small enough to ensure that multiple events in any of the time intervals $(t, t + h)$ have negligible probability of occurrence; yet not so small that it renders the method totally inefficient in terms of computing time.
     Applying this technique to model (4.18), for which

$$B[X(t)] = X(t) \quad \text{and} \quad D[X(t)] = X(t)X(t - 2)/50,$$

produces realizations which quickly give rise to extinction. Not only do the troughs of the stochastic cycles run perilously close to the $X = 0$ axis, but since $B[X(t)] < D[X(t)]$ for $X(t - 2) > 50$ a large population peak automatically results in extinction in the following trough. Thus $t_D = 2$ is too large for stochastic cycles to be sustained. Figure 4.4 shows a realization with the smaller value $t_D = 1.8$ starting from $X(t) = 50$ ($0 \leqslant t \leqslant 1.8$). The time increment h is taken to be 0.001, with $X(t)$ being plotted at times $t = 0, 0.1, 0.2, \ldots, 100.0$. Even though this approach is clearly inefficient at very low values of $B + D$, where most of the simulated events will be 'no change', the computation still took only 8.8 seconds of c.p.u. time on an ICL 2988 machine.

After a short initial period of random wandering, the stochastic cycles (Figure 4.4) soon build up in a manner similar to the deterministic cycles (Figure 4.3). The major difference between them is the widely varying amplitude of the stochastic cycles, with both small ($X(27.3) = 66$) and large ($X(79.4) = 139$) peaks occurring. We see from Table 4.1 that the corresponding deterministic limit cycle period is $4.18t_D = 7.52$, which is close to the average period of 7.96 for the last ten stochastic cycles. Note the variation in the stochastic period, being longer between two large peaks than between two small ones. Even with this reduced value of $t_D = 1.8$, $X(t)$ equals 1 from $t = 82.4$ to 84.5 and so extinction is still a real threat. Thus whilst deterministic limit cycles exist for all values of $t_D > \pi/2 = 1.57$ (since $r = 1$),

Figure 4.4. Stochastic development of the time-delayed logistic model $dN(t)/dt = N(t)[1 - 0.02N(t - 1.8)]$.

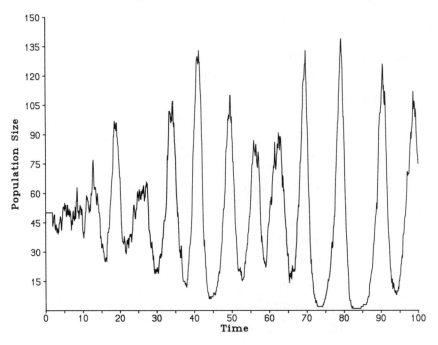

the value $t_D = 1.8$ here represents an upper limit for obtaining sustained stochastic cycles.

4.4   More general deterministic models

In the second type of process considered in Section 4.1, developmental delay is assumed to affect the birth rate but not the death rate. For example, the logistic model corresponding to the general process (4.4) has the form

$$dN(t)/dt = rN(t - t_D) - (r/K)[N(t)]^2. \tag{4.21}$$

Linearizing this equation about the steady state K, with $N(t) = K[1 + n(t)]$, then leads to

$$dn(t)/dt = -2rn(t) + rn(t - t_D) \tag{4.22}$$

which is slightly different from (4.9).

A similar result is obtained even when we work with the most general single time-delay model (4.5), namely

$$dN(t)/dt = H(N, N_D) \tag{4.23}$$

where $N_D(t) = N(t - t_D)$. Let N^* denote the steady state (assuming one exists). Then on expanding the function $H(N, N_D)$ about N^* we have

$$\frac{dN}{dt} \simeq H(N^*, N^*) + (N - N^*)\left(\frac{\partial H}{\partial N}\right)_{N = N_D = N^*}$$
$$+ (N_D - N^*)\left(\frac{\partial H}{\partial N_D}\right)_{N = N_D = N^*}.$$

Write $N(t) = N^*[1 + n(t)]$ and $N_D(t) = N^*[1 + n_D(t - t_D)]$. Then since $H(N^*, N^*) = 0$, this equation reduces to the general form

$$dn/dt = -an - bn_D \tag{4.24}$$

where

$$a = -\left(\frac{\partial H}{\partial N}\right)_{N = N_D = N^*} \quad \text{and} \quad b = -\left(\frac{\partial H}{\partial N_D}\right)_{N = N_D = N^*}. \tag{4.25}$$

Thus with the above logistic model $a = 2r$ and $b = -r$.

Conditions for stability/instability and underdamping/overdamping of solutions to the linearized equation (4.24) may be developed by the same method as used in Section 4.2.1. The only effective difference is that equations (4.11) and (4.12) are replaced by

$$c = a + b \exp\{ct_D\} \cos(wt_D) \tag{4.26}$$

$$w = b \exp\{ct_D\}\sin(wt_D), \tag{4.27}$$

i.e. there is an additional constant (a).

The nature of the solution of these equations is determined by the values of the dimensionless quantities at_D and bt_D; regions corresponding to the four possible types of behaviour for values of at_D and bt_D in the range -4 to 4 are illustrated in Figure 2.8 of Nisbet and Gurney (1982). In summary, two main points emerge.

(1) If a and b are both positive, then a time-delay has a destabilizing effect similar to that seen in Section 4.2.2. Increasing t_D gives rise first to damped oscillations, and then (if b/a is large enough) to instability and divergent oscillations.

(2) If in the absence of a time-delay the steady state would be unstable (i.e. $c < 0$), then adding a time-delay will not change this situation. However, as t_D increases c moves towards zero, thus slowing down the rate of divergence from the steady state and so reducing the strength of the instability (for details see Beddington and May, 1975).

*4.4.1 *Distributed time-delay*

The above model involves idealizing the time-delay as a fixed single value t_D. Now we have already seen through result (4.7) that the birth rate may involve a weighted average of population size taken over all previous times (s). This is called a distributed delay, and has the general form

$$dN/dt = F(N, N_Z) \tag{4.28}$$

where F is any appropriate function. The weighted average

$$N_Z(t) = \int_0^\infty z(s)N(t - s)\, ds \quad \text{where} \quad \int_0^\infty z(u)\, du = 1. \tag{4.29}$$

Linearizing (4.28) about the assumed steady state N^*, in exactly the same way as was done for equation (4.23), yields

$$dn/dt = -an - bn_Z \tag{4.30}$$

where $N_Z(t) = N^*[1 + n_Z(t)]$,

$$a = -\left(\frac{\partial F}{\partial N}\right)_{N = N_Z = N^*} \quad \text{and} \quad b = -\left(\frac{\partial F}{\partial N_Z}\right)_{N = N_Z = N^*}. \tag{4.31}$$

Note that the steady state N^* (assuming there is one) is given by the solution of the equation

$$F(N^*, N^*) = 0. \tag{4.32}$$

Although equations (4.24), (4.25) and (4.30), (4.31) have the same structure, a solution of the form

$$n(t) = \exp\{-ct + iwt\}$$

now involves the far more complicated simultaneous equations

$$c = a + b \int_0^\infty z(u)\, e^{cu} \cos(wu)\, du \qquad (4.33)$$

$$w = b \int_0^\infty z(u)\, e^{cu} \sin(wu)\, du. \qquad (4.34)$$

Choice of the weighting function $\{z(u)\}$ is dictated both by the nature of the delay and the feasibility of solving this pair of equations. Since there are no known experimental results which lead to an automatic choice for $\{z(u)\}$, the sensible approach is to select functions which not only have appropriate shapes but which are also sufficiently simple to enable the integrals in (4.33) and (4.34) to be evaluated directly. A similar situation arises in the study of time-series analysis, in which $z(u)$ is regarded as a 'lag-window' and gives rise to the subject of 'window-carpentry' (see, for example, Chatfield, 1980). Figure 4.5 illustrates four basic shapes for $z(u)$.

Figure 4.5. Four examples of distributed delay functions $z(u)$, each having mean value $t_D = 1$: (a) $z(u)$ is a spike at $u = t_D$, (b) $z(u) = (1/t_D) \exp(-u/t_D)$, (c) $z(u) = (4u/t_D^2) \exp(-2u/t_D)$ and (d) $z(u) = (\pi/4t_D) \sin(\pi u/2t_D)$.

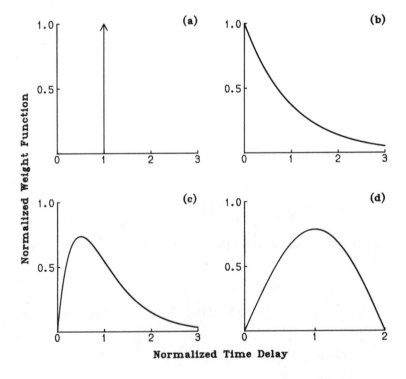

(a) An infinitely large spike at t_D corresponds to the fixed time-delay situation, and has $z(u) \neq 0$ only when $u = t_D$.

(b) A time-delay effect which decreases at a constant rate corresponds to a weighting function which decays exponentially.

(c) Multiplying this exponential by u ensures that $z(0) = 0$ (i.e. the current time t plays no role in the birth rate), whence $z(u)$ rises swiftly to a maximum and then gradually decays to zero.

(d) A half-sine wave is zero at both $u = 0$ and $2t_D$, and provides an example of a symmetric function operating over the finite range $0 \leqslant u \leqslant 2t_D$.

Although equations (4.33) and (4.34) look rather intimidating, simple weights $\{z(u)\}$ like the ones above lead to fairly easy mathematics. For example, in the exponential case (b), namely

$$z(u) = (1/t_D) \exp(-u/t_D), \tag{4.35}$$

we may combine the equations to form

$$c + iw = a + (b/t_D) \int_0^\infty e^{-u/t_D} e^{cu} e^{iwu} \, du \tag{4.36}$$

since $\exp(iwu) = \cos(wu) + i \sin(wu)$. This integral exists provided $c < 1/t_D$, when it integrates to give

$$c + iw = a - b/[(c + iw)t_D - 1]. \tag{4.37}$$

Now this quadratic equation in $(c + iw)$ has the solution

$$c + iw = (1/2t_D)[(1 + at_D) \pm \sqrt{\{(1 + at_D)^2 - 4(a + b)t_D\}}]. \tag{4.38}$$

So we get oscillations $(w \neq 0)$ if and only if (4.38) has an imaginary component, i.e.

$$(1 - at_D)^2 < 4bt_D.$$

For example, we see from (4.30) that if $a = 0$ then the rate of change dn/dt is affected only by the weighted average $n_Z(t)$ (i.e. not by $n(t)$ directly). In this simpler situation (4.38) becomes

$$c + iw = (1/2t_D)[1 \pm \sqrt{\{1 - 4bt_D\}}],$$

so if $b > 0$ then $c > 0$ and the system is stable. If, in addition, $4bt_D < 1$ then $c + iw$ is real so $w = 0$ and the system is overdamped; whilst if $4bt_D > 1$ then $w \neq 0$ and the system is underdamped. Thus although an exponentially distributed time-delay with $a = 0$ can cause damped oscillations about the steady state value, it cannot completely destabilize the process. However, it follows from (4.38) that this restriction does not hold in general. If $a + b < 0$

then $c + iw$ can be real and negative, that is $w = 0$ and $c < 0$, and so $n(t)$ grows exponentially fast.

As a second illustration, the logistic model (4.21) has $a = 2r$ and $b = -r$, whence (4.38) becomes

$$c + iw = (1/2t_D)[(1 + 2rt_D) \pm \sqrt{\{1 + 4r^2t_D^2\}}].$$

Since $c > 0$ and $w = 0$ for all t_D, we have an overdamped stable system regardless of the value of t_D. This is in direct contrast to the fixed reaction time model (Section 4.2.1) for which increasing t_D affects both stability and the type of damping.

MacDonald (1978) provides a comprehensive account of such properties for a wide range of time-lag models drawn from single-species dynamics (logistic growth, the chemostat), interacting pairs of species (predation, mutualism), cell population dynamics (haemopoiesis), and biochemical kinetics (the Goodwin oscillator).

4.5   Periodic and chaotic solutions

So far we have studied processes whose deterministic properties have been described in terms of solutions of differential equations. For example, the logistic equation

$$dN(t)/dt = rN(t)[1 - N(t)/K] \tag{4.1}$$

may be solved either theoretically (solution (3.22)), or numerically by successively evaluating

$$N(t + h) = N(t) + hrN(t)[1 - N(t)/K] \tag{4.39}$$

over $t = 0, h, 2h, \ldots$ for appropriately small h. The question now arises of what happens if h is not small, since in some biological situations (e.g. 13-year periodical cicadas) population growth takes place at discrete intervals of time.

4.5.1   *Three simple deterministic models*

When $h = 1$, equation (4.39) becomes

(A) $$N_{t+1} = N_t[(1 + r) - (r/K)N_t] \qquad (t = 0, 1, 2, \ldots), \tag{4.40}$$

and it by no means automatically follows that (4.39) and (4.40) will have similar properties. Note that (4.40) simplifies, for on writing

$$M_t = [r/(1 + r)K]N_t$$

Table 4.2. *Dynamic properties of models A and B in terms of the growth rate* r

Dynamical behaviour	Model A	Model B
Stable equilibrium point	$0 < r < 2$	$0 < r < 2$
Stable 2-point cycle	$2 < r < 2.5$	$2 < r < 2.526$
Stable 4-point cycle	$2.5 < r < 2.55$	$2.526 < r < 2.656$
Stable cycles, period 8, then 16, 32, etc.	$2.55 < r < 2.57$	$2.656 < r < 2.692$
Chaos	$2.57 < r$	$2.692 < r$

it transforms into

(A') $\qquad M_{t+1} = aM_t(1 - M_t)$ where $a = 1 + r.$ $\hfill$ (4.41)

The study of such non-linear difference equations having the general form

$$N_{t+1} = f(N_t) \tag{4.42}$$

forms a major branch of mathematics, and here we can do no more than scratch the surface. Some of the simplest non-linear difference equations exhibit a remarkable range of dynamic behaviour; from *stable* equilibrium points, to *stable cyclic* oscillations between several points, through to a completely *chaotic* regime in which cycles of any period, or even totally aperiodic fluctuations, can occur. Only after the appearance of May's (1974c, 1975) papers did biologists really begin to take note of this rich dynamic structure.

May (1975) considers the simple non-linear equation

(B) $\qquad N_{t+1} = N_t \exp\{r(1 - N_t/K)\},$ $\hfill$ (4.43)

which may be thought of as being a difference equation analogue of the logistic differential equation (4.1), with r and K being the usual growth rate and carrying capacity, respectively. This model had already been proposed as a rough description for some fish populations by Ricker (1954). Model (A) is sometimes thought to be less satisfactory than (B) since the population becomes negative if N_t exceeds $K(1 + r)/r$. However, the stability characteristics of both models, set out in Table 4.2, are strikingly similar.

Figure 4.6 illustrates *some* of the possible changes in the dynamic behaviour of $\{N_t\}$ under model A, written in the normalized form

$$(N_{t+1}/K) = (N_t/K)[(1 + r) - r(N_t/K)],$$

for $r = 1.90, 2.40, 2.55, 2.90$ and 3.00 with a starting value of $N_1/K = 0.01$. Since K is just a normalizing constant, it may be taken as being 1.

(a) $r = 1.90$: N_t quickly rises to the vicinity of the stable point 1, and then undergoes damped oscillations about it.

(b) $r = 2.40$: After a swift initial rise, N_t rapidly settles down to a 2-point cycle.

(c) $r = 2.55$: Now the cycle is of length $2^3 = 8$; note that as r increases between 2.55 and 2.57 the cycles have increasing period 2^n.

(d) $r = 2.90$: For r greater than 2.57 these cycles give way to chaotic behaviour in the sense that trajectories look like the sample functions of a random process. In this example, if $N_t > 1$ then $N_{t+2} > 1$ or $N_{t+3} > 1$, giving rise to pseudo-cycles of period 2 or 3.

(e) $r = 3.00$: N_t can now drop very close to zero, from which position it can take some time to recover. Thus the pseudo-cycles are longer than in (d). Slightly higher values of r can cause N_t to diverge quickly to minus infinity; the biological implication is that population extinction occurs as soon as N_t becomes negative.

This scenario of order into chaos is a direct consequence of non-linearity, yet it took more than one hundred years following Verhulst's discovery of the

Figure 4.6. Solutions of the difference equation $N_{t+1} = N_t + rN_t(1 - N_t)$ for: (a) $r = 1.90$, stable equilibrium point; (b) $r = 2.40$, stable 2-point cycle; (c) $r = 2.55$, stable 8-point cycle; (d) $r = 2.90$ and (e) $r = 3.00$, chaos.

(a)

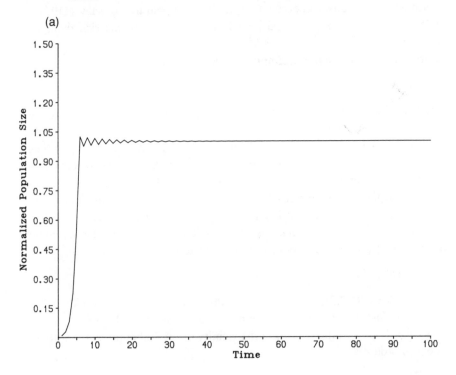

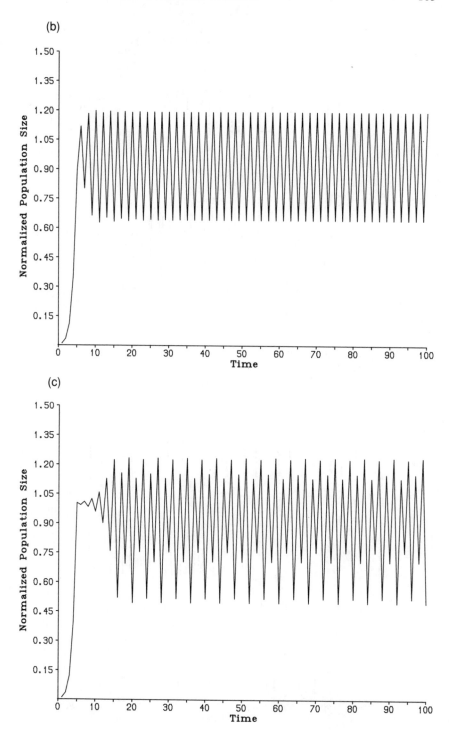

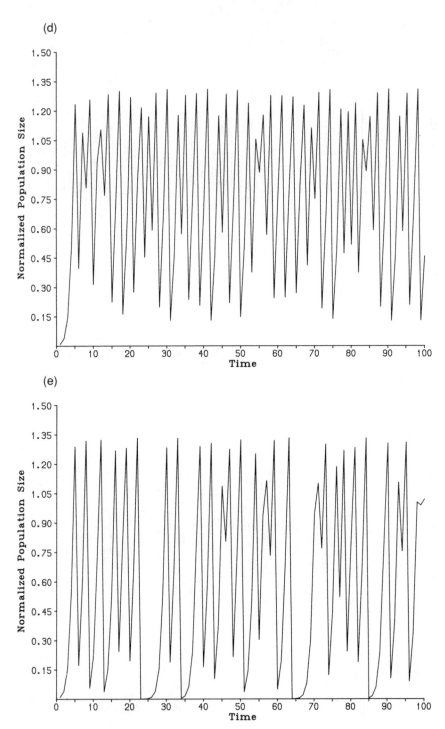

non-linear logistic process before the ensuing complications were explained. As we have just seen, for small growth rates nothing spectacular happens, it is when r increases beyond 2.57 that we are in for surprises. In simple terms, chaos means that the system has gone out of control. There is no way to predict its long-term behaviour even though the process is completely determined by its initial value.

The most exciting aspect of this work lies not so much in its application to population dynamics, but in the scenario by which order turns into chaos. Recent developments in this area have led to the staggeringly beautiful art form of fractals; the excellent description by Peitgen and Richter (1986) is worth acquiring for the pictures alone!

The phenomenon of a three-part regime consisting of a stable point, followed first by stable cycles of period 2^n and then by chaotic behaviour, is likely to occur in many models involving discrete generation, density-dependent, population growth. May (1975) cites five such models from the entomological literature alone. An interesting one is

(C) $$N_{t+1} = \lambda N_t / [1 + a N_t]^b \tag{4.44}$$

(Hassell, 1975). This provides a two-parameter representation for a wide range of field and laboratory data, and exhibits the following behaviour:

>   small b or λ – globally stable point
>   moderate b and λ – stable cycles
>   large b and λ – chaos.

Bearing in mind the general engineering experience that incorporating long-term delays into what are otherwise stable models can lead to stable limit cycles, the movement from stability to cyclic behaviour is not surprising. What may be considered remarkable is that simple, purely deterministic, models give rise to non-predictable behaviour patterns once r is large enough (e.g. $r > 2.57$ and 2.692 for models A and B, respectively).

4.5.2   *More general deterministic results*

Let us briefly mention some basic mathematical results (May, 1975) for the general model

$$N_{t+1} = N_t f(N_t), \tag{4.45}$$

since this contains as special cases all three models A, B and C above. Equilibrium points N^*, where $N_{t+1} = N_t = N^*$, are clearly the solutions of the equation

$$f(N^*) = 1.$$

A standard perturbation argument in which N_t is replaced by $N^* + u_t$ for small u_t, and μ denotes $-N^*[df/dN]$ evaluated at $N = N^*$, then yields for

$0 < \mu < 1$: straight damping (i.e. overdamping)
$1 < \mu < 2$: oscillatory damping (i.e. underdamping).

To study what happens when $\mu > 2$ May (1975) introduces the trick of working with

$$N_{t+2} = N_{t+1}f(N_{t+1}) = N_t f(N_t) f[N_t f(N_t)] = N_t g(N_t) \qquad \text{(say)}$$

to examine conditions for 2-point cycles; whilst looking at

$$N_{t+4} = N_t h(N_t) \qquad \text{(say)}$$

yields conditions for 4-point cycles, etc. Moreover, use of a general theorem shows that if system (4.45) has a 3-point cycle (i.e. $N_t = N_{t+3}$ but $N_t \neq N_{t+1}$ or N_{t+2}) then there also exist cycles of period n where n is *any* positive integer, and that there are an uncountable number of initial points N_1 from which the system does not eventually settle into any of these cycles. The starting point is clearly fundamental in determining the future behaviour of the system.

In model B the maximum of the function

$$N \exp\{r(1 - N/K)\}$$

occurs at $N = K/r$, so regardless of the initial value N_1 the maximum possible size cannot exceed the value of N_{s+1} following $N_s = K/r$, namely

$$N_+ = (K/r) \exp(r - 1). \qquad (4.46)$$

Excluding extremely small starting values, the smallest N_t occurs just one step beyond $N_{s+1} = N_+$, namely

$$N_- = N_+ \exp\{r(1 - N_+/K)\}. \qquad (4.47)$$

Particular values of N_+ and N_- for $K = 1$ are:

r	2.1	3	4	5	6
N_+	1.43	2.46	5.02	10.92	24.74
N_-	0.58	0.031	5.2×10^{-7}	3.2×10^{-21}	3.5×10^{-61}

Clearly N_- is effectively zero for r much beyond 3.

May (1975) shows that for model A the average value of $\{N_t\}$ over a very long run is equal to K, and that as r becomes large $\{N_t\}$ increasingly comprises a few large population values together with long sequences of very low values. These large fluctuations will on average be spaced roughly $(1/r) \exp(r - 1)$ time units apart, and this increase with r explains the difference in appearance between Figures 4.6d and 4.6e. He claims that the transition, as r increases, into a regime of apparent chaos with cycles of essentially arbitrary period is a result with many ecological implications. In particular, that it could be relevant to temperate insect populations where the natural description is in terms of non-linear, i.e. density-dependent, equations often with relatively large r.

Whilst the *mathematical* analysis of non-linear systems is an extremely important field of study, if it is not possible to distinguish between different types of behaviour in real populations then it is not possible to test predictions based on these mathematical systems and so their *ecological* value will be severely limited (see Poole, 1977). Thus a deterministic model which gives rise to different types of behaviour for varying parameter values will only be ecologically useful if these behavioural differences are still in evidence when stochastic effects are introduced.

4.5.3   *Stochastic results*

Simulation of the stochastic counterpart $\{X_t\}$ to the general model

$$N_{t+1} = N_t f(N_t) \tag{4.45}$$

is in principle easily accomplished by computing

$$X_{t+1} = X_t f(X_t) + Z_{t+1}, \tag{4.48}$$

where $\{Z_t\}$ is a sequence of independent random variables with mean 0 and variance σ_t^2. The only difficulty is to determine σ_t^2, for unlike our earlier stochastic analyses we now have to contend with possible dramatic changes in $\{X_t\}$ right across its admissible range.

Smith and Mead (1980) tackle this problem by treating the change in population size between the discrete times t and $t + 1$ as though it resulted from a continuous-time simple birth–death process $\{Y(s)\}$ with rates λ and μ, respectively, and starting value $Y(0) = X_t$. As X_{t+1} then equals $Y(1)$, we see from results (2.49) and (2.50) that this birth–death approximation implies that

$$\text{mean}[X_{t+1}] = \text{mean}[Y(1)] = X_t \exp(\lambda - \mu) \tag{4.49}$$

and

$$\sigma_{t+1}^2 = \text{Var}[X_{t+1}] = \text{Var}[Y(1)]$$
$$= X_t[(\lambda+\mu)/(\lambda-\mu)]e^{(\lambda-\mu)}(e^{(\lambda-\mu)}-1). \tag{4.50}$$

Under model B, for example, on putting $N_{t+1} = \text{mean}(X_{t+1})$ where $N_t = X_t$, we have from (4.43) that

$$X_t \exp(\lambda - \mu) = X_t \exp\{r(1 - X_t/K)\},$$

whence

$$\lambda = r \quad \text{and} \quad \mu = rX_t/K. \tag{4.51}$$

Thus if we take $\{Z_t\}$ to be Normally distributed, simulation involves cycling over the following successive steps:

   (i) evaluate $N_{t+1} = X_t f(X_t)$;
   (ii) compute a Normal pseudo-random variable Z_{t+1} with mean 0 and variance σ_{t+1}^2 given by (4.50);
   (iii) put $X_{t+1} = N_{t+1} + Z_{t+1}$.

To see how appropriate this scheme is under model B for small X_{t+1}, consider $r = 3$. Then with $K = 1000$ (say) and X_t equal to the deterministic

Figure 4.7. Realization of the stochastic difference equation $X_{t+1} = X_t \exp\{r(1 - X_t/1000)\} + Z_{t+1}$ (model B) for: (a) $r = 2.68$, (deterministic) 8-point cycle; and (b) $r = 3.30$, chaos.

(a)

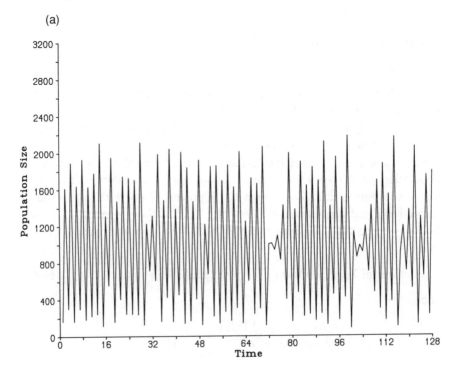

maximum (4.46), i.e. $N_+ = 2463$, we have from (4.47) that $X_{t+1} \simeq N_- = 31$. Hence as (4.50) gives $\sigma_{t+1} = 8.45$, extinction is 3.7 standard deviations away from N_-, far enough for the purpose of simulating long runs. Unlike model A the deterministic part $X_t f(X_t)$ cannot go negative, so extinction results solely from the stochastic term Z_t.

Figure 4.7 shows two realizations of this model for: (a) $r = 2.68$, corresponding to a deterministic 8-point cycle, and (b) $r = 3.30$, corresponding to deterministic chaos (see Table 4.2). A striking feature is that for (a) random variation obliterates the pattern within the four peaks and the four troughs of the corresponding deterministic cycle, namely

$$\begin{array}{cccc} 1544 & 2001 & 1384 & 1916 \\ 165 & 359 & 137 & 495 \quad (165, \text{etc.}). \end{array}$$

All that remains is a saw-tooth structure with a variable amplitude. Thus the deterministic 8-point cycle has been transformed into a stochastic 2-point cycle. Moreover, when $0 < r < 2$ (corresponding to deterministic stable equilibrium) random variation will cause stochastic fluctuations about the value K, so sustaining the oscillations initially present in Figure 4.6a. Thus it may not even be possible to discriminate between stable equilibrium and

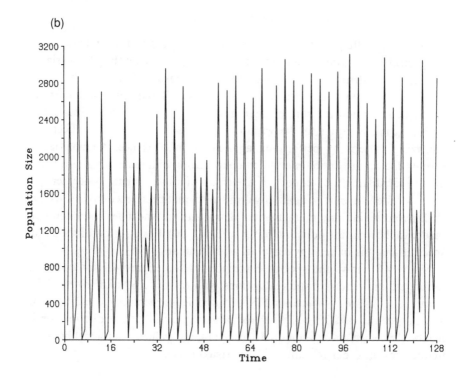

stable cycles, never mind within the class of stable cycles ($2 < r < 2.692$). The effect on the chaotic regime (b) is less serious, with 3-point quasi-cycles occurring both with and without the presence of random variation.

May (1986) points out that the chaotic regime contains an exquisite and delicate fine structure, including infinitely many cascades of bifurcations; he cites a situation in which 93 different intrinsic 11-point cycles each cascade down through their harmonics of stable cycles with periods 11×2^n, yet the entire process occupies an exceedingly small range of the parameters a, r or the like. Whilst this is indeed 'beautiful mathematics', superimposed noise will destroy such fine structure. Thus for practical ecological purposes the chaotic regime can generally be regarded as giving apparently random dynamics.

4.5.4   *Spectral representation*

Precise comparisons between different types of behaviour can be made by using a time-series technique called spectral analysis. Consider the series consisting of the 12 points

$$1 \quad 0 \quad 0 \quad 1 \quad 0 \quad 0 \quad 1 \quad 0 \quad 0 \quad 1 \quad 0 \quad 0.$$

This has 4 complete cycles of length 3, i.e. the series has *frequency* 4 and *period* 3. Clearly

$$\text{frequency} \times \text{period} = \text{length of series } (n),$$

so knowing the frequency is equivalent to knowing the period, and vice versa.

Spectral analysis determines the presence of cycles by splitting the variance of a series into distinct parts, each part being a measure of the contribution of a specific frequency of occurrence to the overall pattern. An important consequence is that the sum of these measures over all possible frequencies $1, 2, \ldots, n/2$ is equal to the variance of the series, so we are effectively analyzing pattern by performing an analysis of variance. An analogy is turning the tuning knob on a radio, identification of a scale of pattern being akin to identifying the frequency/wavelength of a radio signal. Since here we are interested in purely qualitative and not quantitative comparisons, we shall avoid giving any mathematical detail and instead refer the interested reader to the clear introduction by Chatfield (1980). For a review of the use of spectral analysis in ecology see Platt and Denman (1975).

Figure 4.8 shows the spectra corresponding to Figure 4.7 for both the simulated and deterministic realizations with (a) $r = 2.68$ and (b) $r = 3.30$. Since the series are of length $n = 128$, permissible frequencies are $1, 2, \ldots, 64$; and as spectral values usually behave rather erratically (their distribution is highly skew) we have smoothed (by averaging values over neighbouring frequencies) in order to gain a clearer visual effect.

In case (a) (deterministic 8-point cycle) the spectrum shows four peaks centred around frequencies $f = 16, 32, 48$ and 64; analysis of the corresponding unsmoothed spectrum shows four spikes situated exactly at these values with contributions to overall variance of

$$0.02\% \ (f = 16), \ 3.80\% \ (f = 32), \ 0.28\% \ (f = 48), \ 95.90\% \ (f = 64).$$

The corresponding periods (n/f) are 8, 4, 2.67 and 2, which, on disregarding the value 2.67 as a rogue harmonic, points to an 8-point cycle as expected. A particularly interesting feature is that most (95.9%) of the variance is associated not with the whole cycle (period 8), or even the half-cycle (period 4), but with the basic up-down pattern (period 2); indeed the amount of variation associated with period 8 (0.02%) is extremely small. It is therefore not surprising that when stochastic fluctuations are introduced the variation corresponding to $f = 16$ and $f = 32$ is completely drowned out, leaving only the 2-point structure behind.

Figure 4.8. Smoothed spectra ($\times \ 10^{-5}$) for the simulated stochastic (———) and deterministic (– – –) realizations corresponding to Figure 4.7, namely (a) $r = 2.68$ and (b) $r = 3.30$.

(a)

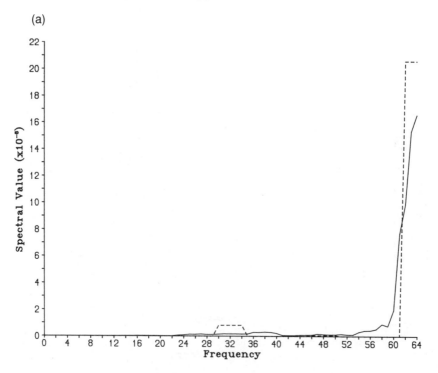

In case (b), i.e. $r = 3.30$, we are now in the chaotic regime in the sense that the deterministic realization *appears* to move randomly about a pseudo-cycle of length 3. Adding true (i.e. stochastic) variation on top of apparent (i.e. chaotic) variation still gives variation, no matter how we define it, and so one might intuitively expect far less qualitative difference between stochastic and deterministic behaviour than exists in case (a). This is confirmed by the close similarity between the deterministic and stochastic spectra (Figure 4.8b), with peaks centred around $f = 43$ (i.e. an average period of $128/43 = 3$) in both cases.

In practice random variation will always be present, so the relevance of using deterministic models which have substantially different behaviour from their stochastic counterparts is highly questionable. Poole (1977), for example, constructs three different deterministic realizations corresponding to stable equilibrium, stable periods, and chaos, and shows that the corresponding stochastic realizations are effectively indistinguishable!

If it is not possible to distinguish between different types of behaviour in simulated situations (where we actually know the precise form of the underlying model), in real-life situations discrimination becomes even more hopeless. The value of simple deterministic representations of biological

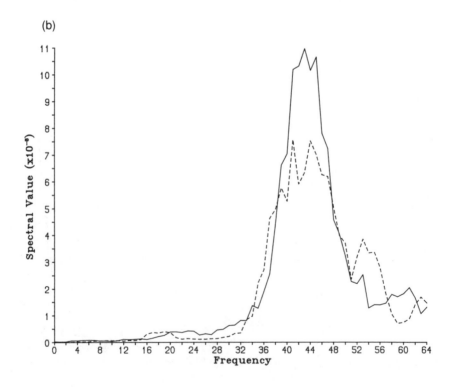

(b)

development can therefore be severely limited. Thus not only should deterministic and stochastic behaviour patterns be examined together, but if their qualitative behaviour differs substantially then far more importance should be placed on the stochastic model than has previously been acknowledged.

4.5.5   *Final comments*

The widespread prevalence of deterministic models is due mainly to their mathematical convenience; they are often 'justified' by the implicit assumption that in large populations stochastic deviations are small enough to be ignored. Unfortunately there is no simple rule which says how large a population must be before deterministic approximations become reasonable; the only effective test is to conduct a simulation exercise. Smith and Mead (1980) go even further by claiming that: 'Many of the details of the dynamics of deterministic models are artefacts of the deterministic approximation and have little or no importance in ecology'. Moreover, we have already seen examples in which stochastic behaviour can be adequately described in terms of fluctuations around a point equilibrium, 2-point and 3-point cycles. Thus a stochastic system may actually represent a considerable *reduction* in behavioural complexity from its deterministic counterpart, and thereby provide a far more realistic benchmark.

A totally opposing point of view is made by modellers who place importance in the fact that apparently chaotic behaviour can arise in a wholly deterministic system (e.g. Beddington, Free and Lawton, 1975). They not only argue that the existence of high-period cycles or chaotic behaviour may be of considerable benefit in interpreting the patterns of fluctuation shown by many arthropod populations in the field, but they also suggest that the description of such behaviour as being due to environmental fluctuation is a temptation to be avoided.

Yet further complications arise when attempts are made to fit a model to specific data sets, since, as Smith and Mead (1980) point out, the choice of model or fitting technique can lead to quite different interpretations. For example, Hassell, Lawton and May (1976) analyzed data of Fujii (1967) on a laboratory population of the cowpea weevil *Callosobruchus maculatus* (F.) and found that their parameter estimates put the population on the boundary between damped oscillations and stable limit cycles when one model was used, but with another model the population seemed to be well within the region of chaos. If one takes the view that simple mathematical models can reflect only basic qualitative phenomena of complicated biological structures, such as chaos or limit cycles, then use of quantitative fitting techniques in non-linear systems may well be over-ambitious. Nevertheless, parameter estimates (such as for r) are still likely to contain valuable ecological information.

To conclude, we note that since most biological regulatory processes involve a time-delay, the simple models we have considered above provide a very useful means of understanding different types of behaviour. Indeed, without such mathematical insight it is unlikely that biologists would have become quite so aware that certain insect populations may exhibit chaotic behaviour if density-dependent effects are very severe.

4.6   Analysis of field and laboratory data

Let us now see how our theoretical analyses relate to real-world situations. Hassell, Lawton and May (1976) provide a substantial survey of a variety of data sets from field and laboratory populations which have approximately discrete generations, and fit Hassell's simple three-parameter model

(C) $$N_{t+1} = \lambda N_t (1 + aN_t)^{-b} \tag{4.44}$$

to them. The estimated parameters are then used to compare the kinds of dynamic behaviour that are found in real populations with those known to be theoretically possible.

Note that although laboratory studies have the advantage that they can provide good approximations to single-species development, they are not subject to hazards of the outside world which can result in boom growth closely followed by collapse. Conversely, field populations involve dynamic interaction with food supply, competitors, predators, etc., so the single-species assumption can give rise to a misleading impression of dynamic behaviour.

Estimates of b and λ are provided for 24 types of insect, together with the ratio of population maxima to minima in a variety of 56 populations, the theoretical value for this ratio being

$$N_{\max}/N_{\min} = (1/\lambda)[1 + \lambda(b-1)^{b-1}/b^b]^b. \tag{4.52}$$

The type of behaviour depends on both b and λ, but damped oscillations are only possible if $\lambda > 2.7$, stable limit cycles if $\lambda > 7.4$, and chaos if $\lambda > 15$. The minimum values of λ for these different stability regions correspond to infinite b. As b decreases, the minimum λ-values increase until when $b < 3$ chaos is no longer possible for any sensible choice of λ. Similarly, stable limit cycles cannot occur if $b < 2$, and not even damped oscillations if $b < 1$.

Model B is a limiting case of model C. Since on putting $\lambda = \exp(r)$ and

$ab = r/K$ in (4.44) and then letting $a \to 0$ and $b \to \infty$, we obtain

$$N_{t+1} = e^r N_t \lim_{b \to \infty} \{1 + (rN_t/bK)\}^{-b}$$

$$= N_t \exp\{r(1 - N_t/K)\}. \tag{4.43}$$

This simpler two-parameter model gives a poorer fit to population data than the more detailed three-parameter model C, and it is interesting to compare them. The size of the above minimum λ-values for model C will preclude almost all natural populations from the chaos regime, and most from the cycle regime, regardless of the value of b. Indeed, Hassell, Lawton and May's analysis of field and laboratory populations yields the following classification:

	Stable point	Stable cycles	Chaos
Model C	26	1	1
Model B	15	7	6

So five populations that are classified as stable under model C are chaotic under model B.

By focussing on the more general model C we may conclude that some types of dynamic behaviour that are possible in theory rarely occur in real populations. Following a disturbance, most of the 28 populations featured show monotonic damping with only an occasional case of oscillatory damping or a low-order limit cycle; high-order limit cycles and chaos appear to be relatively rare phenomena. The discrepancy in classification between models B and C is disturbing, since it demonstrates that conclusions reached for a particular data set may well be sensitive to model choice. A particularly interesting observation is that the most *complex* behaviour (chaos) is far more often predicted by the *simpler* of the two models!

Experiments which study change in dynamic behaviour are of two distinct types, involving:

(i) different strains of a particular organism; or,
(ii) the same strain in different conditions (e.g. temperature).

An example of the first type is provided by combining the data from Fujii (1967) and Utida (1967) on the southern cowpea weevil (*Callosobruchus maculatus* (F.)) with that of Fujii (1968) on the azuki bean weevil (*C. chinensis* (F.)). Three selected examples of population fluctuation are shown in Figure

4.9, and portray (a) stable overdamping, (b) stable underdamping, and (c) a 2-point limit cycle. Experiments of the second type include studies on: *Daphnia magna* (Pratt, 1943), in which the population appears stable and overdamped at 18 °C and cyclic at 25 °C; and *Folsomia candida* (Beddington, 1974), with stable underdamping at 10 °C and overdamping at 16 °C. The effect of increasing temperature is to increase the parameter r in the first example, but to decrease it in the second.

4.7   Nicholson's blowflies

Data sets commonly used to demonstrate the existence of large, self-sustaining, quasi-cyclic fluctuations when a laboratory population receives a regular food supply are those obtained by Nicholson (1954, 1957) on the Australian sheep blowfly *Lucilia cuprina*. More recently, Taylor and Sokal (1976) observed similar large amplitude oscillations in cultures of the housefly *Musca domestica*. In spite of their popularity as examples of sustained cyclic behaviour, such data sets have received little attention in the theoretical literature. May (1974b) and Maynard Smith (1974) suggest two

Figure 4.9. Population dynamics of three laboratory cultures of *Callosobruchus*: (a) *C. chinensis* (Fujii, 1968) (○), (b) *C. maculatus* (Utida, 1967) (△) and (c) *C. maculatus* (Fujii, 1967) (∗) (data points taken from Hassell, Lawton and May, 1976).

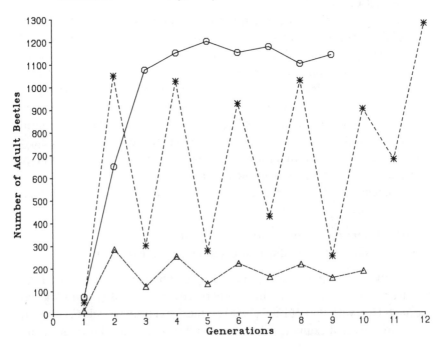

possible models, but it is left to Nisbet and Gurney (1982) to provide a detailed mathematical study.

4.7.1 *Description of the data*

We shall examine just one set of data (Nicholson, 1954, Figure 3), for a population maintained under as nearly constant conditions as possible. A blowfly culture was held at 25° C, and the governing requisite was ground liver (which was available to the adults alone); each day 0.5 g was placed in the breeding cage as food supply. The outstanding characteristic of this culture is the maintenance of violent and fairly regular oscillations in the density of the adult population (see Figure 4.10).

Nicholson observed that significant egg generation occurred only when the adult population was very low; at higher densities competition for food amongst the adults was so severe that few, if any, blowflies ate enough to enable them to develop eggs. Natural mortality therefore caused the population to decay until it was sufficiently small for some blowflies to eat enough liver to lay eggs. Since it takes more than two weeks for eggs to

Figure 4.10. A population of *Lucilia cuprina* (———) governed by a constant daily food supply, and the fitted time-delayed logistic equation (4.3) with $rt_D = 2.1$ (---) (data are reproduced from Figure 3 of Nicholson, 1954).

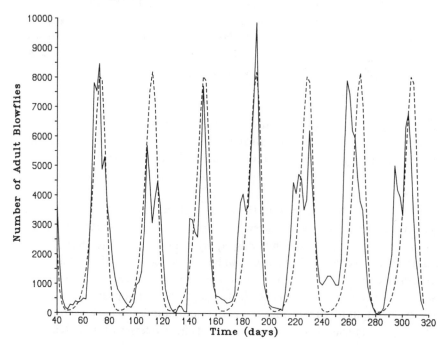

develop into adults the population decayed still further during this period, so enabling even more eggs to be generated. The emergence of new adults then led to a rapid rise in the adult population, and thereby to the ending of egg production and the beginning of a new cycle.

4.7.2   *A simple time-delay model*

If acquisition of food caused the immediate birth of fully mature adults instead of merely producing eggs which develop into adults later on, then the system would be non-oscillatory. It would simply wander around its carrying capacity (K) as a stochastic logistic process (Chapter 3). The time-lag (t_D) between the appearance of eggs and adults is therefore fundamental to the observed process. However, whilst the underlying biology may be described fairly easily, translation of these biological ideas into mathematical equations is not so simple. May (1974b) therefore used the basic equation

$$dN(t)/dt = rN(t)[1 - N(t - t_D)/K] \qquad (4.3)$$

as 'an extremely crude first approximation, incorporating the minimum amount of essential biological information about the system'. The justification for using a differential, rather than a difference, equation is that the data are measured continuously (in days), rather than discretely (in generations).

May argues that the single parameter rt_D is completely determined by Nicholson's data because it depends sensitively on the ratio N_{max}/N_{min}. He takes $rt_D = 2.1$, possibly because the largest (i.e. fourth) cycle in Figure 4.10 has $N_{max}/N_{min} = 84$ (see Table 4.1). The associated cycle period is $4.54t_D$. So as six full cycles occur in 232 days, we have

$$6 \times (4.54t_D) = 232$$

giving t_D as 8.5 days (which May took to be roughly 9 days).

Figure 4.10 shows the comparison between Nicholson's data and the limit cycle obtained by solving equation (4.3) numerically. The carrying capacity K is crudely estimated as the mean population size 2500 (approx.), whilst since $t_D \simeq 8.5$ we may estimate r by $2.1/t_D = 0.25$.

May's claim that, 'in view of the crudity of the model, the agreement is surprisingly good', has to be placed in perspective. First, we see that the peaks occur at almost equal time intervals (apart from cycle 6 which has an early start), so *any regular* cyclic process can be positioned to ensure that its peaks generally coincide with the observed ones. Second, since K is effectively a scale parameter, it can always be chosen to provide an appropriate value for the cycle amplitude. Thus once the period and amplitude have been chosen there is not much left to fit! Note that 'fit' here refers to the matching of general qualitative features, and is not used in the statistical (i.e. regression) sense. A least-squares analysis might well provide 'better' estimates of r, K and t_D, but the exercise would be pointless since it would not lead to an increase in biological understanding.

Clearly the simple time-delay equation (4.3) can provide no more than a summary of the data, since we know that it does not mimic the underlying biology. It also fails as a realistic model, for the implied time-delay t_D of 8.5 days is far too small when compared with the true value of approximately 15 days. Moreover, if we assume that $t_D = 15$ then

$$(\text{cycle period})/t_D = (232/6 \times 15) = 2.6,$$

and we see from Table 4.1 that this does not conform to limit cycle behaviour.

4.7.3 *Nisbet and Gurney's time-delay model*

In an attempt to incorporate biological realism, Maynard Smith (1974) developed the linear time-delay differential equation

$$dN(t)/dt = \tfrac{1}{2}ksw - CN(t) - \tfrac{1}{2}mksN(t - t_D). \tag{4.53}$$

Here C denotes the adult death rate, s is the constant probability that an egg will survive to become an adult, and $\tfrac{1}{2}k[w - mN(t - t_D)]$ is argued to be the rate at which eggs are laid by the population. In this last term k is a constant, whilst w and m respectively denote the actual and required rate of food supply. Though he shows that use of the scale transformation $N(t) = N^*[1 + n(t)]$ about the assumed equilibrium point N^* leads to the general local form (4.24), no assessment is made of limit cycle behaviour.

A serious attempt at a full mathematical study was made by Nisbet and Gurney (1982) who developed their own model by first returning to the general birth–death scenario

$$dN(t)/dt = B[N(t)] - D[N(t)]. \tag{3.2}$$

They assume that the rate of recruitment at time t depends only on the size of the population at time $t - t_D$, so that B may be written as $B[N(t - t_D)]$; whilst deaths occur according to a simple death process with rate δ. Then (3.2) becomes

$$dN(t)/dt = B[N(t - t_D)] - \delta N(t). \tag{4.54}$$

On assuming further that the average per capita fecundity drops exponentially with increasing population size, and that each egg has a constant probability of becoming an adult, equation (4.54) takes the final form

$$dN(t)/dt = \gamma N(t - t_D)\exp\{-N(t - t_D)/N_0\} - \delta N(t) \tag{4.55}$$

for some constants γ, δ and N_0.

Analysis of this equation follows a now familiar path. Scaling the variables $N(t)$ and t by denoting

$$N(t) = N_0 M(s) \quad \text{with} \quad s = t/t_D \tag{4.56}$$

leads to

$$dM(s)/ds = \gamma t_D M(s-1) \exp\{-M(s-1)\} - \delta t_D M(s), \tag{4.57}$$

so the qualitative behaviour of the system is defined solely by γt_D and δt_D. On placing $dM/ds = 0$ we see that there is just one equilibrium point, namely

$$M^* = \log_e(\gamma/\delta).$$

This will be positive (and therefore meaningful) provided that $\gamma > \delta$, i.e. provided that the maximum possible per capita reproduction rate is greater than the per capita death rate.

To investigate the stability of this steady state M^*, we write

$$M(s) = M^*[1 + m(s)]$$

in equation (4.57) and then ignore terms in $m^2, m^3, \ldots$. This yields the local equation

$$dm(s)/ds = -\delta t_D \{m(s) + m(s-1)[\log_e(\gamma/\delta) - 1]\}. \tag{4.58}$$

The next step is to assume a solution of the form

$$m(s) = \exp\{-cs + iws\}, \tag{4.10}$$

and then to show that such a form is possible if and only if the damping constant c and frequency w satisfy the equations

$$c = \delta t_D \{1 + [\log_e(\gamma/\delta) - 1] e^c \cos(w)\} \tag{4.59}$$

$$w = \delta t_D \{\log_e(\gamma/\delta) - 1\} e^c \sin(w). \tag{4.60}$$

Note that this pair of simultaneous equations is identical in form to the earlier pair (4.11 and 4.12) apart from an added constant term in (4.59).

Since analytic solutions to (4.59) and (4.60) are not available, numerical solutions must be obtained over an appropriate set of values of γt_D and δt_D. As usual, general qualitative behaviour is categorized by

$$c > 0 \text{ (stable)}, \qquad c < 0 \text{ (unstable)}$$
$$w = 0 \text{ (overdamped)}, \qquad w \neq 0 \text{ (underdamped)};$$

and Nisbet and Gurney provide a diagram (Figure 4.11) which segments these characteristics over the biologically meaningful parameter range of $1 \leqslant \gamma t_D \leqslant 1000$ and $0.01 \leqslant \delta t_D \leqslant 10$. They present two hypotheses as explanations for the quasi-periodic fluctuations observed by Nicholson: a locally stable and underdamped system in which stochasticity gives rise to quasi-cyclic oscillations (H1); and a locally unstable system which gives rise to self-sustaining deterministic limit cycles (H2).

They consider the period to delay ratio (T/t_D) and the maximum to minimum population size ratio (N_{max}/N_{min}) under these two hypotheses, and then make a comparison with the values $T/t_D = 2.6 \pm 0.1$ and $N_{max}/N_{min} = 36 \pm 17$ derived from the data. On this basis they obtain the estimates

H1: $\gamma t_D = 23.5 \pm 4.5$ and $\delta t_D = 3.0 \pm 0.7$
H2: $\gamma t_D = 150 \pm 70$ and $\delta t_D = 2.9 \pm 0.5,$ (4.61)

so γt_D alone provides the means for discriminating between H1 and H2.

One may argue that during that part of each cycle where $N(t)$ approaches its minimum value N_{min}, the rate of adult recruitment $B[N(t - t_D)]$ is essentially zero. Equation (4.54) then tells us that

$$dN(t)/dt \simeq -\delta N(t),$$

which integrates to give

$$\log_e[N(t)] \simeq \text{constant} - \delta t.$$ (4.62)

Figure 4.11. Qualitative behaviour of the more realistic model (4.55). (Reproduced from Nisbet and Gurney, 1982, by permission of John Wiley & Sons.)

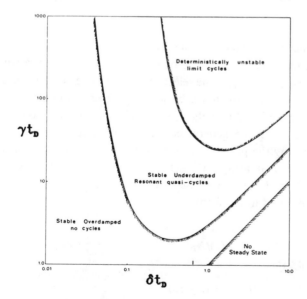

Plots of $\log_e[N(t)]$ against t in these minimum regions do indeed yield straight lines, the slopes of which provide an estimate for δ of 0.27 ± 0.025 days^{-1}, and hence of

$$\delta t_D = 4.0 \pm 0.5.$$

Whilst neither of the δt_D-values (4.61) agree closely with this value, they are near enough to maintain plausibility in both H1 and H2.

An estimate of γt_D may be obtained by noting that in equation (4.55) the future recruitment rate

$$B(t + t_D) = \gamma N(t) \exp\{-N(t)/N_0\}$$

takes the minimum value $\gamma N_0\, e^{-1}$ at $N(t) = N_0$. Thus

$$\frac{\text{maximum rate of future recruitment}}{\text{population at time of maximum production}} = \gamma\, e^{-1}. \qquad (4.63)$$

Since Nicholson's original data also includes the number of eggs generated each day, and since it is known that egg to adult survival is a constant (91%), the ratio (4.63) may be used to deduce

$$100 < \gamma t_D < 160$$

(for details see Nisbet and Gurney, 1982). This range of values is clearly compatible with H2 ($80 < \gamma t_D < 220$) and totally incompatible with H1 ($19 < \gamma t_D < 28$), which leads Nisbet and Gurney to state categorically that 'the fluctuations observed by Nicholson in this case are quite clearly of limit cycle type'.

4.8   Simulation of the two blowfly models

The comparison between the two hypotheses H1 (underdamping) and H2 (limit cycles) resulting in the selection of H2 has been made purely on *deterministic* grounds; that *stochastic* cycles might be sustainable under H1 still needs to be examined. For the simple delay model (Section 4.3) we initially proposed using the general continuous-time simulation program GENONE by storing the population size $\{X(t)\}$ at the discrete time points $t = 0, 0.01, 0.02, \ldots$ in an array $\{A(j)\}$ for $j = 0, 1, 2, \ldots$. With this scheme the stochastic birth and death rates corresponding to the deterministic model (4.55) can be computed in the form

$$\begin{aligned}
B(t) &= \gamma X(t - t_D) \exp\{-X(t - t_D)/N_0\} \\
&\simeq \gamma A(j) \exp\{-A(j)/N_0\} \\
D(t) &= \delta X(t),
\end{aligned} \qquad (4.64)$$

where j is the integer part of $100(t - t_D)$.

The alternative discrete-time procedure (4.20), in which time is segmented into small intervals of length h, works well provided that the constraint $\{B[X(ih)] + D[X(ih)]\}h < 1$ does not require h to be unreasonably small. Unfortunately, with Nicholson's data, $\{X(t)\}$ comes close to 10^4, which means that h should be at most 10^{-5} (say). During much of a cycle $B + D$ will be considerably less than its maximum value, whence it follows from (4.20) that many of the resulting 32 000 000 discrete time points (the experiment lasted 320 days) would simply record 'no change' and thereby give rise to considerable inefficiency.

Returning to (4.64), if we simulate over the range $t = 0$ to 300 (say) then the array $\{A(j)\}$ will be of size 30 000. With many computers this will not be a problem, but if necessary the array size can be drastically reduced either by storing $\{X(t)\}$ at the larger time increment of $h = 0.1$, which sacrifices a little accuracy, and/or by treating $\{A(j)\}$ as a circular array. This circularity is easily achieved by putting

$j =$ integer part of t/h (modulus t_D/h).

For example, if $h = 0.1$ and $t_D = 15$ then at $t = 33.61$

$j =$ integer part of 336.1 (modulus 150) $= 36$.

Note that here the size of the required store is a mere $t_D/h = 150$.

The program GENDEL (see Appendix to this chapter) incorporates the rates (4.64) into GENONE, and we shall now use GENDEL with Nisbet and Gurney's time-delay estimate of $t_D = 14.8$, result (4.61) providing the values

H1: $\gamma = 1.59$ and $\delta = 0.20$
H2: $\gamma = 10.14$ and $\delta = 0.20$.

We have already seen from (4.56) that N_0 is purely a scale parameter, and so the values

$N_0 = 3125$ (under H1)   and $N_0 = 500$ (under H2)

are chosen to ensure that the simulated peaks are of a similar size to the data peaks (about 8000).

Figure 4.12 shows a simulation run over $0 \leqslant t \leqslant 280$ under each hypothesis, $\{X(t)\}$ initially being held constant at $X(t) = 100$ for $0 \leqslant t \leqslant t_D$ before being released. The corresponding deterministic realizations are obtained by writing the differential equation (4.55) in the discrete form

$$N[(i+1)h] = N[ih] + h\gamma N[(i - i_D)h] \exp\{-N[(i - i_D)h]/N_0\}$$
$$- \delta N[ih] \qquad (4.65)$$

where $t = ih$ and $t_D = i_D h$, and then solving (4.65) sequentially with $h = 0.01$ over $i_D \leqslant i \leqslant 280/h$, i.e. over $i = 1480, \ldots, 28\,000$.

In both cases the deterministic and stochastic plots are virtually identical. In particular, we see that hypothesis H1 (Figure 4.12a) is immediately ruled out since the supposition that stochasticity can cause large sustained cycles is blatantly untrue. Indeed, the stochastic population size approaches the deterministic steady state $N_0 \log_e(\gamma/\delta) = 6479$ quite quickly with successive amplitudes decaying at rate 0.66. Hypothesis H2 (Figure 4.12b) behaves far better. Not only does it lead to sustained cycles which are similar to those in Nicholson's data (Figure 4.10), but the double peaks so evident in the data also feature in the simulated realization.

The fundamental difference between the data and simulated realization is that the data cycles exhibit far more variation, both in the range of cycle peak and trough values and in the form of the double peaks. Thus the model, in both its stochastic and deterministic forms, produces realizations which are too regular. However, H2 does reproduce the basic qualitative structure

Figure 4.12. Simulated (———) and deterministic (−−−) realizations of Nisbet and Gurney's time-delayed blowfly model with (a) underdamped (H1) and (b) limit cycle (H2) behaviour.

(a)

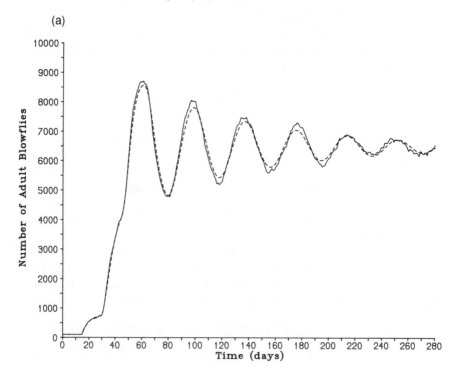

shown in the data, and to this extent Nisbet and Gurney's limit cycle model is highly successful.

Appendix

The following FORTRAN program 'GENDEL' simulates the general time-delay birth and death process. In this example the birth and death rates, *B* and *D*, are those of (4.64); they are easily replaced by any other suitable set of rates.

```
C    THIS PROGRAM IS USED TO SIMULATE A GENERAL TIME-
C    DELAY BIRTH AND DEATH PROCESS WITH RATES B AND
C    D, RESPECTIVELY, SEED S, INITIAL POPULATION SIZE N0,
C    DURATION TIME MAXTIME, TIME-DELAY TD, BIRTH AND
C    DEATH CONSTANTS GAMMA, CON AND DELTA. N(T-TD)
C    IS RECORDED IN TIME INCREMENTS OF LENGTH H.
C
C
     REAL*8 G05CAF,Y
     REAL MAXTIME
```

(b)

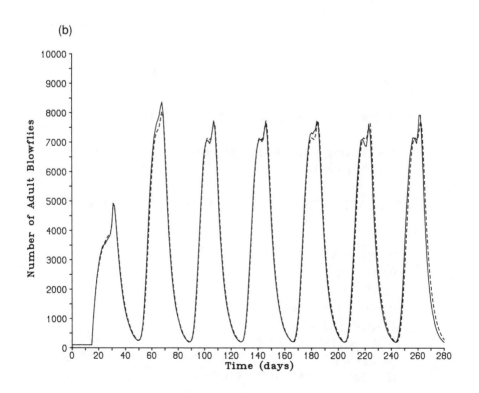

```
        DIMENSION A(30000)
        INTEGER COUNT,SEED
C
C       READ IN PARAMETER VALUES
C
        READ(01,8000) SEED,N0,MAXTIME,TD,GAMMA,DELTA,
     1  CON,H
        CALL G05CBF(SEED)
C
C       SET INITIAL VALUES
C
        MAX = 1000000
        TIME = 0.0
        TINC = 1.0
        N = N0
        WRITE(02,8001) TIME,N
        TIME = TD
        TIMEOUT = 14.0
        M = NINT(TD/H)
        M1 = M + 1
        DO 10 J = 1,M1
10      A(J) = N0
C
        J = 1
        DO 20 COUNT = 1,MAX
C
C       SET BIRTH AND DEATH RATES B AND D
C
        B = GAMMA*A(J)*EXP( - A(J)/CON)
        D = DELTA*N
C
C       SIMULATE INTER-EVENT TIMES AND NEW POPULATION
C       SIZE
C
        Y = G05CAF(Y)
        RATE = B+D
        TEST = B/RATE
        REXP = - (DLOG(Y))/RATE
        TIMEX = TIME
        TIME = TIME + REXP
        NOLD = N
        IF(TIME.GT.MAXTIME) STOP
```

```
C
        Y = G05CAF(Y)
        N = N−1
        IF(Y.LE.TEST) N=N+2
C
C     UPDATE TIME-DELAY ARRAY
C
        J1 = NINT((TIME−TD)/H) + 1
        IF((J1−J).EQ.0) GO TO 500
        MA = M+J+1
        MB = M+J1
        DO 30 L=MA,MB
  30    A(L) = NOLD
        IF(TIME.LE.TIMEOUT) GO TO 500
        WRITE(02,8001) TIMEX,NOLD
        TIMEOUT = TIMEOUT + TINC
 500    CONTINUE
        J = J1
C
  20    CONTINUE
C
8000    FORMAT(2I8,6F8.3)
8001    FORMAT(F10.4,6X,I8)
        STOP
        END
```

5

Competition processes

So far we have just considered single-species population dynamics. However, in nature organisms do not generally exist in isolated populations but they live alongside organisms from many other species. Whilst a large number of these species will be unaffected by the presence or absence of one another, in some cases two or more species will interact competitively. Such competition may either be for common resources that are in short supply, such as food or space, or it may be that organisms from different species attack each other directly.

Now there is considerable evidence to suggest that species population stability is typically greater in communities with many interacting species than in simple ones. For example, it has been noted that simple laboratory predator–prey populations characteristically undergo violent oscillations; cultivated land and orchards have shown themselves to be fairly unstable; whilst the rain forest, a highly complex structure, appears to be very stable. On closer examination, however, the issue clouds over since species integration in a complex community is a highly non-linear affair, and quite remarkable instabilities can ensue from the introduction or removal of a single species (May, 1971b). We shall therefore ignore the difficult world of three or more interacting populations and concentrate on just two (an extremely important field of study in its own right).

Before we begin it is worthwhile repeating Park's (1954) warning that the functional existence of inter-species competition may be inferred from a body of data even when no such inter-species dependence exists. If two species are correlated in such a way that the population increase of one tends to be associated with the decrease of the other, then competition between the populations can be suspected as a causal agent. But other possibilities must also be considered. For example, if each species eats entirely different food and the availability of each food type alternates temporally, then it might be expected that the two population sizes would respond with similar alternations. Observed correlations can therefore be environmental as well as competitive in nature, and this must be borne in mind by the investigator working with field observations.

5.1 Introduction

This said, consider the simple case of two species living together, and assume that the growth rate of each is inhibited by members both of its own species and of the other species. Denote by $N_1(t)$ and $N_2(t)$ the (deterministic) number of individuals of species 1 and 2, respectively. Then the simplest mathematical approach is to extend the one-species logistic model

$$dN/dt = N(r - sN) \tag{3.20}$$

to the two-species representation

$$dN_1/dt = N_1(r_1 - s_{11}N_1 - s_{12}N_2)$$
$$dN_2/dt = N_2(r_2 - s_{21}N_1 - s_{22}N_2). \tag{5.1}$$

Here between- and within-species competitive effects are contained in the same factor controlling the rate of growth of the individual species.

The obvious extension for m-species competition is

$$dN_i/dt = N_i(r_i - s_{i1}N_1 - \cdots - s_{im}N_m) \qquad (i = 1, \ldots, m). \tag{5.2}$$

Each population would grow logistically if it were alone, with parameters r_i and s_{ii}; whilst s_{ij} $(j \neq i)$ measures the extent to which the presence of species j affects the growth of species i.

Although the single-species equation (3.20) can be solved explicitly, in general the same does not hold true for the multi-species simultaneous differential equations (5.1) and (5.2). There is, however, one particular situation in which a partial solution to (5.1) may be obtained quite easily. Suppose that the inhibitory effects of species 1 and 2 are the same for both populations, that is we can write

$$dN_1/dt = N_1[r_1 - s_1(N_1 + pN_2)]$$
$$dN_2/dt = N_2[r_2 - s_2(N_1 + pN_2)]. \tag{5.3}$$

Then an individual of either species behaves as though it were competing with a population of size $N = N_1 + pN_2$, where the constant p allows each species to differ in its inhibitory effect. If individuals of species 2 make smaller demands on available resources then $p < 1$, otherwise $p > 1$. Comparison of (5.1) and (5.3) gives $s_{11} = s_1$, $s_{12} = ps_1$, $s_{21} = s_2$ and $s_{22} = ps_2$. Hence as

$$p = s_{12}/s_{11} = s_{22}/s_{21}$$

i.e.

$$s_{11}s_{22} = s_{12}s_{21}, \tag{5.4}$$

the product of the within-species inhibitory growth rates equals the product of the between-species inhibitory growth rates.

Now equations (5.3) may be written as

$$dN_1/dt = N_1(r_1 - s_1N)$$
$$dN_2/dt = N_2(r_2 - s_2N),$$

(5.5)

and these can be partially solved by eliminating N, viz.

$$N = \frac{1}{s_1}\left[r_1 - \left(\frac{1}{N_1}\frac{dN_1}{dt}\right)\right] = \frac{1}{s_2}\left[r_2 - \left(\frac{1}{N_2}\frac{dN_2}{dt}\right)\right].$$

Thus

$$s_2\left(\frac{1}{N_1}\frac{dN_1}{dt}\right) - s_1\left(\frac{1}{N_2}\frac{dN_2}{dt}\right) = s_2r_1 - s_1r_2$$

which integrates to give

$$s_2[\log_e\{N_1(t) - N_1(0)\}] - s_1[\log_e\{N_2(t) - N_2(0)\}] = t(s_2r_1 - s_1r_2).$$

Taking the exponential of both sides then yields

$$\frac{[N_1(t)]^{s_2}}{[N_2(t)]^{s_1}} = \frac{[N_1(0)]^{s_2}}{[N_2(0)]^{s_1}} \exp\{t(s_2r_1 - s_1r_2)\}.$$

(5.6)

Whilst this is not a complete solution, in that it does not detail the separate development of $N_1(t)$ and $N_2(t)$, it does provide valuable information on their relative magnitudes. Noting from (5.5) that dN_1/dt and dN_2/dt are both negative if $N = N_1 + pN_2$ exceeds r_1/s_1 and r_2/s_2, we see that $N_1(t)$ and $N_2(t)$ have maximum permissible sizes which are not greater than their individual carrying capacities $K_1 = r_1/s_1$ and $K_2 = r_2/s_2$. Thus (5.6) shows that

   (i) if $s_2r_1 > s_1r_2$ then $N_2(t) \to 0$,
   (ii) if $s_2r_1 < s_1r_2$ then $N_1(t) \to 0$,
   (iii) if $s_2r_1 = s_1r_2$ then neither $\to 0$.

So unless the exceptional case (iii) occurs, only one of the species will persist and the other will eventually die out. If $s_2r_1 > s_1r_2$ then species 1 will ultimately 'win', and conversely if $s_2r_1 < s_1r_2$; the speed at which these events happen is governed by the term $\exp\{t(s_2r_1 - s_1r_2)\}$. Once the losing species has become extinct the winner will grow in accordance with the single-species logistic process.

On a final point, we have already seen that single-species logistic growth can be described either by the differential equation

$$dN(t)/dt = N(t)[r - sN(t)]$$

(3.20)

or by the difference equation

$$N(t+1) = e^r N(t)/[1 + N(t)(e^r - 1)/K].$$

(3.25)

This suggests that the differential equations (5.1) might correspond to a pair of difference equations

$$N_i(t + 1) = a_i N_i(t)/[1 + b_i N_i(t) + c_i N_j(t)] \qquad (5.7)$$

for some constants a_i, b_i and c_i ($i, j = 1, 2; j \neq i$). Pielou (1977) provides the equivalence

$$a_i = \exp(r_i), \quad b_i = s_{ii}(a_i - 1)/r_i \quad \text{and} \quad c_i = s_{ij}(a_i - 1)/r_i,$$

and since equations (5.1) do not have a known solution this discrete representation is especially useful when sketching the trajectories of $\{N_1(t)\}$ and $\{N_2(t)\}$ against t.

5.2   Experimental background

Not only is expression (5.6) an incomplete solution, but assumption (5.4) on which it is based will not usually hold true. Thus further pursuit of more general mathematical solutions is, at least for ecological purposes, pointless. We shall therefore now look at some biological systems, and then see whether equations (5.1) can be used to provide a partial representation of biological behaviour.

5.2.1   *Gause's yeast experiments*

The Russian microbiologist Gause conducted one of the earliest and most important investigations into competing systems. Gause (1932) studied in detail the mechanism of competition between two species of yeast, *Saccharomyces cerevisiae* and *Schizosaccharomyces kefir*. He began with four series of experiments on the growth of the two species taken separately, and fitted the logistic curves

$$\begin{aligned} N_1(t) &= 13.0/[1 + \exp\{3.328\ 16 - 0.218\ 27t\}] \quad (Sacch.) \\ N_2(t) &= \ 5.8/[1 + \exp\{2.475\ 50 - 0.060\ 69t\}] \quad (Schiz.). \end{aligned} \qquad (5.8)$$

Figure 5.1a shows these curves together with data from the first two series of experiments, and the fit is (visually) reasonable. Gause notes that other experiments have produced growth curves which are 'somewhat asymmetrical', but such asymmetry is not marked here. *Schizosaccharomyces* clearly grows more slowly than *Saccharomyces*, and the carrying capacity of the former is smaller.

In contrast, experiments on the growth of a mixed population of the two species showed that the growth of the combined population (Figure 5.1a) lies in between that of the two individual species taken separately. The individual species in the mixed population (Figure 5.1b) still appear to exhibit logistic growth with evidence of possible stabilization around new carrying capacity levels. Gause found that the influence of *Schizosaccharomyces* upon the unutilized growth opportunity of *Saccharomyces* is decidedly marked (estimated $s_{12}/s_{11} = 3.15$), though *Saccharomyces* influences the growth of *Schizosaccharomyces* only slightly (estimated $s_{21}/s_{22} = 0.439$). He retained the parameters

$$r_1 = 0.218\,27, \quad s_{11} = 0.016\,79 \quad (Sacch.)$$
$$r_2 = 0.060\,69, \quad s_{22} = 0.010\,46 \quad (Schiz.) \tag{5.9}$$

estimated from the separate series of experiments, whence

$$s_{12} = 3.15s_{11} = 0.052\,89$$

Figure 5.1. (a) Development of two separate populations of *Saccharomyces* (□) and *Schizosaccharomyces* (◇), the two associated logistic curves (5.8), and the total amount of yeast (○) in the combined population (from data of Gause, 1932). (b) Development of the populations of *Saccharomyces* (□) and *Schizosaccharomyces* (◇) in the combined experiment, and the two associated curves (5.11) with parameter values (5.9) and (5.10) (from data of Gause, 1932). (c) Development of the two deterministic curves of Figure 5.1b for *Sacch.* (———) and *Schiz.* (– – –) over the time period $t = 0$ to 360 (hours).

(a)

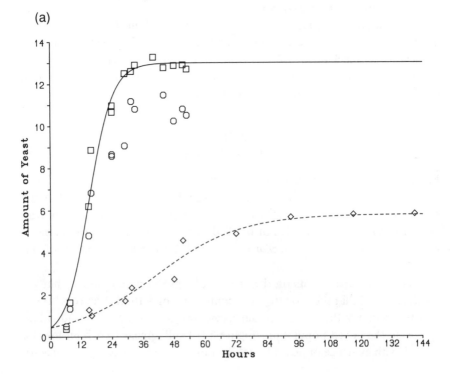

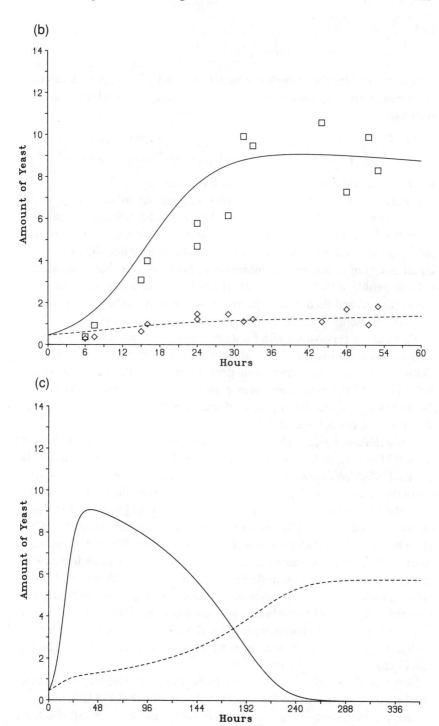

and

$$s_{21} = 0.439s_{22} = 0.004\ 59. \qquad (5.10)$$

In order to plot the growth curves $\{N_1(t)\}$ and $\{N_2(t)\}$ it is sufficient to solve equations (5.1) numerically through the simple procedure of successively evaluating

$$N_1[(i+1)h] = N_1(ih) + hN_1(ih)[r_1 - s_{11}N_1(ih) - s_{12}N_2(ih)]$$
$$N_2[(i+1)h] = N_2(ih) + hN_2(ih)[r_2 - s_{21}N_1(ih) - s_{22}N_2(ih)] \quad (5.11)$$

where i takes the values $1, 2, 3, \ldots$ and h is a suitably small time increment. For example, we might choose Gause's starting values of $N_1(0) = N_2(0) = 0.45$, select $h = 0.01$, and output when $i = 0, 50, 100, \ldots$, i.e. at times $t = 0, 0.5, 1, \ldots$. Figure 5.1b includes the result of such a computation using the parameter values (5.9) and (5.10). Careful inspection shows that our initial assertion of eventual stabilization around two possible asymptotes is not compatible with this model; $N_1(t)$ reaches a maximum value of 9.1 at $t = 41$ (hours) and then decays thereafter. A plot over the larger time scale of $t = 0$ to 360 hours (Figure 5.1c) shows this effect quite clearly.

Note that this cross-over effect is a prediction over 360 hours based purely on a model fitted over the much shorter time scale of 53 hours. It does not follow that the underlying biological process would generate similar behaviour had the experiment been performed for the same length of time. Nevertheless, it is still instructive to discuss the reasoning which lies behind this model-based behaviour.

Whilst *Saccharomyces* (N_1) has a much greater initial rate of increase ($r_1 = 0.218$) than *Schizosaccharomyces* ($r_2 = 0.061$), *Schiz.* (N_2) has a far stronger effect on *Sacch.* ($s_{12}/s_{11} = 3.15$) than *Sacch.* has on *Schiz.* ($s_{21}/s_{22} = 0.439$). So although $N_1(t)$ initially increases substantially faster than $N_2(t)$, once the total competitive effect on $N_1(t)$, namely $s_{11}[N_1(t) + 3.15N_2(t)]$, exceeds r_1, then $dN_1(t)/dt$ goes negative and $N_1(t)$ subsequently decays. Note that $N_2(t)$ has 3.15 times the effect on $N_1(t)$ than $N_1(t)$ exerts on itself. After reaching the maximum value $N_1(41) = 9.1$, $N_1(t)$ changes from a state of rapid growth to moderate decline towards zero ($N_1(280) = 0.001$). The initial growth of $N_2(t)$ is severely restricted by the rapid rise of $N_1(t)$, even though $s_{21}/s_{22} = 0.439$ is substantially less than one. However, once $N_1(t)$ starts to decay then the competitive effect of $N_1(t)$ on $N_2(t)$ decays with it and so $N_2(t)$ is governed more and more by its own carrying capacity. The two curves cross at $t = 180$ hours.

Note that the accuracy of this crude numerical procedure may be partially checked by comparing the asymptotic value of $N_2(t)$ (i.e. 5.8021) with its known carrying capacity of 5.8 – a mere 0.04% too high. A full check on

Table 5.1. Results of four competition experiments (from Birch, 1953)

Expt.	Species	Grain	Temp. (°C)	Moisture (%)	r	Surv. spec.	Ext. times (weeks)
1	A & B	Maize	29.1	13	0.417 0.436	B	74–150
2	A & B	Wheat	29.1	14	0.772 0.564	A	46–95
3	A & C	Wheat	29.1	14	0.772 0.578	A	38–190+
4	A & C	Wheat	32.3	14	0.501 0.686	C	26–47

computed values can be obtained by comparing them with exact values generated from the discrete representation (5.7).

5.2.2   *Birch's grain beetle experiments*

Shortly after completing this work Gause (1934) conducted experiments over much longer time periods in which extinction of one of the species was almost always achieved. He studied competition between *Paramecium aurelia* and *Paramecium caudatum*, and found that whilst each species would grow well separately, when they were raised in the same tube *P. aurelia* always drove out *P. caudatum*. Note that this competitive situation is not as simple as the yeast one; not only did the values of s_{12}/s_{11} and s_{21}/s_{22} vary with the age of the culture, but also the bacteria that were the food source produced a toxin that harmed the *P. caudatum* more than the *P. aurelia*.

Similar experiments by Birch (1953) showed that a relatively slight change in environment could alter which of the two competing species became extinct. Mixed populations of grain beetles were developed within either maize or wheat at two different temperatures and relative humidities. The results are summarized in Table 5.1; *r* denotes the innate capacity for increase (i.e. r_1, r_2 in equations (5.1)), and A, B and C denote *Calandra oryzae* (small strain), *C. oryzae* (large strain) and *Rhizopertha dominica*, respectively.

Each of the four experiments was replicated 15 times. In Experiment 1, B had a higher *r*-value than A and so had an advantage, with the converse holding in Experiment 2. Similarly, A had the advantage over C in Experiment 3, and C over A in Experiment 4. In each of the 60 cases the species which held the advantage *always* won, though care must be taken not to claim this as a universal principle.

Birch found that the results from different replicates within the same experiment were surprisingly uniform. With the exception of Experiment 3,

one outcome of which is shown in Figure 5.2, one species always became
extinct though there was wide variation in the times to extinction (see Table
5.1). In Experiment 3, although *R. dominica* became extinct in only 7 of the 15
replicates by the 190th week (when the experiment had to be terminated),
numbers of *R. dominica* in the remaining replicates were then very low and so
the species was deemed to have lost.

Inspection of Figure 5.2 shows that both *C. oryzae* (small) and *R. dominica*
exhibit logistic growth for the first 10 weeks, but then competition affects
both species with *R. dominica* reaching a maximum population size at 22
weeks and decaying thereafter. Natural variation clearly plays a major role in
both populations after about 30 weeks have elapsed, the variation for *C.
oryzae* (small) being comparable to that of the South Australian sheep
population (Figure 3.3). Unlike the deterministic behaviour exhibited in
Figure 5.1c, the population size of *R. dominica* does not decay steadily to zero
after reaching its peak, but appears from week 70 onwards to wander
randomly in the range 7 to 50. Consideration of stochastic behaviour is
therefore just as important for this two-species competition process as it is for

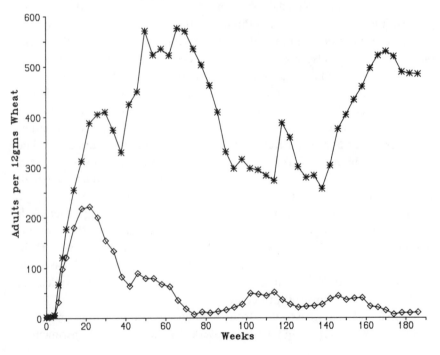

Figure 5.2. Birch's competition experiment between *C. oryzae* (small) (∗)
and *R. dominica* (◊) (from data of Birch, 1953).

the one-species logistic process, and raises the whole question of population size stability. Is one species always destined to become extinct (as is suggested by Experiments 1, 2 and 4), or can two competing species successfully coexist with eventual extinction of either species being due solely to natural variation (as might be the case in Experiment 3)?

5.3 Stability

To answer this question let us return to the deterministic equations (5.1). For an equilibrium solution (N_1^*, N_2^*) to exist we require $dN_1/dt = 0 = dN_2/dt$ at $N_1 = N_1^*$, $N_2 = N_2^*$, i.e.

$$0 = r_1 - s_{11}N_1^* - s_{12}N_2^*$$
$$0 = r_2 - s_{21}N_1^* - s_{22}N_2^*, \qquad (5.12)$$

whence

$$N_1^* = \frac{r_1 s_{22} - r_2 s_{12}}{s_{11}s_{22} - s_{12}s_{21}} \quad \text{and} \quad N_2^* = \frac{r_2 s_{11} - r_1 s_{21}}{s_{11}s_{22} - s_{12}s_{21}} \qquad (5.13)$$

provided that $N_1^* > 0$ and $N_2^* > 0$. We now have to determine whether an equilibrium point (N_1^*, N_2^*) is *stable* or *unstable*. That is, if the population values N_1 and N_2 are slightly displaced from this point, will (N_1, N_2) return to (N_1^*, N_2^*) or will it move away?

Consider the geometry of equations (5.12). From the first, $N_1^* = 0$ when $N_2^* = r_1/s_{12}$ and $N_2^* = 0$ when $N_1^* = r_1/s_{11}$; whilst from the second, $N_1^* = 0$ when $N_2^* = r_2/s_{22}$ and $N_2^* = 0$ when $N_1^* = r_2/s_{21}$. Four different types of behaviour are therefore possible depending on whether (i) r_1/s_{12} is greater or less than r_2/s_{22} and (ii) r_1/s_{11} is greater or less than r_2/s_{21}. Each is illustrated in Figure 5.3, and we see via equations (5.1) that

$$\text{above line (a)} - dN_1/dt < 0, \text{ so } N_1(t) \text{ decreases}$$
$$\text{on line (a)} \quad - dN_1/dt = 0, \text{ so } N_1(t) \text{ is constant}$$
$$\text{below line (a)} - dN_1/dt > 0, \text{ so } N_1(t) \text{ increases.}$$

Similarly, above, on, or below line (b) $N_2(t)$ decreases, is constant, or increases. Sketching these directions onto Figure 5.3 then provides a pictorial representation of the following four possible outcomes.

(i) If $r_1/r_2 > s_{12}/s_{22}$ and s_{11}/s_{21}, then species 1 wins.
(ii) If $r_1/r_2 < s_{12}/s_{22}$ and s_{11}/s_{21}, then species 2 wins.

In neither of these cases does a positive equilibrium solution (N_1^*, N_2^*) exist.

(iii) If $s_{12}/s_{22} < r_1/r_2 < s_{11}/s_{21}$, then a stable equilibrium exists, i.e. as t increases $N_1(t)$ and $N_2(t)$ approach N_1^* and N_2^*, respectively.

(iv) If $s_{11}/s_{21} < r_1/r_2 < s_{12}/s_{22}$, then there is an unstable equilibrium; if N_1 and N_2 are displaced from it then they continue to move away, and one of the species will ultimately become extinct (which one depends on the starting point $(N_1(0), N_2(0))$.

These simple graphical arguments show that a stable equilibrium can exist if and only if

$$r_1 s_{22} > r_2 s_{12} \quad \text{and} \quad r_2 s_{11} > r_1 s_{21}. \tag{5.14}$$

One way of interpreting this condition is to denote $t_{ij} = s_{ij}/r_i$ $(i, j = 1, 2)$ as the competitive effect of species j on species i relative to the natural growth rate of species i. Then (5.14) is equivalent to

$$t_{22} > t_{12} \quad \text{and} \quad t_{11} > t_{21}, \tag{5.15}$$

Figure 5.3. Stability diagrams for two-species competition: (i) N_1 wins – line (a) lies wholly above line (b); (ii) N_2 wins – line (b) lies wholly above line (a); (iii) stable equilibrium – line (a) crosses line (b) and has steeper gradient; (iv) unstable equilibrium – line (b) crosses line (a) and has steeper gradient.

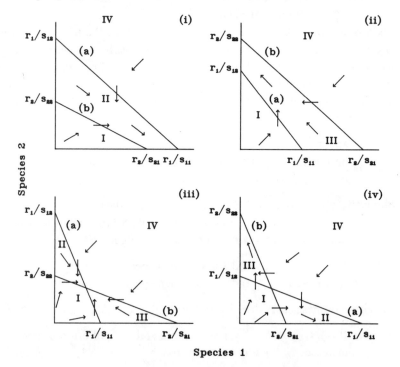

i.e. the relative competitive effect of a species on itself is greater than its effect on the other species.

Comparison of Figures 5.3(iii) and (iv) shows that condition (5.14) is also equivalent to the requirement that the equilibrium point (N_1^*, N_2^*) lies above the line joining K_1 on the N_1-axis and K_2 on the N_2-axis, where $K_i = r_i/s_{ii}$ $(i = 1, 2)$ denotes the carrying capacity of species i in the absence of species j.

*5.3.1 *Local stability*

To effect a more mathematically rigorous approach, suppose we restrict our attention to *local* stability, i.e. any *small* deviation from (N_1^*, N_2^*) must ultimately decay to zero. Then on writing

$$N_i(t) = N_i^*[1 + n_i(t)] \qquad (i = 1, 2) \tag{5.16}$$

where $n_i(t)$ is assumed to be small, equations (5.1) become

$$N_1^* \, dn_1/dt = N_1^*(1 + n_1)[r_1 - s_{11}N_1^*(1 + n_1) - s_{12}N_2^*(1 + n_2)]$$
$$N_2^* \, dn_2/dt = N_2^*(1 + n_2)[r_2 - s_{21}N_1^*(1 + n_1) - s_{22}N_2^*(1 + n_2)].$$

On ignoring terms in n_1^2, n_2^2 and $n_1 n_2$, and using (5.12), these equations simplify to

$$dn_1/dt = -(s_{11}N_1^*)n_1 - (s_{12}N_2^*)n_2$$
$$dn_2/dt = -(s_{21}N_1^*)n_1 - (s_{22}N_2^*)n_2. \tag{5.17}$$

Writing the first equation in the form

$$n_2 = -[dn_1/dt + (s_{11}N_1^*)n_1]/(s_{12}N_2^*), \tag{5.18}$$

and substituting this expression into the second, then gives

$$\frac{d^2 n_1}{dt^2} + \frac{dn_1}{dt}(s_{11}N_1^* + s_{22}N_2^*)$$
$$+ n_1(s_{11}s_{22} - s_{12}s_{21})N_1^* N_2^* = 0. \tag{5.19}$$

This is a standard second-order differential equation, and to solve it we first have to construct the associated 'characteristic equation', namely

$$w^2 + w(s_{11}N_1^* + s_{22}N_2^*) + (s_{11}s_{22} - s_{12}s_{21})N_1^* N_2^* = 0. \tag{5.20}$$

The roots of this quadratic equation are

$$w_1, w_2 = \tfrac{1}{2}[-(s_{11}N_1^* + s_{22}N_2^*) \pm \sqrt{\{(s_{11}N_1^* + s_{22}N_2^*)^2}$$
$$- 4(s_{11}s_{22} - s_{12}s_{21})N_1^* N_2^*\}] \tag{5.21}$$

$$= \tfrac{1}{2}[-(s_{11}N_1^* + s_{22}N_2^*) \pm \sqrt{\{(s_{11}N_1^* - s_{22}N_2^*)^2}$$
$$+ 4s_{12}s_{21}N_1^* N_2^*\}], \tag{5.22}$$

whence the general solution to the differential equation (5.19) has the form

$$n_1(t) = A \exp(w_1 t) + B \exp(w_2 t) \tag{5.23}$$

for some constants A and B. Substituting (5.23) into (5.18) then gives

$$n_2(t) = -[A(w_1 + s_{11}N_1^*) \exp(w_1 t) \\ + B(w_2 + s_{11}N_1^*) \exp(w_2 t)]/(s_{12}N_2^*). \tag{5.24}$$

The unknown A and B are determined from the initial values

$$n_1(0) = A + B \\ n_2(0) = -[A(w_1 + s_{11}N_1^*) + B(w_2 + s_{11}N_1^*)]/(s_{12}N_2^*). \tag{5.25}$$

The trajectories of $n_1(t)$ and $n_2(t)$, and hence of $N_1(t)$ and $N_2(t)$, can now be plotted directly from (5.23)–(5.25).

As the square-root part of expression (5.22) consists of two positive terms, w_1 and w_2 are both real, so spontaneous cycles cannot arise. Moreover, since local stability requires that $n_1(t)$ and $n_2(t)$ decay to zero as t increases, it follows from (5.23) and (5.24) that both $w_1 < 0$ and $w_2 < 0$. Direct examination of (5.21) shows that this can only occur if

$$s_{11}s_{22} - s_{12}s_{21} > 0. \tag{5.26}$$

For equilibrium to exist we also require $N_1^* > 0$ and $N_2^* > 0$, so if (5.26) holds then expressions (5.13) show that

$$r_1 s_{22} - r_2 s_{12} > 0 \quad \text{and} \quad r_2 s_{11} - r_1 s_{21} > 0.$$

These inequalities combine with (5.26) to produce a rigorous proof of the *local stability* condition (iii) above.

Conversely, if $s_{11}s_{22} - s_{12}s_{21} < 0$ so that $w_1 > 0$, i.e. $n_1(t)$ and $n_2(t)$ grow exponentially, then the equilibrium solution (5.13) with $N_1^* > 0$ and $N_2^* > 0$ gives $r_1 s_{22} - r_2 s_{12} < 0$ and $r_2 s_{11} - r_1 s_{21} < 0$, i.e.

$$s_{11}/s_{21} < r_1/r_2 < s_{12}/s_{22}.$$

This proves condition (iv) above for *local instability*.

5.3.2 *Global stability*

The sketching technique (Figure 5.3) yields valuable qualitative information on the behavioural characteristics of a model, whilst the subsequent mathematical treatment provides formal conditions for the existence of locally stable and unstable equilibria. However, the only effective way of studying global behaviour is to compute (N_1, N_2)-trajectories via the numerical procedure (5.11) for various parameter values r_i, s_{ij} and initial population sizes $N_i(0)$ ($i, j = 1, 2$). We have already used this approach to model competition between *Saccharomyces* and *Schizosaccharomyces* (Figure 5.1c), and observed that in spite of *Saccharomyces* having the larger initial rate of growth it eventually loses. Note that this outcome agrees with that predicted by local condition (ii) above; i.e. species 2 wins over species 1 and no equilibrium exists, since $r_1/r_2 = 3.60$ is less than both $s_{12}/s_{22} = 5.06$ and $s_{11}/s_{21} = 3.66$.

In contrast, Figures 5.4a and 5.4b show plots of $N_2(t)$ against $N_1(t)$ for various initial population sizes when an equilibrium solution does exist. In both cases $r_1 = 2.0$ and $r_2 = 1.5$, with:

Figure 5.4. (a) Convergence of trajectories towards the stable equilibrium point $N_1^* = 50$, $N_2^* = 25$ for the competition model

$dN_1/dt = N_1(2.0 - 0.03N_1 - 0.02N_2)$
$dN_2/dt = N_2(1.5 - 0.01N_1 - 0.04N_2)$

and initial population sizes $(1, 1)$, $(1, 70)$, $(120, 1)$ and $(150, 70)$.
(b) Divergence of trajectories away from the unstable equilibrium point $N_1^* = 20$, $N_2^* = 45$ for the competition model

$dN_1/dt = N_1(2.0 - 0.01N_1 - 0.04N_2)$
$dN_2/dt = N_2(1.5 - 0.03N_1 - 0.02N_2)$

and initial population sizes $(20, 20)$, $(22, 43)$, $(200, 75)$, $(3, 15)$, $(18, 47)$ and $(30, 75)$.

(a)

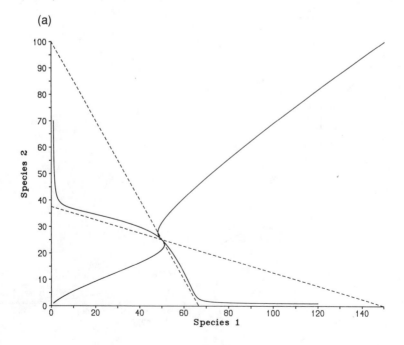

Species 1

(a) $s_{11} = 0.03$, $s_{12} = 0.02$, $s_{21} = 0.01$ and $s_{22} = 0.04$, so as $s_{12}/s_{22} = 0.5 < r_1/r_2 = 1.33 < s_{11}/s_{21} = 3$ this corresponds to a stable equilibrium at $N_1^* = 50$, $N_2^* = 25$ (condition (iii));

(b) $s_{11} = 0.01$, $s_{12} = 0.04$, $s_{21} = 0.03$ and $s_{22} = 0.02$, so as $s_{11}/s_{21} = 0.33 < r_1/r_2 = 1.33 < s_{12}/s_{22}$ this corresponds to an unstable equilibrium at $N_1^* = 20$, $N_2^* = 45$ (condition (iv)).

In Figure 5.3 the positive quadrant of the (N_1, N_2)-plane is partitioned into four zones:

Zone I $- dN_1/dt > 0$ and $dN_2/dt > 0$;
Zone II $- dN_1/dt > 0$ and $dN_2/dt < 0$;
Zone III $- dN_1/dt < 0$ and $dN_2/dt > 0$;
Zone IV $- dN_1/dt < 0$ and $dN_2/dt < 0$.

Figure 5.4a shows the convergence to stable equilibrium from four starting points, one in each zone, namely:

(I) $N_1(0) = N_2(0) = 1$, both populations start very small so initial growth is approximated by two separate birth processes.

(II) $N_1(0) = 1$ and $N_2(0) = 70$, $N_2(0)$ is above its carrying capacity and quickly drops down to the lower equilibrium line, the trajectory then follows this closely to the equilibrium point.

(b)

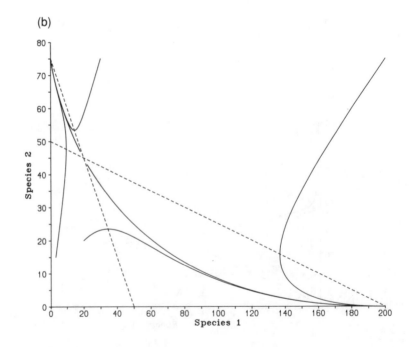

(III). $N_1(0) = 120$ and $N_2(0) = 1$, as for II but with $N_1(t)$ and $N_2(t)$ interchanged.

(IV) $N_1(0) = 150$ and $N_2(0) = 70$, both populations start well above their respective carrying capacities and rapidly drop down to equilibrium.

Although the trajectories from zones I and IV travel almost directly towards the equilibrium point, note that they enter zones II and III before actually reaching it. Suppose that $N_1(0) < N_1^*$ and $N_2(0) < N_2^*$. Then if equilibrium is reached via zone II, since $dN_2/dt < 0$ there, it follows that $N_2(t)$ must *overshoot* N_2^* before reaching it. Similarly for arrival at equilibrium via zone III, $N_1(t)$ must overshoot N_1^*. Thus an overshoot for one of the populations is inevitable.

Figure 5.4b illustrates the converse phenomenon, where trajectories move away from an unstable equilibrium position towards an axis. Trajectories starting in zones I or IV cross over into zones II or III (see Figure 5.3iv). Once a trajectory is in zone III, $N_1(t)$ decays towards zero and $N_2(t)$ converges towards its carrying capacity K_2; once in zone II, $N_2(t)$ decays towards zero and $N_1(t)$ converges towards its carrying capacity K_1. The six illustrated trajectories have the following initial population values $(N_1(0), N_2(0))$:

> extinction of species 1:   (3, 15) – moves from I to III
> (18, 47) – remains in III
> (30, 75) – moves from IV to III;
> extinction of species 2:   (20, 20) – moves from I to II
> (22, 43) – remains in II
> (200, 75) – moves from IV to II.

Note that if the initial and equilibrium values are close together, then a small change to the former can radically alter the (deterministic) outcome of the competition.

*5.3.3   *General stability conditions*

Model (5.1) is clearly the simplest model possible which allows for natural growth and competition both within and between species. An obvious generalization is to replace it by the equations

$$dN_1/dt = N_1 F_1(N_1, N_2)$$
$$dN_2/dt = N_2 F_2(N_1, N_2) \tag{5.27}$$

where F_1 and F_2 are two given functions of N_1 and N_2. Rescigno and Richardson (1967) develop the corresponding stability arguments to those above, where several equilibrium points, stable and unstable, may now exist simultaneously. These points are given by the solutions (N_1^*, N_2^*) of the equations

$$F_1(N_1, N_2) = 0 = F_2(N_1, N_2).$$
(5.28)

Zones I to IV correspond to values of N_1 and N_2 such that: (I) $F_1 > 0$, $F_2 > 0$; (II) $F_1 > 0$, $F_2 < 0$; (III) $F_1 < 0$, $F_2 > 0$; and (IV) $F_1 < 0$, $F_2 < 0$. If the curves (5.28) intersect in the positive (N_1, N_2)-quadrant, then Figures 5.3iii and 5.3iv may be generalized as in Figure 5.5. Direct comparison of Figures 5.3iii and 5.3iv with the areas local to the points of intersection P and Q in Figure 5.5 shows that P and Q correspond to stable and unstable equilibrium points, respectively. When the curves $F_1 = 0$ and $F_2 = 0$ have many such intersections, then various fates are possible for the two species depending upon the initial conditions. In general, the points of intersection where the curve $F_1 = 0$ has a steeper gradient (in magnitude) than the curve $F_2 = 0$ are points of local stable coexistence.

Once generalizations such as (5.27) are allowed we have to consider whether an equilibrium state is *locally* or *globally* stable. With the basic competition equations (5.1) any perturbation from a stable equilibrium point, no matter how large, results in the deterministic trajectory returning to that point

Figure 5.5. Stable (P) and unstable (Q) equilibria for a general competition process.

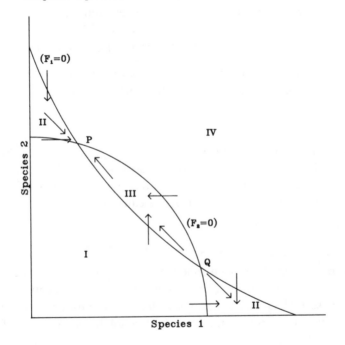

(Figure 5.3iii). Thus here local stability automatically implies global stability. However, if in Figure 5.5 a perturbation is large enough for lower zone II to be reached, then the resulting deterministic trajectory will not return to the stable point P but will hit the species 1 axis instead. So P is stable only for small, i.e. local, perturbations. An equilibrium point may therefore be locally stable without necessarily being globally stable.

Realistic competition models usually involve complicated mathematical forms for the birth and death rates, and although specific solutions may be computed numerically it is notoriously difficult to gain a deep understanding of a process from a numerical approach. However, Nisbet and Gurney (1982) argue that whilst *non-linear* models are fundamental to population dynamics there are only a few occasions when significant progress cannot be made with the aid of purely *linear* mathematics. We have already seen how to construct locally linear approximations from non-linear models which are valid near a steady state, an example being the linear approximation (5.17) to the quadratic equations (5.1). Though in principle such approximations are valid only if perturbations from the steady state are small, in practice they generally work with perturbations that are surprisingly large. Moreover, for models that possess just one equilibrium state, although use of local linear approximations can only prove local stability, this in turn usually strongly suggests global stability.

Local stochastic linearization can also lead both to theoretical estimates of the pseudo-equilibrium distribution and to qualitatively acceptable estimates of extinction probabilities (and thereby to estimates of ecological stability), but more of this later.

To demonstrate the widespread applicability of the linearization approach suppose we replace equations (5.27) by the totally general form

$$dN_1/dt = G_1(N_1, N_2)$$
$$dN_2/dt = G_2(N_1, N_2) \tag{5.29}$$

where G_1 and G_2 are two functions of N_1 and N_2. Then if (N_1^*, N_2^*) is an equilibrium solution of these equations, we have

$$G_1(N_1^*, N_2^*) = 0 = G_2(N_1^*, N_2^*). \tag{5.30}$$

Writing $n_1 = N_1 - N_1^*$ and $n_2 = N_2 - N_2^*$, and expanding (5.29) as two Taylor's series, then gives

$$\frac{dN_1}{dt} = G_1 + n_1 \frac{\partial G_1}{\partial N_1} + n_2 \frac{\partial G_1}{\partial N_2} + \text{higher-order terms} \tag{5.31}$$

together with a similar equation for G_2, where the G_i and their partial derivatives are evaluated at (N_1^*, N_2^*). These higher-order terms involve products such as n_1^2, $n_1 n_2^4$, etc., and for small n_i they should be negligible in comparison with G_i and $n_j \, \partial G_i / \partial n_j$ ($i, j = 1, 2$). Thus (5.31) leads to the linear approximation

$$dn_1/dt = A_{11}n_1 + A_{12}n_2$$
$$dn_2/dt = A_{21}n_1 + A_{22}n_2 \qquad (5.32)$$

where $A_{ij} = \partial G_i / \partial n_j$ are evaluated at $N_i = N_i^*$ ($i, j = 1, 2$). Equations (5.17) and (5.32) are identical provided we denote

$$s_{ij} = -A_{ij}/N_j^*,$$

and so the previous stability analysis immediately extends to the general process (5.29). For example, condition (5.26), i.e. $s_{11}s_{22} - s_{12}s_{21} > 0$, becomes $A_{11}A_{22} - A_{12}A_{21} > 0$.

This is really as far as we can go, even though in theory global stability occurs if we can find a Lyapunov function $L(N_1, N_2)$ with the following properties:

(i) $L(N_1^*, N_2^*) = 0$;
(ii) $L(N_1, N_2) > 0$ for all $(N_1, N_2) \neq (N_1^*, N_2^*)$;
(iii) $dL/dt \leqslant 0$ with equality only if $(N_1, N_2) = (N_1^*, N_2^*)$.

Unfortunately there is no universal principle to help us choose an appropriate Lyapunov function. The usefulness of this technique as a practical means of demonstrating global stability is therefore severely limited, and even when such functions have been successfully constructed their ecological interpretation has often been obscure.

5.4 Stochastic behaviour

We have already noted that in Gause's (1934) experiments on competition between *Paramecium aurelia* and *P. caudatum* the former always won (Section 5.2.2). Moreover, whilst Birch's (1953) experiments showed that different environmental conditions led to different eventual winners, for any given set of experimental conditions all fifteen replicates had the same outcome. However, it does *not* follow that repeated 'identical' experiments *must* necessarily produce the same outcome.

5.4.1 *Park's flour beetle experiments*

Several laboratory examples illustrating variable outcomes are provided by Krebs (1985); here we shall consider a set of classic experiments developed by Park (1954). Park studied the effect of competition between two types of flour beetle, *Tribolium castaneum* and *Tribolium confusum*, under six regimes of temperature and humidity, where the two types were initially

Table 5.2. Competition results between *T. castaneum* and *T. confusum* (from Park, 1954)

Regime	Temp. (°C)	Rel. hum. (%)	Climate	No. reps.	% wins conf. (f)	% wins cast. (s)	Winning species	Median no. days for elim. (f/s)
I	34	70	hot–wet	30	0	100	s	330
II	34	30	hot–dry	30	90	10	f	210/150
III	29	70	temp.–wet	30	14	86	s	645/405
IV	29	30	temp.–dry	30	87	13	f	675/300
V	24	70	cold–wet	30	69	31	f	825/585
VI	24	30	cold–dry	20	100	0	f	270

present in equal amounts. The results (Table 5.2) show that in intermediate climates (i.e. the middle four) the outcome was uncertain; sometimes *T. confusum* won and sometimes *T. castaneum* won. However, the median age for elimination of *T. confusum* was substantially higher than that for *T. castaneum* under each of these regimes.

Regime V (cold–wet) presents a particularly interesting situation, since when living apart *T. castaneum* maintained a mean density of 45 beetles per gram of flour and *T. confusum* 28 beetles per gram. Thus *T. confusum* is the stronger survivor in spite of being in the presence of a competitor which appears to have the advantage when judged purely on single-species performance. This situation parallels the (presumed) deterministic prediction for competition between *Saccharomyces* and *Schizosaccharomyces* (Figure 5.1c).

Park (1954) graphs the population size of *T. confusum* ($N_1(t)$) and *T. castaneum* ($N_2(t)$) against t for ten particular experimental runs which between them cover all six regimes and possible winning combinations. Neyman, Park and Scott (1956) take two of these examples (regime V) and plot the trajectories of $N_2(t)$ against $N_1(t)$ (Figure 5.6); *T. confusum* becomes extinct in one and *T. castaneum* in the other. We see that in the early stages of the competition the two trajectories cross each other several times, but once two (arbitrarily placed) 'barriers' are reached such intersections no longer occur. Within the central zone the trajectories move more or less randomly with no preferred direction towards either barrier. However, once a barrier has been crossed there is an ever-increasing tendency to move further and further away from it until the declining species becomes extinct.

5.4.2   *Gause's competitive exclusion principle*

Although in all the 170 cases covered by Table 5.2 one of the species always became extinct, it must be stressed that in certain situations it is perfectly possible for two species to live together in harmony even if they differ only slightly in their requirements. At first sight this seems to contradict Gause's (1934) competitive exclusion principle, namely that 'as a result of competition two similar species scarcely ever occupy similar niches, but displace each other in such a manner that each takes possession of certain peculiar kinds of food and modes of life in which it has an advantage over its competitor'. He takes *niche* to be defined (Elton, 1927) as 'the niche of an animal means its place in the biotic environment, its relations to food and enemies'. However, for some species such differences in habitat or requirements need only be *very* slight for their niches to be effectively different.

For example, Crombie (1945) found that the grain beetles *Rhizopertha* and *Oryzaephilus* would coexist indefinitely in wheat. The larvae of one species live and feed inside the wheat grain, and those of the other species outside, and this small larval difference is sufficient to allow coexistence to occur.

Figure 5.6. Two competition trajectories for *T. confusum* and *T. castaneum*; one species winning in each case (redrawn from Neyman, Park and Scott, 1956).

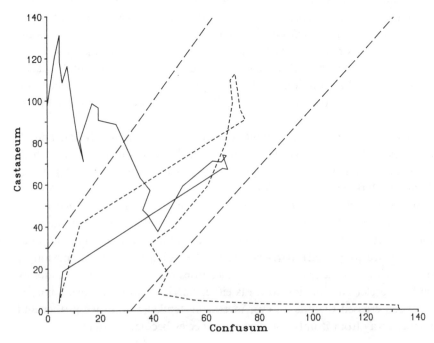

Gause (1935) studied *Paramecium aurelia* and *Paramecium bursaria* populations living in a tube containing yeast. Whilst the former fed on the yeast suspended in the upper layers of the fluid medium, the latter fed on the lower layers, and this difference in feeding behaviour was sufficient to allow the two species to coexist. Krebs (1985, chapter 12) provides further examples, and he also summarizes the wide range of opinion on both the importance of the competitive exclusion principle and the concept of niche.

5.4.3   *Simulation of two-species competition*

In spite of these early experimental developments it took the pioneering papers of Leslie (1958), Leslie and Gower (1958) and Barnett (1962) to stimulate interest in the simulation of two interacting species, though in fairness such research was really dependent on the emergence of sufficiently powerful computers.

To illustrate the type of computational procedure involved, suppose that species $i = 1, 2$ has individual birth rate r_i and individual death rate $s_{ii}X_i(t) + s_{ij}X_j(t)$ $(j \neq i)$ where $X_i(t)$ denotes stochastic population size at time t. Thus death is assumed to be due to competitive effects both within and between species. Then in successive small time intervals $(t, t + h)$ we have

$$
\begin{aligned}
\Pr[X_1 \to X_1 + 1] &\simeq X_1 r_1 h & &= B_1(X_1, X_2; t)h \\
\Pr[X_1 \to X_1 - 1] &\simeq X_1[s_{11}X_1 + s_{12}X_2]h & &= D_1(X_1, X_2; t)h \\
\Pr[X_2 \to X_2 + 1] &\simeq X_2 r_2 h & &= B_2(X_1, X_2; t)h \\
\Pr[X_2 \to X_2 - 1] &\simeq X_2[s_{21}X_1 + s_{22}X_2]h & &= D_2(X_1, X_2; t)h.
\end{aligned}
\tag{5.33}
$$

On denoting

$$
R = B_1 + D_1 + B_2 + D_2,
$$

we see that the probability that the next event is a birth or death of a member of species i is B_i/R or D_i/R, respectively. So event types may be simulated by using the following procedure:

  (i) generate a uniform pseudo-random number $0 \leqslant Y \leqslant 1$;
 (ii) use (5.33) to evaluate B_i, D_i and R for $i = 1, 2$;
(iii) if $0 \leqslant Y \leqslant B_1/R$ then $X_1 \to X_1 + 1$;
 (iv) if not, but $Y \leqslant (B_1 + D_1)/R$ then $X_1 \to X_1 - 1$;
  (v) if neither, but $Y \leqslant (B_1 + D_1 + B_2)/R$ then $X_2 \to X_2 + 1$;
 (vi) otherwise, $X_2 \to X_2 - 1$.

Whilst to simulate inter-event times (s):

 (vii) generate a new Y;
(viii) evaluate $s = -[\log_e(Y)]/R$ (as in (3.40));

(ix) update t to $t + s$;

then

(x) return to (i).

To illustrate the type of stochastic trajectories which correspond to the situations shown in Figure 5.3, we shall now use this simulation procedure on three specific processes, namely:

(A) stable equilibrium: $r_1 = 2.0$, $s_{11} = 0.03$, $s_{12} = 0.02$ and $r_2 = 1.5$, $s_{21} = 0.01$, $s_{22} = 0.04$ (as in Figure 5.4a);
(B) unstable equilibrium: $r_1 = 2.0$, $s_{11} = 0.01$, $s_{12} = 0.04$ and $r_2 = 1.5$, $s_{21} = 0.03$, $s_{22} = 0.02$ (as in Figure 5.4b);
(C) non-equilibrium: $r_1 = 2.0$, $s_{11} = 0.03$, $s_{12} = 0.01$ and $r_2 = 1.5$, $s_{21} = 0.04$, $s_{22} = 0.02$.

Figure 5.7 shows the simulated development of the first 2000 events of the stable equilibrium process (A) starting from $X_1(0) = X_2(0) = 1$. Although the trajectory initially wiggles its way fairly directly towards the deterministic equilibrium point $(50, 25)$, once $X_1(t)$ reaches the value 32 it then performs a

Figure 5.7. Simulation of the first 2000 events of the stable equilibrium competition process (A) starting from $X_1(0) = X_2(0) = 1$.

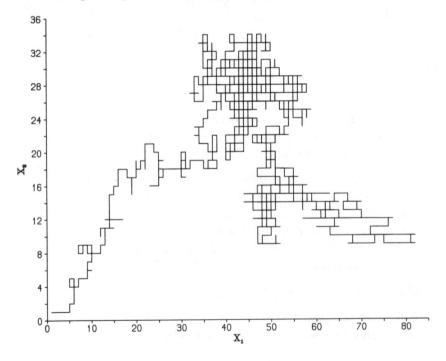

random walk within a roughly polygon-shaped region bounded by $32 \leqslant X_1 \leqslant 81$ and $9 \leqslant X_2 \leqslant 34$. Simulation over a much larger number of events would enable this central occupied region to be better defined. However, here we are interested not in the determination of space-filling properties, but in the simple illustration that the behaviour of stochastic and deterministic trajectories can be totally different. In particular, once the stochastic trajectory reaches the vicinity of the deterministic equilibrium point, further convergence towards it effectively ceases and random wandering begins.

In the unstable equilibrium process (B) the deterministic trajectories first move into regions II or III (see Figure 5.3iv) before moving directly away from the equilibrium point (N_1^*, N_2^*). Figures 5.8a and 5.8b show stochastic trajectories for two simulations which start at the unstable equilibrium point $N_1^* = 20$, $N_2^* = 45$, and they clearly parallel the biological trajectories shown in Figure 5.6. For the first few events of run 1 the process remains close to (N_1^*, N_2^*), but then it gradually drifts diagonally down towards the X_1-axis

Figure 5.8. Simulation of the unstable equilibrium process (B) starting from the equilibrium point $X_1(0) = 20$, $X_2(0) = 45$; (a) species 1 wins, (b) species 2 wins.

(a)

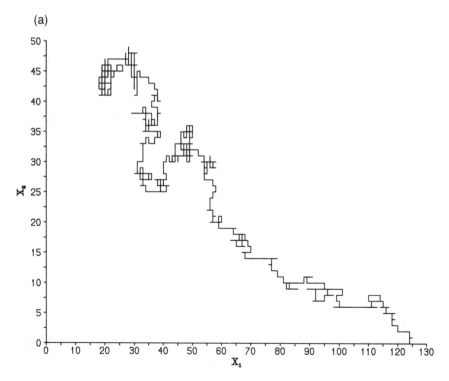

which it intersects at $X_1 = 124$. In contrast, run 2 immediately moves away from (N_1^*, N_2^*), but then spends a considerable time randomly wandering above it. However, once X_1 finally drops below 10 the trajectory drifts towards the X_2-axis which it meets at $X_2 = 64$. Thus not only can either species win from the same starting point, but also the trajectory may, or may not, exhibit random wandering before drift towards an axis takes place. Note that the winning population values of $X_1 = 124$ and $X_2 = 64$ are quite different from the deterministic end-points of $r_1/s_{11} = 200$ and $r_2/s_{22} = 75$, respectively.

To investigate this last result further, case B was simulated 100 times. Species 1 won in 39 runs and species 2 won in the remaining 61 runs, so here the probabilities of winning are approximately 0.4 and 0.6, respectively. Now the limits of zones II and III (Figure 5.3iv) are $r_2/s_{21} = 50$, $r_1/s_{11} = 200$ (species 1) and $r_1/s_{12} = 50$, $r_2/s_{22} = 75$ (species 2), respectively. Hence as the winning population values over the 100 simulations were in the range $108 \leqslant X_1 \leqslant 198$ and $49 \leqslant X_2 \leqslant 84$, it appears that stochastic trajectories of the winning species are likely to intersect the axes anywhere within, or near to, zones II and III.

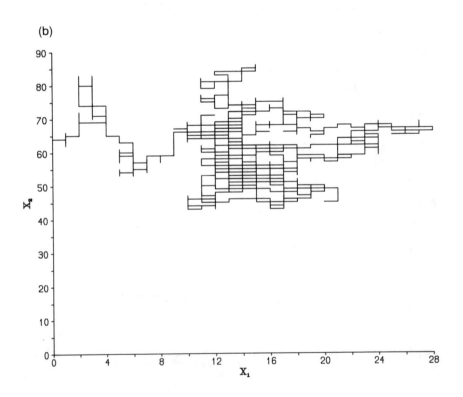

(b)

Figure 5.9 shows a simulation run for the non-equilibrium case C in which species 1 ultimately wins even though species 2 has the apparent initial advantage of a high starting value ($X_2(0) = 200$, whereas $X_1(0) = 5$). Though X_2 drops rapidly in the opening stages of the process there is little movement away from the X_2-axis until X_2 reaches 100. Thereafter X_1 drifts fairly rapidly upwards until X_2 reaches 10, at which point X_1 starts to wander.

Simulating case C 100 times (starting from the point (10, 200)) *always* led to species 1 winning, so here the general prediction of the deterministic analysis is correct. However, the lack of upwards drift of X_1 for large X_2 suggests that if $X_1(0)$ is very small then random excursions of X_1 might cause species 1 to become extinct in the early stages of the competition. Indeed, out of 100 simulations with the start point (1, 200) a total of 61 resulted in a win for species 2. Thus in non-equilibrium one species can be a certain winner only if its initial population size is not too small.

If we wish to make more precise statements about probabilities of extinction, mean times (or number of steps) to extinction, etc., then experiments need to be conducted which comprise considerably more than 100 simulation runs. Computational efficiency then becomes an important

Figure 5.9. Simulation of the non-equilibrium competition process (C) starting from $X_1(0) = 5$, $X_2(0) = 200$.

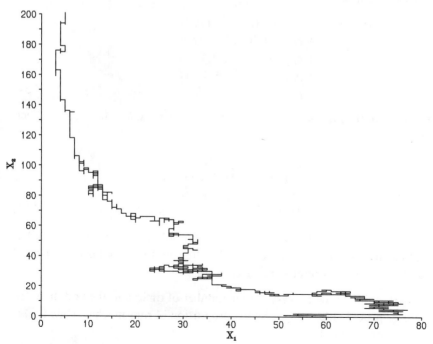

factor, so it is worth mentioning a simple variance-reduction idea proposed by Barnett (1962). Suppose that a simulation is started at the point $(X_1(0) = i, X_2(0) = j)$ and that it hits the X_1-axis (i.e. species 2 becomes extinct) after n steps. Then if, in the meantime, it passes through the point (i', j') (say) after r steps, this single simulation run also provides an example of a competition process which starts from (i', j') and leads to the extinction of species 2 after $n - r$ steps. Thus a simulation which starts from any point (i, j) also provides information on runs starting from points other than (i, j). Though this method clearly does not use independent realizations, the gain in computational efficiency can be enormous.

*5.5   Probability equations

Construction of the two-species equations for the probabilities $p_i(t) = \Pr\{X_1(t) = i, X_2(t) = j\}$ directly parallels that for the one-species growth process (Section 3.1.1). Given the general transition probabilities (5.33), in the small time interval $(t, t + h)$

$$
\begin{aligned}
\Pr\{\text{no event in } (t, t+h)\} &= 1 - [B_1(i, j; t) + D_1(i, j; t) \\
&\quad + B_2(i, j; t) + D_2(i, j; t)]h \\
&= 1 - R(i, j; t)h \quad \text{(say)},
\end{aligned}
$$

whence it follows that

$$
\begin{aligned}
p_{ij}(t + h) = {}&p_{i-1,j}(t)B_1(i - 1, j; t)h \\
&+ p_{i,j-1}(t)B_2(i, j - 1; t)h \\
&+ p_{i+1,j}(t)D_1(i + 1, j; t)h \\
&+ p_{i,j+1}(t)D_2(i, j + 1; t)h \\
&+ p_{ij}(t)[1 - R(i, j; t)h].
\end{aligned} \tag{5.34}
$$

Dividing both sides of this equation by h and letting $h \to 0$, as before, then leads to

$$
\begin{aligned}
dp_{ij}(t)/dt = {}&B_1(i - 1, j; t)p_{i-1,j}(t) \\
&+ B_2(i, j - 1; t)p_{i,j-1}(t) \\
&+ D_1(i + 1, j; t)p_{i+1,j}(t) \\
&+ D_2(i, j + 1; t)p_{i,j+1}(t) - R(i, j; t)p_{ij}(t).
\end{aligned} \tag{5.35}
$$

Except in the most trivial of cases direct solution of equation (5.35) is mathematically intractable. We are therefore forced back on to numerical solutions and two options are open to us:

(i) simulate the process a large number of times and then estimate the probabilities $\{p_{ij}(t)\}$ by the proportion of runs in which state (i, j) is occupied at time t;

(ii) solve equations (5.35) numerically – the easiest (though not the most accurate) way is to evaluate (5.34) at successive time points $t = 0, h, 2h, \ldots$ for suitably small h.

Unfortunately, unless t is very small the computer time needed to simulate sufficiently precise probability estimates can be horrendous. Option (ii) also has computational problems, since we have to choose not only an appropriately small value for h but also a suitable rectangular region $0 \leqslant i \leqslant I, 0 \leqslant j \leqslant J$ (say) within which equation (5.35) is to be solved. If I and J are too large then $\{p_{ij}(t)\}$ will be effectively zero well within the region and substantial computing time will be spent unnecessarily; whilst if I and J are too small then some of the $p_{ij}(t)$ for $i > I$ or $j > J$ will be non-negligible giving rise to substantial numerical error.

To illustrate option (ii) consider the symmetric competition process with parameter values $r_1 = r_2 = 0.075$, $s_{11} = s_{22} = 0.015$ and $s_{12} = s_{21} = 0.01$. Then (5.13) shows that there is a deterministic equilibrium point $(N_1^*, N_2^*) = (3, 3)$, and since

$$r_1/s_{12} = r_2/s_{21} = 7.5 > r_2/s_{22} = r_1/s_{11} = 5 \qquad (5.36)$$

we see from Figure 5.3iii that this equilibrium is stable. Choose h by putting the probability of an event in $(t, t + h)$ when $i = N_1^*$ and $j = N_2^*$ to be 0.1 (say), i.e.

$$R(3, 3)h = [3(r_1 + 3s_{11} + 3s_{12}) + 3(r_2 + 3s_{21} + 3s_{22})]h \simeq 0.1.$$

This gives $h \simeq 0.1$. Also select $I = J = 15$ as trial boundary values, since they are double the deterministic intercept values (5.36). Finally, let the population start at equilibrium.

Table 5.3 shows the resulting probability values at times $t = 1$ and 20. By $t = 1$ the probability distribution is starting to spread out from the equilibrium point (3,3), though the probability of a species becoming extinct by this time $(p_{\text{ext}}(t))$ is only 0.0002. By $t = 20$ this probability has risen to 0.2764, and conditional on extinction not having occurred the probabilities $p_{ij}(t)/[1 - p_{00}(t)]$ are beginning to approach their limiting values.

With $h = 0.1$ and $t = 20$ the program cycles 200 times over the 256 values of $i, j = 0, \ldots, 15$, giving a total of 51 200 operations. On an ICL 2988 machine this took 9 seconds of c.p.u. time which is large considering the trivially small size of the equilibrium value $(3, 3)$. More realistic values of $I = J = 200$ and $h = 0.01$ would involve around $200 \times 200 \times 2000 = 8 \times 10^7$ operations, equivalent to nearly 4 hours of c.p.u. time. If the required computing resource is acceptable then fine, but if not then the simulation

Table 5.3. Competition probabilities $\{p_{ij}(t)\} \times 10^3$ for $i, j = 0, \ldots, 10$ and (a) $t = 1$, (b) $t = 20$; parameter values are $r_1 = r_2 = 0.075$, $s_{11} = s_{22} = 0.015$ and $s_{12} = s_{21} = 0.01$

i						j					
	0	1	2	3	4	5	6	7	8	9	10
(a) 0	0	0	0	0	0	0	0	0	0	0	0
1	0	0	1	7	1	0	0	0	0	0	0
2	0	1	23	114	24	3	0	0	0	0	0
3	0	7	114	438	94	12	1	0	0	0	0
4	0	1	24	94	18	2	0	0	0	0	0
5	0	0	3	12	2	0	0	0	0	0	0
6	0	0	0	1	0	0	0	0	0	0	0
7	0	0	0	0	0	0	0	0	0	0	0
8	0	0	0	0	0	0	0	0	0	0	0
9	0	0	0	0	0	0	0	0	0	0	0
10	0	0	0	0	0	0	0	0	0	0	0
(b) 0	2	12	22	27	26	20	14	8	4	2	1
1	12	32	40	39	31	21	12	7	3	1	1
2	22	40	44	37	26	16	9	4	2	1	0
3	27	39	37	28	18	10	5	2	1	0	0
4	26	31	26	18	11	6	3	1	1	0	0
5	20	21	16	10	6	3	1	1	0	0	0
6	14	12	9	5	3	1	1	0	0	0	0
7	8	7	4	3	1	1	0	0	0	0	0
8	4	3	2	1	1	0	0	0	0	0	0
9	2	1	1	0	0	0	0	0	0	0	0
10	1	1	0	0	0	0	0	0	0	0	0

approach (i) (preferably combined with Barnett's (1962) variance-reduction procedure) will at least produce ball-park estimates fairly quickly.

5.6   Extinction

If competition goes on indefinitely then under both equilibrium and non-equilibrium conditions one of the species is certain to become extinct. In principle, the probability that it is species 1 (say) may be evaluated by solving equation (5.35) numerically for very large t. However, unless the population sizes $\{X_1(t), X_2(t)\}$ remain small this procedure is both extremely inefficient and computationally suspect.

*5.6.1   *Extinction probabilities*

To attack the extinction probabilities directly denote

$$q_{ij} = \text{Pr}\{\text{species 1 becomes extinct first given}$$
$$\text{that } X_1(0) = i, X_2(0) = j\}. \quad (5.37)$$

Then considering the *first* move away from (i, j) (in contrast to (5.34) which involves the *last* move) gives

$$q_{ij} = \Pr\{i \to i+1, j \to j\}q_{i+1,j} + \Pr\{i \to i-1, j \to j\}q_{i-1,j}$$
$$+ \Pr\{i \to i, j \to j+1\}q_{i,j+1} + \Pr\{i \to i, j \to j-1\}q_{i,j-1},$$

yielding the general equation

$$q_{ij} = B_1'(i, j)q_{i+1,j} + D_1'(i, j)q_{i-1,j} + B_2'(i, j)q_{i,j+1}$$
$$+ D_2'(i, j)q_{i,j-1}. \tag{5.38}$$

Here we have used (5.33) to form

$$B_1'(i, j) = \Pr\{\text{next event is a species 1 birth}\}$$
$$= B_1(i, j)/R(i, j),$$

and similarly $D_1' = D_1/R$, $B_2' = B_2/R$ and $D_2' = D_2/R$.

Although equation (5.38) is valid over all $i, j = 0, 1, 2, \ldots$, it is only computationally feasible to obtain a numerical solution within a finite region, say $0 \leqslant i \leqslant I$, $0 \leqslant j \leqslant J$. Let us therefore adopt the following procedure.

(i) Select appropriate values for I and J.

(ii) Put $q_{0j} = 1$ (species 1 loses) for $j = 1, 2, \ldots, J$, and $q_{i0} = 0$ (species 1 wins) for $i = 1, 2, \ldots, I$.

(iii) Set trial starting values $q_{ij}^{(0)} = 0.5$ (for example), for $1 \leqslant i \leqslant I + 1$, $1 \leqslant j \leqslant J + 1$.

(iv) Solve equations (5.38) in the iterative form

$$q_{ij}^{(n+1)} = B_1'(i, j)q_{i+1,j}^{(n)} + D_1'(i, j)q_{i-1,j}^{(n)}$$
$$+ B_2'(i, j)q_{i,j+1}^{(n)} + D_2'(i, j)q_{i,j-1}^{(n)} \tag{5.39}$$

for $n = 0, 1, 2, \ldots$ until the $\{q_{ij}^{(n)}\}$-values 'settle down'. In practice this happens when the maximum difference $|q_{ij}^{(n+1)} - q_{ij}^{(n)}|$ over $1 \leqslant i \leqslant I$, $1 \leqslant j \leqslant J$ is less than some suitably low value δ (e.g. 10^{-6}).

Since the probabilities $q_{I+1,j}^{(n)}$ and $q_{i,J+1}^{(n)}$ are not included in the above scheme they have to be estimated separately. One easy way is to use the linear approximation

$$q_{I+1,j}^{(n+1)} = 2q_{I,j}^{(n)} - q_{I-1,j}^{(n)} \qquad (j = 1, \ldots, J)$$
$$q_{i,J+1}^{(n+1)} = 2q_{i,J}^{(n)} - q_{i,J-1}^{(n)} \qquad (i = 1, \ldots, I) \tag{5.40}$$

(being careful to ensure that these values are constrained to lie between 0 and 1). When n is large $q_{I,j}^{(n)}$ is then the average of $q_{I-1,j}^{(n)}$ and $q_{I+1,j}^{(n+1)} \simeq q_{I+1,j}^{(n)}$ (and similarly for $q_{i,J}^{(n)}$).

Table 5.4 shows the resulting extinction probabilities $\{q_{ij}\}$ corresponding to: (a) the previous stable equilibrium example (used for Table 5.3) with $r_1 = r_2 = 0.075$, $s_{11} = s_{22} = 0.015$ and $s_{12} = s_{21} = 0.01$; and (b) the cor-

Table 5.4. Extinction probabilities {q_{ij}} × 10² for species 1 with $r_1 = r_2 = 0.075$ for (a) stable equilibrium ($s_{11} = s_{22} = 0.015$, $s_{12} = s_{21} = 0.01$) and (b) unstable equilibrium ($s_{11} = s_{22} = 0.01$, $s_{12} = s_{21} = 0.015$). **Bold** values correspond to probabilities in the range 0.45–0.55

(a)

j \ i	0	1	2	3	4	5	6	7	8	9	10	11	12	13	14	15	16	17	18	19
0	0	100	100	100	100	100	100	100	100	100	100	100	100	100	100	100	100	100	100	100
1	0	**50**	65	72	76	79	82	83	85	86	87	87	88	89	89	90	90	90	91	91
2	0	35	**50**	58	64	68	70	73	75	76	77	79	80	80	81	82	83	83	84	84
3	0	28	42	**50**	56	60	63	65	68	69	71	72	73	74	75	76	77	78	78	79
4	0	24	36	44	**50**	**54**	57	60	62	64	66	67	69	70	71	72	72	73	74	75
5	0	21	32	40	**46**	**50**	**53**	56	58	60	62	63	65	66	67	68	69	70	70	71
6	0	18	30	37	43	**47**	**50**	**53**	**55**	57	59	60	62	63	64	65	66	67	67	68
7	0	17	27	35	40	44	**47**	**50**	**52**	**54**	56	58	59	60	61	62	63	64	65	66
8	0	15	25	32	38	42	**45**	**48**	**50**	**52**	**54**	**55**	57	58	59	60	61	62	63	63
9	0	14	24	31	36	40	43	**46**	**48**	**50**	**52**	**53**	**55**	56	57	58	59	60	61	62
10	0	13	23	29	34	38	41	44	**46**	**48**	**50**	**52**	**53**	**54**	**55**	56	57	58	59	60
11	0	13	21	28	33	37	40	42	**45**	**47**	**49**	**50**	**51**	**53**	**54**	**55**	56	57	58	58
12	0	12	20	27	31	35	38	41	43	**45**	**47**	**49**	**50**	**51**	**52**	**53**	**54**	**55**	56	57
13	0	11	20	26	30	34	37	40	42	44	**46**	**47**	**49**	**50**	**51**	**52**	**53**	**54**	**55**	56
14	0	11	19	25	29	33	36	39	41	43	**45**	**46**	**48**	**49**	**50**	**51**	**52**	**53**	**54**	**55**
15	0	10	18	24	28	32	35	38	40	42	44	**45**	**47**	**48**	**49**	**50**	**51**	**52**	**53**	**53**
16	0	10	17	23	28	31	34	37	39	41	43	44	**46**	**47**	**48**	**49**	**50**	**51**	**52**	**52**
17	0	10	17	22	27	30	33	36	38	40	42	43	**45**	**46**	**47**	**48**	**49**	**50**	**51**	**51**
18	0	9	16	22	26	30	33	35	37	39	41	42	44	**45**	**46**	**47**	**48**	**49**	**50**	**51**
19	0	9	16	21	25	29	32	34	37	38	40	42	43	44	**45**	**47**	**47**	**48**	**49**	**50**

(b)

j	0	1	2	3	4	5	6	7	8	9	10	11	12	13	14	15	16	17	18	19
0	0	100	100	100	100	100	100	100	100	100	100	100	100	100	100	100	100	100	100	100
1	0	**50**	68	78	83	87	90	91	93	94	95	95	96	96	97	97	97	97	98	98
2	0	32	**50**	62	70	75	80	83	85	87	89	90	91	92	93	93	94	95	95	95
3	0	22	38	**50**	59	66	71	75	78	81	83	85	86	88	89	90	91	92	92	93
4	0	17	30	41	**50**	57	63	68	72	75	78	80	82	84	85	86	87	88	89	90
5	0	13	25	34	43	**50**	56	61	65	69	72	75	77	79	81	83	84	85	86	87
6	0	10	20	29	37	44	**50**	55	60	64	67	70	73	75	77	79	80	82	83	84
7	0	9	17	25	32	39	45	**50**	55	59	62	66	68	71	73	75	77	79	80	81
8	0	7	15	22	28	35	40	45	**50**	54	58	61	64	67	69	72	74	75	77	78
9	0	6	13	19	25	31	36	41	46	**50**	54	57	60	63	66	68	70	72	74	76
10	0	5	11	17	22	28	33	38	42	46	**50**	53	57	60	62	65	67	69	71	73
11	0	5	10	15	20	25	30	34	39	43	47	**50**	53	56	59	61	64	66	68	70
12	0	4	9	14	18	23	27	32	36	40	43	47	**50**	53	56	58	61	63	65	67
13	0	4	8	12	16	21	25	29	33	37	40	44	47	**50**	53	55	58	60	62	64
14	0	3	7	11	15	19	23	27	31	34	38	41	44	47	**50**	53	55	57	60	62
15	0	3	7	10	14	17	21	25	28	32	35	39	42	45	47	**50**	53	55	57	59
16	0	3	6	9	13	16	20	23	26	30	33	36	39	42	45	47	**50**	52	55	57
17	0	3	5	8	12	15	18	21	25	28	31	34	37	40	43	45	48	**50**	52	54
18	0	2	5	8	11	14	17	20	23	26	29	32	35	38	40	43	45	48	**50**	**52**
19	0	2	5	7	10	13	16	19	22	24	27	30	33	36	38	41	43	46	**48**	**50**

responding unstable process with s_{ii} and s_{ij} interchanged. Note that both equilibrium points are at $N_1^* = N_2^* = 3$. The cut-off point δ was chosen to be 10^{-8} which resulted in (a) 238 and (b) 255 iterations, the c.p.u. time for both being around 13 seconds. In case (b), for example, the number of iterations corresponding to the following δ-values were:

δ	10^{-2}	10^{-3}	10^{-4}	10^{-5}	10^{-6}	10^{-7}	10^{-8}
No. iterations	34	86	121	154	188	222	255

Thus once $\delta = 0.001$ had been reached, i.e. the start values had lost their effect, each extra decimal place in δ required 34 additional iterations.

As far as extinction probabilities are concerned, the main difference between stability and instability lies in the shape of the region within which trajectories are equally likely to hit either axis. For example, if we define this region as being between the $q_{ij} = 0.45$ and 0.55 contour lines, then Table 5.4 shows that whereas the region diverges quite widely in the stable case (a), it forms a much narrower wedge in the unstable case (b).

*5.6.2 *Times to extinction*

Only a marginal change is required to convert this approach into one that determines the expected time to extinction, where 'time' here denotes number of events and not absolute (i.e. clock) time. Let

$$t_{ij} = \text{expected time to extinction of either species}$$
$$\text{given } X_1(0) = i,\, X_2(0) = j, \tag{5.41}$$

and consider the first move away from the starting point (i, j). If the first event is a move from (i, j) to $(i + 1, j)$ then the expected number of events from then on is $t_{i+1,j}$, whence the total expected duration of the competition is $1 + t_{i+1,j}$. Thus on considering all four possible first moves we obtain the general equation

$$t_{ij} = B_1'(i,j)(t_{i+1,j} + 1) + D_1'(i,j)(t_{i-1,j} + 1)$$
$$+ B_2'(i,j)(t_{i,j+1} + 1) + D_2'(i,j)(t_{i,j-1} + 1),$$

i.e.

$$t_{ij} = B_1'(i,j)t_{i+1,j} + D_1'(i,j)t_{i-1,j}$$
$$+ B_2'(i,j)t_{i,j+1} + D_2'(i,j)t_{i,j-1} + 1 \tag{5.42}$$

since $B_1' + D_1' + B_2' + D_2' = 1$.

Apart from the extra '1' on the right-hand side, equation (5.42) is identical to (5.38). Only the boundary conditions

$$t_{i0} = t_{0j} = 0 \qquad \text{for all } i, j = 0, 1, 2, \ldots \tag{5.43}$$

are substantially different: if only one species is initially present then the time to extinction of the other must be zero. Operations (ii) to (iv) above (Section 5.6.1) are therefore modified as follows:

(ii)′ Put $t_{i0} = t_{0j} = 0$ for $i = 1, \ldots, I$ and $j = 1, \ldots, J$.
(iii)′ Set start values, $t_{ij}^{(0)} = 1$ (for example), for $1 \leqslant i \leqslant I$ and $1 \leqslant j \leqslant J$.
(iv)′ Solve equations (5.42) in the iterative form

$$t_{ij}^{(n+1)} = B_1'(i,j) t_{i+1,j}^{(n)} + \cdots + D_2'(i,j) t_{i,j-1}^{(n)} + 1 \tag{5.44}$$

for $n = 0, 1, 2, \ldots$.

Table 5.5 shows the expected times to extinction computed for two examples with the same equilibrium point $N_1^* = N_2^* = 10$:

(a) $r_1 = r_2 = 0.25$, $\quad s_{11} = s_{22} = 0.015$, $\quad s_{12} = s_{21} = 0.01$ (stable equilibrium);
(b) $r_1 = r_2 = 0.25$, $\quad s_{11} = s_{22} = 0.01$, $\quad s_{12} = s_{21} = 0.015$ (unstable equilibrium).

In both cases the two t_{ij}-surfaces quickly rise to an almost flat plateau. Thus unless either i or j is so small that there is a non-negligible probability of the opening events leading directly to extinction, both processes can evidently wander around for some considerable time before extinction occurs. One might well expect the high value $t_{10, 10} = 882$ in case (a), since there is always a driving force towards the equilibrium point $(10, 10)$ and so extinction involves a substantial journey *against* it. However, the size of $t_{10, 10} = 306$ in case (b) is rather surprising, since trajectories are now subject to a drift *towards* the axes which leads one to expect a mean time to extinction that is much smaller.

*5.7   Quasi-equilibrium probabilities

To develop the corresponding set of equations for the quasi-equilibrium probabilities $\{\pi_{ij}^{(Q)}\}$ consider the *last* move made by the process, as distinct from equations (5.38) for $\{q_{ij}\}$ and (5.42) for $\{t_{ij}\}$ which were constructed by considering the *first* move. Then

$\pi_{ij}^{(Q)} = \Pr[\text{trajectory is currently at } (i, j)]$
$\quad = \Pr[\text{previously at } (i - 1, j) \text{ followed by a species-1 birth}]$
$\quad + \Pr[\text{previously at } (i + 1, j) \text{ followed by a species-1 death}]$

Table 5.5 Expected times to extinction $\{t_{ij}\} \times 10^{-1}$ for either species with $r_1 = r_2 = 0.25$ for (a) stable equilibrium ($s_{11} = s_{22} = 0.015$, $s_{12} = s_{21} = 0.01$) and (b) unstable equilibrium ($s_{11} = s_{22} = 0.01$, $s_{12} = s_{21} = 0.015$)

(a)

j											i									
	0	1	2	3	4	5	6	7	8	9	10	11	12	13	14	15	16	17	18	19
0	0	0	0	0	0	0	0	0	0	0	0	0	0	0	0	0	0	0	0	0
1	0	50	55	53	51	48	46	44	42	41	39	38	37	36	35	34	33	33	32	31
2	0	55	67	69	68	67	65	64	62	60	59	58	56	55	54	53	52	51	51	50
3	0	53	69	74	75	75	74	73	72	71	70	69	68	67	66	65	64	63	63	62
4	0	51	68	75	78	79	79	79	78	78	77	76	75	74	73	73	72	71	70	70
5	0	48	67	75	79	81	82	82	82	81	81	80	80	79	79	78	77	77	76	76
6	0	46	65	74	79	82	83	84	84	84	84	83	83	83	82	82	81	81	80	80
7	0	44	64	73	79	82	84	85	85	86	86	85	85	85	85	84	84	84	83	83
8	0	42	62	72	78	82	84	85	86	87	87	87	87	87	87	86	86	86	86	85
9	0	41	60	71	78	81	84	86	87	87	88	88	88	88	88	88	88	88	87	87
10	0	39	59	70	77	81	84	86	87	88	88	89	89	89	89	89	89	89	89	89
11	0	38	58	69	76	80	83	85	87	88	89	89	90	90	90	90	90	90	90	90
12	0	37	56	68	75	80	83	85	87	88	89	90	90	90	91	91	91	91	91	91
13	0	36	55	67	74	79	83	85	87	88	89	90	90	91	91	91	91	91	92	92
14	0	35	54	66	73	79	82	85	87	88	89	90	91	91	91	92	92	92	92	92
15	0	34	53	65	73	78	82	84	86	88	89	90	91	91	92	92	92	93	93	93
16	0	33	52	64	72	77	81	84	86	88	89	90	91	91	92	92	93	93	93	93
17	0	33	51	63	71	77	81	84	86	88	89	90	91	91	92	93	93	93	93	94
18	0	32	51	63	70	76	80	83	86	87	89	90	91	92	92	93	93	93	94	94
19	0	31	50	62	70	76	80	83	85	87	89	90	91	92	92	93	93	94	94	94

(b)

j	i																			
	0	1	2	3	4	5	6	7	8	9	10	11	12	13	14	15	16	17	18	19
0	0	0	0	0	0	0	0	0	0	0	0	0	0	0	0	0	0	0	0	0
1	0	14	16	15	14	12	12	11	10	9	9	9	8	8	8	7	7	7	7	7
2	0	16	20	21	20	19	18	17	16	16	15	14	14	13	13	13	12	12	12	11
3	0	15	21	23	23	23	22	22	21	20	19	19	18	17	17	16	16	16	15	15
4	0	14	20	23	25	25	25	25	24	23	23	22	21	21	20	20	19	19	18	18
5	0	12	19	23	25	26	27	26	26	26	25	25	24	24	23	22	22	21	21	21
6	0	12	18	22	25	27	27	28	28	27	27	27	26	26	25	25	24	24	23	23
7	0	11	17	22	25	26	28	28	29	29	29	28	28	27	27	27	26	26	25	25
8	0	10	16	21	24	26	28	29	29	29	30	29	29	29	29	28	28	27	27	27
9	0	9	16	20	23	26	27	29	29	30	30	30	30	30	30	30	29	29	28	28
10	0	9	15	19	23	25	27	29	30	30	31	31	31	31	31	31	30	30	30	29
11	0	9	14	19	22	25	27	28	29	30	31	31	32	32	32	32	31	31	31	31
12	0	8	14	18	21	24	26	28	29	30	31	32	32	32	32	32	32	32	32	32
13	0	8	13	17	21	24	26	27	29	30	31	32	32	32	33	33	33	33	32	32
14	0	8	13	17	20	23	25	27	29	30	31	32	32	33	33	33	33	33	33	33
15	0	7	13	16	20	22	25	27	29	30	31	32	32	33	33	33	34	34	34	34
16	0	7	12	16	19	22	24	26	28	29	31	32	32	33	343	34	34	34	34	34
17	0	7	12	16	19	21	24	26	28	29	30	31	32	33	33	34	34	34	35	35
18	0	7	12	15	18	21	23	25	27	28	30	31	32	33	33	34	34	35	35	35
19	0	7	11	15	18	21	23	25	27	28	29	31	32	32	33	34	34	35	35	35

Table 5.6. Quasi-equilibrium probabilities $\{\pi_{ij}^{(Q)}\} \times 10^4$ for the stable competition process (a) of Table 5.5. **Bold** values correspond to probabilities greater than 0.005

										i										
j	1	2	3	4	5	6	7	8	9	10	11	12	13	14	15	16	17	18	19	20
1	0	0	0	0	0	1	2	4	6	10	14	19	23	26	28	28	27	24	20	16
2	0	0	0	0	1	2	4	7	11	16	21	27	31	34	35	34	31	26	21	16
3	0	0	0	1	2	4	7	11	17	23	30	36	40	42	41	38	33	27	22	16
4	0	0	1	2	4	7	12	18	25	33	40	46	48	48	46	41	34	27	21	15
5	0	1	2	4	7	12	19	27	35	44	50	**54**	**55**	**53**	48	41	33	26	19	13
6	1	2	4	7	12	19	27	37	46	**54**	**59**	**61**	**59**	**55**	48	39	31	23	17	11
7	2	4	7	12	19	27	37	47	**56**	**63**	**66**	**65**	**61**	**53**	45	36	27	20	14	9
8	4	7	11	18	27	37	47	**57**	**64**	**69**	**69**	**65**	**58**	50	40	31	23	16	11	7
9	6	11	17	25	35	46	**56**	**64**	**70**	**71**	**68**	**62**	**53**	44	35	26	19	13	8	5
10	10	16	23	33	44	**54**	**63**	**69**	**71**	**69**	**64**	**56**	47	37	28	20	14	9	6	4
11	14	21	30	40	50	**59**	**66**	**69**	**68**	**64**	**57**	48	39	30	22	15	10	7	4	3
12	19	27	36	46	**54**	**61**	**65**	**65**	**62**	**56**	48	39	31	23	16	11	7	5	3	2
13	23	31	40	48	**55**	**59**	**61**	**58**	**53**	47	39	31	23	17	12	8	5	3	2	1
14	26	34	42	48	**53**	**55**	**53**	50	44	37	30	23	17	12	8	5	3	2	1	1
15	28	35	41	46	48	48	45	40	35	28	22	16	12	8	5	3	2	1	1	0
16	28	34	38	41	41	39	36	31	26	20	15	11	8	5	3	2	1	1	0	0
17	27	31	33	34	33	31	27	23	19	14	10	7	5	3	2	1	1	0	0	0
18	24	26	27	27	26	23	20	16	13	9	7	5	3	2	1	1	0	0	0	0
19	20	21	22	21	19	17	14	11	8	6	4	3	2	1	1	0	0	0	0	0
20	16	16	16	15	13	11	9	7	5	4	3	2	1	1	0	0	0	0	0	0
21	12	12	11	10	9	7	6	4	3	2	1	1	1	0	0	0	0	0	0	0
22	8	8	7	6	5	4	3	3	2	1	1	1	0	0	0	0	0	0	0	0
23	5	5	4	4	3	2	2	1	1	1	0	0	0	0	0	0	0	0	0	0
24	3	3	2	2	2	1	1	1	0	0	0	0	0	0	0	0	0	0	0	0
25	1	1	1	1	1	0	0	0	0	0	0	0	0	0	0	0	0	0	0	0

+ Pr[previously at $(i, j - 1)$ followed by a species-2 birth]
+ Pr[previously at $(i, j + 1)$ followed by a species-2 death],

giving

$$\pi_{ij}^{(Q)} = \pi_{i-1,j}^{(Q)} B_1'(i - 1, j) + \pi_{i+1,j}^{(Q)} D_1'(i + 1, j)$$
$$+ \pi_{i,j-1}^{(Q)} B_2'(i, j - 1) + \pi_{i,j+1}^{(Q)} D_2'(i, j + 1). \quad (5.45)$$

Remember that $B_1' = B_1/R$, etc., where $R = B_1 + D_1 + B_2 + D_2$ and

$$B_1(i, j) = ir_1; \qquad\qquad B_2(i, j) = jr_2$$
$$D_1(i, j) = i(is_{11} + js_{12}); \qquad D_2(i, j) = j(is_{21} + js_{22}).$$

To use our iterative procedure on equations (5.45) we set $\pi_{ij}^{(Q)} = 0$ for (i, j) outside the rectangular region $1 \leqslant i \leqslant I, 1 \leqslant j \leqslant J$ and evaluate $\{\pi_{ij}^{(Q)}\}$ within it. Ideally, I and J should be chosen large enough to ensure that the probability of being on either of the edges (I, j) or (i, J) is negligible. However, some leakage of probability from the rectangle is bound to occur, either by deaths onto the axes $(i, 0)$ and $(0, j)$, or else by births into the interior states $(i, J+1)$ and $(I+1, j)$. At the end of each iteration the $\{\pi_{ij}^{(Q)}\}$ must therefore be rescaled to ensure that they sum to one over $1 \leqslant i \leqslant I$, $1 \leqslant j \leqslant J$.

Table 5.6 gives the solution for the stable equilibrium example (a) of Section 5.6.2, and shows contours that are roughly elliptical in shape (illustrated by the **bold** values). The maximum values $\pi_{10,9}^{(Q)} = \pi_{9,10}^{(Q)} = 0.0071$ occur very close to the deterministic equilibrium point $(10, 10)$, but here the similarity between deterministic and stochastic behaviour ends. Since $\pi_{1,16}^{(Q)} = \pi_{16,1}^{(Q)} = 0.0028$, a stochastic trajectory is only $2\frac{1}{2}$ times more likely to be at the equilibrium point than at either of these two edge points, which shows the extent to which trajectories are likely to wander away from (N_1^*, N_2^*). If X_2 is small then its competitive effect on X_1 is negligible and so X_1 behaves as a single-species logistic process centred around the carrying capacity $r_1/s_{11} = 0.25/0.015 \simeq 17$. Similarly, when X_1 is small X_2 is centred around $r_2/s_{22} \simeq 17$.

For processes with much larger equilibrium values than $(10, 10)$ the effective range of the trajectories may lie well away from the axes. In such cases the shape of the $\{\pi_{ij}^{(Q)}\}$-surface may be approximated by a bivariate Normal distribution, though we shall leave discussion of such considerations until the next chapter.

6

Predator–prey processes

Our study of two interacting populations has so far been restricted to direct competition within and between species. That is, in terms of the general equations

$$dN_1/dt = N_1(r_1 + a_{11}N_1 + a_{12}N_2)$$
$$dN_2/dt = N_2(r_2 + a_{21}N_1 + a_{22}N_2) \tag{6.1}$$

we have taken

$$r_1 > 0 \quad \text{with} \quad a_{11} = -s_{11} < 0, \quad a_{12} = -s_{12} < 0$$
$$r_2 > 0 \quad \text{with} \quad a_{21} = -s_{21} < 0, \quad a_{22} = -s_{22} < 0. \tag{6.2}$$

Such a configuration of the r_i and a_{ij} may be denoted in an obvious notation by

$$\begin{pmatrix} + & | & - & - \\ + & | & - & - \end{pmatrix},$$

and examples of other possible configurations include:

(i) two non-interacting populations

$$\begin{pmatrix} + & | & - & 0 \\ + & | & 0 & - \end{pmatrix} \quad \text{(independent logistics);}$$

(ii) the second population lives on the waste products of the first, but otherwise does it neither harm nor good

$$\begin{pmatrix} + & | & - & 0 \\ - & | & + & - \end{pmatrix} \quad \text{(scavenging);}$$

(iii) the second population not only lives on the first but also cultivates it (e.g. humans farming cows)

$$\begin{pmatrix} - & | & 0 & + \\ - & | & + & 0 \end{pmatrix} \quad \text{(symbiosis).}$$

Since each of the six parameters may be either + (enhancement),

– (impedance) or 0 (no effect), the total number of possible configurations for the basic system (6.1) is $3^6 = 729$! However, few of these will have any biological relevance, and since only one truly interactive two-species process, namely *predation*, has as much biological importance as *competition*, we shall restrict our attention in this chapter to predation alone. Less specific studies of the general representation (6.1) are given in texts such as Keyfitz (1977); whilst Hallam (1986) develops theoretical results for a wide class of models involving predation, competition and cooperation.

6.1   The Lotka–Volterra model

Predation occurs when members of one species eat those of another species. Four types of biological predation may be distinguished, each of which is equivalent when expressed in their simplest mathematical form. *Herbivores* are animals that prey on plants or their fruits or seeds, and although the plants eaten are often not killed they may nevertheless be damaged. *Carnivores* prey on herbivores or other carnivores, and such behaviour is the common concept of predation. *Parasitism* is a variant on predation, and involves the parasite laying eggs on or near the host, which is then subsequently eaten. *Cannibalism* is a particular form of predation involving just one species, with predator and prey often being the adults and young, respectively.

The earliest representations of predator–prey behaviour were constructed independently by Lotka (1925) and Volterra (1926), but although their point of departure was much the same Volterra probed far more deeply. He discussed several possible cases, the simplest being that of two associated species. One species, finding sufficient food in its environment, would multiply indefinitely when left to itself, whilst the other would perish for lack of nourishment if left alone; but the second feeds upon the first, and so the two species can coexist together. (See Scudo, 1971, for a concise and non-technical description of Volterra's major contributions to ecology.)

Denote

$$N_1(t) = \text{number of prey (or hosts)}$$

and

$$N_2(t) = \text{number of predators (or parasites)}$$

at time t. Suppose that in the absence of predators prey increase at rate r_1, whilst in the absence of prey predators die at rate r_2. Then under the deterministic scheme (6.1) the simplest representation is

$$dN_1/dt = N_1(r_1 - b_1 N_2)$$
$$dN_2/dt = N_2(-r_2 + b_2 N_1), \tag{6.3}$$

corresponding to the configuration

$$\begin{pmatrix} + & 0 & - \\ - & + & 0 \end{pmatrix}.$$

Here within-species competition has been ignored ($a_{11} = a_{22} = 0$). The constant b_1 measures the death rate of prey due to being eaten by predators; the greater the number of predators the quicker the prey population will be depleted. The constant b_2 measures the skill of the predator in catching prey; the greater the number of prey the greater the availability of predator food resource.

Equations (6.3) combine to form the single equation

$$\frac{dN_1}{dN_2} = \frac{N_1(r_1 - b_1 N_2)}{N_2(-r_2 + b_2 N_1)}, \tag{6.4}$$

which, on writing in the form

$$[-(r_2/N_1) + b_2]\, dN_1 = [(r_1/N_2) - b_1]\, dN_2,$$

integrates directly to give

$$r_2 \log_e(N_1) - b_2 N_1 + r_1 \log_e(N_2) - b_1 N_2 = \text{constant}. \tag{6.5}$$

This expression represents a family of closed curves in which each member of the family corresponds to a different value of the constant, i.e. a curve is determined by the initial position $(N_1(0), N_2(0))$.

Five such curves are illustrated in Figure 6.1 for the process

$$\begin{aligned} dN_1/dt &= N_1(1.50 - 0.1 N_2) \\ dN_2/dt &= N_2(-0.25 + 0.01 N_1) \end{aligned} \tag{6.6}$$

with start points $N_2(0) = 15$ and $N_1(0) = 1, 5, 10, 15$ and 20. Any trajectory will follow a closed path in an anticlockwise direction indefinitely; there is neither damping towards the equilibrium point $N_1^* = r_2/b_2 = 25$, $N_2^* = r_1/b_1 = 15$, nor an outwards drift towards the axes. The trajectory for $N_1(0) = 20$ is seen to be almost elliptical in shape, that for $N_1(0) = 15$ to be slightly skew, whilst trajectories sited further and further away from the equilibrium point become more and more squashed near the axes since they are not allowed to cross them.

Whilst the predator–prey model (6.6) implies cyclic trajectories, we must caution against the assumption that observed cyclic behaviour automatically results from a predator–prey scenario (Williamson, 1975). Various species in the Canadian coniferous forests show a strong ten-year population cycle, in particular the snow-shoe hare, lynx, fox, marten, fisher, and probably also the musk-rat and mink; a four-year cycle found typically in the tundra north of the forests is associated with lemmings and fox. The temptation to draw (wrong) biological conclusions about possible underlying predator–prey mechanisms can sometimes be irresistible!

6.1.1   *Local deterministic solution*

Suppose we restrict attention to small departures from equilibrium. Then in developing an argument analogous to that used for the competition process (Section 5.3.1), we may write either

$$N_i(t) = N_i^*[1 + n_i(t)] \qquad \text{(multiplicative effect)} \qquad (6.7)$$

or

$$N_i(t) = N_i^* + n_i(t) \qquad \text{(additive effect)}. \qquad (6.8)$$

As these yield the same approximate solutions (to first-order), either can be used. Substituting (6.8) into equations (6.3) yields

$$dn_1/dt = (N_1^* + n_1)[r_1 - b_1(N_2^* + n_2)]$$
$$dn_2/dt = (N_2^* + n_2)[-r_2 + b_2(N_1^* + n_1)].$$

Hence as $N_1^* = r_2/b_2$ and $N_2^* = r_1/b_1$, for small n_1 and n_2 we have

$$dn_1/dt \simeq -(b_1 r_2/b_2)n_2 \qquad (6.9)$$

$$dn_2/dt \simeq (b_2 r_1/b_1)n_1. \qquad (6.10)$$

Figure 6.1. Family of closed curves (6.5) for the Lotka–Volterra process (6.6)

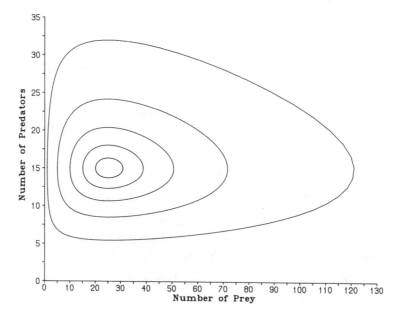

Eliminating n_2 between (6.9) and (6.10) gives

$$d^2 n_1/dt^2 = -r_1 r_2 n_1$$

which has the general solution

$$n_1(t) = (\alpha r_2/b_2) \cos\{t\sqrt{(r_1 r_2)} + \beta\} \tag{6.11}$$

where $(\alpha r_2/b_2)$ and β are constants of integration. Thus

$$\begin{aligned} N_1(t) &= N_1^* + n_1(t) \\ &= (r_2/b_2)[1 + \alpha \cos\{t\sqrt{(r_1 r_2)} + \beta\}]. \end{aligned} \tag{6.12}$$

To determine $n_2(t)$ we substitute for $n_1(t)$ from (6.11) into (6.9), obtaining

$$\begin{aligned} n_2(t) &= -(b_2/b_1 r_2)(dn_1/dt) \\ &= (\alpha/b_1)\sqrt{(r_1 r_2)} \sin\{t\sqrt{(r_1 r_2)} + \beta\}, \end{aligned} \tag{6.13}$$

whence

$$\begin{aligned} N_2(t) &= N_2^* + n_2(t) \\ &= (r_1/b_1)[1 + \alpha\sqrt{(r_2/r_1)} \sin\{t\sqrt{(r_1 r_2)} + \beta\}]. \end{aligned} \tag{6.14}$$

The constants α and β may be determined directly from known initial values

$$N_1(0) = (r_2/b_2)[1 + \alpha\cos(\beta)] \quad \text{and} \quad N_2(0) = (r_1/b_1)[1 + \alpha\sqrt{(r_2/r_1)}\sin(\beta)].$$

The ellipsoidal shape of local trajectories follows immediately. Since $\cos^2(\beta) + \sin^2(\beta) = 1$ for all angles β, solutions (6.11) and (6.13) combine to form

$$(b_2^2 r_1)n_1^2 + (b_1^2 r_2)n_2^2 = \alpha^2 r_2^2 r_1 \tag{6.15}$$

which is the standard equation for an ellipse. Higher-order approximations may be obtained by first writing (6.5) as

$$\begin{aligned} r_2 \log_e[1 + (n_1/N_1^*)] - b_2 n_1 + r_1 \log_e[1 + (n_2/N_2^*)] - b_1 n_2 \\ = \text{(another) constant,} \end{aligned}$$

and then expanding the $\log_e$-terms to obtain

$$\begin{aligned} r_2[\tfrac{1}{2}(n_1/N_1^*)^2 - \tfrac{1}{3}(n_1/N_1^*)^3 + \cdots] \\ + r_1[\tfrac{1}{2}(n_2/N_2^*)^2 - \tfrac{1}{3}(n_2/N_2^*)^3 + \cdots] = \text{constant.} \end{aligned}$$

We can now include terms up to any desired power.

Various deterministic properties of the prey and predator populations follow directly from the local solutions (6.12) and (6.14).

(i) Both populations vary sinusoidally with period $T = 2\pi/\sqrt{(r_1 r_2)}$ (so in example (6.6) $T = 10.26$).

(ii) The populations are always $\frac{1}{2}\pi$ (i.e. 90 degrees) out of phase, prey leading.

(iii) Their respective amplitudes are $(\alpha r_2/b_2)$ and $\alpha\sqrt{(r_1 r_2)}/b_1$, so the ratio of prey to predator amplitude is $(b_1/b_2)\sqrt{(r_2/r_1)}$ (here 0.41). Note that both this ratio and T are independent of the starting point $(N_1(0), N_2(0))$.

(iv) The mean population size taken around a cycle is equal to the equilibrium value.

However, these properties are based purely on deterministic considerations; corresponding stochastic trajectories might meander amongst the entire family of closed curves (6.5) and eventually hit one of the axes. So are these deterministic results due more to mathematical imagination than biological reality?

6.1.2   *Biological investigations*

The first empirical test of the Lotka–Volterra model was made by Gause (1934). The protozoans *Paramecium caudatum* (prey) and *Didinium nasutum* (predator) were reared together in an oat medium, and despite Gause altering the circumstances of his experiment in various ways *Didinium* always exterminated *Paramecium* and then died of starvation. Instead of the classic oscillations predicted by deterministic theory, the experiments all yielded divergent oscillations resulting in extinction of the prey.

Gause believed that the desired theoretical results were not achieved because of the biological peculiarity of predators being able to multiply very rapidly even when prey were scarce. In an attempt to counterbalance this effect he introduced a prey-refuge into the system in which prey were safe from predator attack. The predicted oscillations still did not occur. *Didinium* eliminated *Paramecium*, as before, but only outside the refuge; they then starved to death, after which the *Paramecium* came out from hiding and increased in numbers. Undaunted, Gause tried yet another approach, this time by introducing immigration. Every third day he added one predator and one prey, but although oscillations did now occur he was forced to concede that they resulted only from interference (i.e. immigration) from outside the system.

Note the rather unusual approach. Instead of the usual concept of *mathematical* modelling in which fairly simple mathematical models are developed to describe the salient points of biological phenomena, here *biological* systems are created in an attempt to mimic mathematical phenomena!

Success was achieved some years later by Utida (1957), who ran a series of laboratory experiments in which mixed populations of the azuki bean weevil (*Callosobruchus chinensis*) as host and its larval wasp parasite (*Heterospilus prosopidis*) were developed in Petri dishes maintained at a constant temperature of 30 °C and a constant relative humidity of 75%. Ten grams of azuki beans were placed in each Petri dish every ten days to serve as food for the weevils. Three different experimental runs were started with different initial densities of host and parasite, and population counts were made every ten days. Not only were oscillations observed, but in the case of the first run they continued right up until the 112th generation when the population was accidentally destroyed. Figure 6.2 shows the fluctuation of population density for the third run in which complete extinction was observed as a consequence of the eruptive increase in first host, and then parasite, densities at generations 74, and 81, respectively. In all three runs the densities of host and parasite initially rose and fell broadly in line with deterministic prediction. Thereafter damping occurred, with fluctuations decreasing in amplitude. The populations then passed through several tens of generations in a state of fairly steady density, until relatively violent oscillations suddenly recurred which were then damped in their turn (especially noticeable for the first run).

Figure 6.2. Fluctuations in population density in a host–parasite system of the azuki bean weevil (*Callosobruchus chinensis* – ○) (host) and its larval wasp parasite (*Heterospilus prosopidis* – △) (redrawn from Utida, 1957).

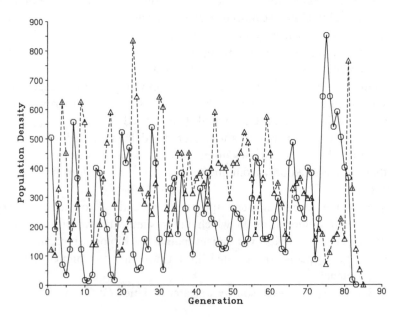

Utida argued that since internal oscillations had been clearly shown in his previous, as well as (then) current work, the existence of predator–prey cycles was beyond doubt. However, since he also acknowledged the delayed effect of the parasite upon the host population, lag effects should really be introduced into the basic differential equations (6.3) before any comparison is made between theory and practice. This apart, we see from Figure 6.2 that over the first 34 generations the hosts and parasites both exhibit 5 full cycles, giving a period of oscillation of about 7 generations. The time gaps between successive host and predator peaks are 3, 2, 4, 3, 2, which averaging at just under 3 generations are more than the 7/4 generations (i.e. $\frac{1}{2}\pi$ phase difference) predicted. Even so, the basic qualitative features of the Lotka–Volterra model are clearly sustained over a reasonably long time period.

An interesting experimental approach was taken by Luckinbill (1973). He realized that the confinement of prey in the small arenas necessary for laboratory study may make it impossible for prey to become so scarce that predators cannot find them. Coexistence might therefore be accomplished in a small system by slowing down the frequency of contact between predator and prey. He proved his point by adding methyl cellulose to interacting laboratory populations of *Paramecium aurelia* and its predator *Didinium nasutum*.

Another consideration is the question of natural selection. Contrary to the constant parameter assumption inherent in equations (6.3), the game of hide-and-seek between host and parasite might intensify the host's ability to conceal itself, or otherwise to avoid attack by the parasite. Indeed, significant evolutionary changes in laboratory predator–prey systems have been demonstrated over as few as 20 generations – see, for example, the wasp parasite (*Nasonia vitripennis*) and house fly host (*Musca domestica*) interaction (Pimental, Nagel and Madden, 1963). The study of related mathematical models which incorporate gradual parameter change clearly has considerable potential.

6.1.3   *Simulation of the stochastic model*

Such is the extent of the natural variation exhibited in Utida's data (Figure 6.2), that the deterministic model (6.3) cannot hope to mimic the development of these two interacting populations. Let us therefore simulate the corresponding stochastic model to see if it can provide us with a more realistic representation. This exercise proceeds exactly as described in Section 5.4.3, where the birth and death rates (5.33) now take the form

$$B_1(X_1, X_2) = r_1 X_1 \quad \text{and} \quad D_1(X_1, X_2) = b_1 X_1 X_2 \quad \text{(prey)}$$
$$B_2(X_1, X_2) = b_2 X_1 X_2 \quad \text{and} \quad D_2(X_1, X_2) = r_2 X_2 \quad \text{(predators)}.$$

$$(6.16)$$

Figure 6.3 shows the trajectories of two runs for the process corresponding to example (6.6), i.e. $r_1 = 1.5, r_2 = 0.25, b_1 = 0.1$ and $b_2 = 0.01$. The first run starts at the deterministic equilibrium point (25, 15), and since near this point B_i and D_i are in approximate balance the trajectory initially wanders randomly about it. Eventually a random excursion takes X_1 past 50, where X_2 experiences a strong upwards pull since $B_2 \simeq 0.5X_2$ is now twice $D_2 \simeq 0.25X_2$. This continues until X_2 reaches 25 at which point X_1 starts to decay towards zero since $D_1 \simeq 2.5X_1$ is substantially larger than $B_1 \simeq 1.5X_1$. Moreover, since $B_1 < D_1$ until X_2 drops below 16, X_1 has a high probability of drifting onto the X_2-axis, resulting in prey extinction.

The further a trajectory moves from its equilibrium point the more force it experiences. The second run demonstrates this effect clearly, showing far more decisive behaviour than the first. The small start value (3, 2) is well away from the equilibrium point (25, 15), and the trajectory moves in a fairly smooth manner almost until its end. Observe that initially B_1 is substantially greater than D_1, B_2 or D_2, and the prey population increases in size to over 110 before the first predator event occurs. That the prey population is almost

Figure 6.3. Two simulations corresponding to the deterministic Lotka–Volterra process (6.6) with start points of (25, 15) and (3, 2).

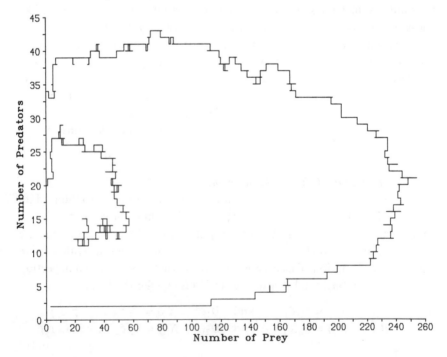

certain to become extinct during the first cycle is self-evident from inspection of the closed deterministic curves shown in Figure 6.1. The deterministic trajectory passing through $(3, 2)$ lies well outside that passing through $(1, 15)$, and opportunity for a slight perturbation onto the X_2-axis is clearly rife.

This contrast between sustained deterministic cycles and early stochastic prey extinction explains Gause's difficulty in modelling the deterministic process biologically in the laboratory.

*6.1.4   A trajectory indicator

Expression (6.5), namely

$$F(X_1, X_2) = r_2 \log_e(X_1) - b_2 X_1 + r_1 \log_e(X_2) - b_1 X_2, \qquad (6.17)$$

provides a useful indicator for distinguishing between stochastic trajectories. Under deterministic theory $F(X_1, X_2)$ remains constant at the value $F(X_1(0), X_2(0))$, so any change in $F(X_1(t), X_2(t))$ as the process develops reflects the way in which the stochastic trajectory wanders across the range of deterministic trajectories. That is, $F(X_1(t), X_2(t))$ tells us which deterministic trajectory the stochastic process $\{X_1(t), X_2(t)\}$ is sitting on at time t.

To determine the range over which F can vary, note that any closed curve (6.17) must cross the horizontal line $X_2 = N_2^*$ between $X_1 = N_1^*$ and $X_1 = 0$. So on fixing $X_2 = N_2^*$ and differentiating (6.17) with respect to X_1, we have

$$dF/dX_1 = (r_2/X_1) - b_2 = 0 \quad \text{at} \quad X_1 = r_2/b_2 = N_1^* \qquad (6.18)$$

which gives the maximum value

$$F_{\max} = r_2[\log_e(N_1^*) - 1] + r_1[\log_e(N_2^*) - 1].$$

The effective minimum value, corresponding to $X_1 = 1$, $X_2 = N_2^*$ ($X_1 = 0$ implies extinction), is

$$F_{\min} = -b_2 + r_1[\log_e(N_2^*) - 1].$$

For the simulations shown in Figure 6.3, and X_2 fixed at $N_2^* = 15$, F therefore ranges from 3.12 ($X_1 = 25$) down to 2.55 ($X_1 = 1$). Intermediate values are:

X_1	1	5	10	15	20	25
F	2.55	2.91	3.04	3.09	3.11	3.12

Note how slowly F falls as X_1 starts to move away from $N_1^* = 25$. In its first 15 moves, simulation 1 of Figure 6.3 travels from $(25, 15)$ to $(26, 11)$, at which point $F = 3.05$. Then F gradually drifts downwards, reaching 2.80 at (26, 27). Since this point has the same F-value as (3, 15), the stochastic trajectory already seems destined to travel close to the predator axis. In simulation 2 the start point (3, 2) has $F = 1.08$, and since this is considerably less than 2.55 the predators seem doomed to extinction from the outset.

6.1.5   *Final comments*

It is far too easy to fall into either the deterministic or the stochastic camps. The 'determinists' may claim, quite justifiably, that they can predict the basic shape and properties of the stochastic trajectories. For example, simulation 2 of Figure 6.3 reaches $X_1 = N_1^*$ at time $t = 1.69$ on its outward journey and $t = 5.71$ on its return, giving an elapsed time for this half-cycle of 4.02. This is predicted quite well by the deterministic value $T/2 = \pi/\sqrt{(r_1 r_2)}$ $= 5.13$, even though the calculations assume that trajectories remain close to the equilibrium point. The 'stochastics' may argue that random wandering across the family of closed curves (6.5) is bound to occur, and that deterministic theory predicts neither extinction nor the movement away from equilibrium experienced by simulation 1.

Surely the correct response is that common sense must prevail. If we take features from *both* analyses then we can predict the overall qualitative shape of the trajectories *and* the virtual certainty of extinction.

6.2   The Volterra model

Though Bartlett (1957a) did manage to obtain a simulation run of the Lotka–Volterra model which displayed three clear full cycles before extinction of predators occurred (he used $r_1 = 1.0$, $b_1 = 0.1$, $r_2 = 0.5$ and $b_2 = 0.02$), in view of the general difficulty of generating sustained stochastic cycles some modification to this model is clearly necessary if results like those displayed in Figure 6.2 are to be modelled successfully by a predator–prey relationship.

In the absence of predators, the prey population grows exponentially at rate r_1 according to a pure birth process. To prevent this potential population explosion a natural first extension to model (6.3) is to introduce a logistic term $(-cN_1^2)$ into the prey equation. The deterministic equations may now be written in the form (developed in considerable generality by Volterra, 1931)

$$dN_1/dt = N_1(r_1 - cN_1 - b_1 N_2) \quad \text{(prey)}$$
$$dN_2/dt = N_2(-r_2 + b_2 N_1) \quad \text{(predators)}. \tag{6.19}$$

Note that introducing a logistic term $(-c'N_2^2)$ into the predator equation offers little new, since it merely strengthens the death term already present.

Though introducing time-lags into the equations does increase the variety of possible behaviour (see Caswell, 1972).

The three basic assumptions underlying equations (6.19) are:

(i) in the absence of predators the prey population develops as a logistic process with intrinsic rate of increase r_1 and carrying capacity r_1/c;

(ii) the rate at which prey are eaten is proportional to the product of the two population sizes;

(iii) the rate at which predators are born is proportional to the product of the two population sizes, and no time-lag is involved.

Whilst assumption (i) is reasonable, (ii) assumes that predators are free-ranging and are not subject to territorial limits within the confines of the study area. Whilst this may be reasonable for laboratory experiments, its relevance to field data is more suspect (we shall introduce territorial considerations in the next chapter). Though assumption (iii) may be considered biologically naive, it does enable interesting conclusions to be drawn about the general development of the process.

6.2.1 *Deterministic trajectories*

To analyze the nature of the deterministic trajectories let us construct a phase-plane diagram analogous to that developed for the competition process (Figure 5.3). From (6.19)

$$dN_1/dt = 0 \quad \text{at} \quad cN_1 + b_1N_2 = r_1$$
$$dN_2/dt = 0 \quad \text{at} \quad b_2N_1 = r_2, \tag{6.20}$$

so there is an equilibrium point at

$$N_1^* = r_2/b_2 \quad \text{and} \quad N_2^* = (1/b_1)(r_1 - cr_2/b_2). \tag{6.21}$$

Here N_1^* is necessarily positive, but N_2^* is positive only if

$$r_1 > cr_2/b_2, \quad \text{i.e. } r_1/c > r_2/b_2. \tag{6.22}$$

Thus equilibrium can exist only if the prey carrying capacity r_1/c is large enough to support the predator population.

The dynamics of the system can be obtained directly from equations (6.19) by noting that:

$dN_1/dt > 0$ for points (N_1, N_2) below the line $cN_1 + b_1N_2 = r_1$;

$dN_1/dt < 0$ for points (N_1, N_2) above the line $cN_1 + b_1N_2 = r_1$;

$dN_2/dt > 0$ for $N_1 > r_2/b_2$;

$dN_2/dt < 0$ for $N_1 < r_2/b_2$.

Inserting the appropriate arrows in the resulting phase-plane diagram

(Figure 6.4), and then joining them up, produces a spiral trajectory converging towards the equilibrium point (N_1^*, N_2^*).

*6.2.2   General local solution

This result may be justified theoretically by paralleling the local stability argument developed in Section 5.3.1. On writing $N_i(t) = N_i^* + n_i(t)$ (i.e. using (6.8) instead of (6.7)), equations (6.19) become

$$dn_1/dt = (N_1^* + n_1)[r_1 - c(N_1^* + n_1) - b_1(N_2^* + n_2)]$$
$$dn_2/dt = (N_2^* + n_2)[-r_2 + b_2(N_1^* + n_1)].$$

From the equilibrium equations (6.20) $r_1 - cN_1^* - b_1N_2^* = 0$ and $b_2N_1^* - r_2 = 0$, so

$$dn_1/dt = (N_1^* + n_1)[-cn_1 - b_1n_2] \simeq -N_1^*(cn_1 + b_1n_2)$$
$$dn_2/dt = (N_2^* + n_2)(b_2n_1) \simeq N_2^*b_2n_1 \qquad (6.23)$$

where terms in n_1^2, n_1n_2 and n_2^2 have been ignored.

Now bearing in mind that we have already developed a similar system (5.17) for the competition process, it is expedient to study the general local

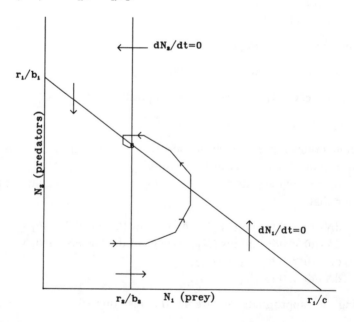

Figure 6.4. Phase-plane diagram for the Volterra predator–prey process (6.19) with $r_1/c > r_2/b_2$.

equations

$$dn_1/dt = k_{11}n_1 + k_{12}n_2 \tag{6.24}$$

$$dn_2/dt = k_{21}n_1 + k_{22}n_2. \tag{6.25}$$

Any associated two-species system can then be analyzed by simply inserting the appropriate constants k_{ij}. On denoting $n' = dn/dt$ and $n'' = d^2n/dt^2$, differentiating (6.24) yields

$$n_1'' = k_{11}n_1' + k_{12}n_2',$$

i.e. using (6.25),

$$n_1'' = k_{11}n_1' + k_{12}(k_{21}n_1 + k_{22}n_2),$$

i.e. using (6.24),

$$n_1'' = k_{11}n_1' + k_{12}k_{21}n_1 + k_{22}(n_1' - k_{11}n_1),$$

whence

$$n_1'' - (k_{11} + k_{22})n_1' + (k_{11}k_{22} - k_{12}k_{21})n_1 = 0. \tag{6.26}$$

To solve this second-order differential equation consider the trial solution

$$n_1(t) = A \exp(\lambda t)$$

for some constants A and λ. For this expression to satisfy (6.26) we need

$$A\lambda^2 e^{\lambda t} - (k_{11} + k_{22})A\lambda e^{\lambda t} + (k_{11}k_{22} - k_{12}k_{21})A e^{\lambda t} = 0.$$

Thus as $A \exp(\lambda t)$ cancels we require λ to be a solution of the 'auxiliary equation'

$$\lambda^2 - (k_{11} + k_{22})\lambda + (k_{11}k_{22} - k_{12}k_{21}) = 0. \tag{6.27}$$

Since this quadratic has the two roots

$$\lambda_1, \lambda_2 = \tfrac{1}{2}[(k_{11} + k_{22}) \pm \sqrt{\{(k_{11} - k_{22})^2 + 4k_{12}k_{21}\}}], \tag{6.28}$$

the general solution to equation (6.26) may be then written as

$$n_1(t) = A_1 \exp(\lambda_1 t) + A_2 \exp(\lambda_2 t) \tag{6.29}$$

for some constants A_1 and A_2. To find $n_2(t)$ we simply substitute (6.29) back into (6.24), obtaining

$$\begin{aligned} n_2(t) = (1/k_{12})[A_1(\lambda_1 - k_{11}) \exp(\lambda_1 t) \\ + A_2(\lambda_2 - k_{11}) \exp(\lambda_2 t)]. \end{aligned} \tag{6.30}$$

If required, A_1 and A_2 can be determined from the initial values

$$n_1(0) = A_1 + A_2$$

and

$$n_2(0) = (1/k_{12})[A_1(\lambda_1 - k_{11}) + A_2(\lambda_2 - k_{11})].$$ (6.31)

The behavioural characteristics of this local deterministic system are determined by the values of λ_1 and λ_2. In particular, on defining

$$\Delta = (k_{11} - k_{22})^2 + 4k_{12}k_{21}$$ (6.32)

we see from (6.28) that if $\Delta < 0$ then λ_1 and λ_2 are complex numbers, which results in damped cycles if $(k_{11} + k_{22}) < 0$ and increasing cycles if $(k_{11} + k_{22}) > 0$. Whilst if $\Delta > 0$ then both λ_1 and λ_2 are real numbers, resulting in straight exponential decay if $\lambda_1 < 0$ and $\lambda_2 < 0$, and straight exponential growth if either $\lambda_1 > 0$ or $\lambda_2 > 0$. For a complete discussion of possible modes of behaviour see Chapter 12 of Keyfitz (1977).

Pielou (1977, Chapter 6) discusses the extension of this characterization to k-species interactions (the Routh–Hurwitz criteria), which is potentially very interesting since controversy reigns on whether increased trophic web complexity leads to increased community stability. On the one hand both mathematical models and laboratory experiments on one-predator–one-prey systems oscillate violently (Elton, 1958); cultivated land and orchards are fairly unstable; whereas the rain forest, the paradigm of trophic web complexity, appears very stable (May, 1971b). Whilst on the other, May considers a simple mathematical model for a many-predator–many-prey deterministic system, and shows it to be in general less stable, and never more stable, than the analogous one-predator–one-prey community. Though this last result seems to caution against any simple belief that increasing population stability is a mathematical consequence of increasing multi-species complexity, it does invoke a deterministic argument. A stochastic analysis might well resolve this apparent dilemma between observation and theory.

6.2.3   *Local Volterra solution*

To see how these general results may be applied to specific processes, consider the Volterra system (6.23) for which

$$k_{11} = -cN_1^*, \quad k_{12} = -b_1 N_1^*, \quad k_{21} = b_2 N_2^* \quad \text{and} \quad k_{22} = 0.$$ (6.33)

Inserting these values into (6.28) gives

$$\lambda_1, \lambda_2 = \tfrac{1}{2}[-cN_1^* \pm \sqrt{\{(cN_1^*)^2 - 4r_2 b_1 N_2^*\}}].$$ (6.34)

Thus if $c > 0$ then $n_1(t)$ and $n_2(t)$ decay exponentially to zero. To examine the nature of this decay, note that (6.32) yields

$$\Delta = (cN_1^*)^2 - 4r_2 b_1 N_2^*.$$ (6.35)

So as c varies, the transition point ($c = c'$, say) between cyclic damping ($\Delta < 0$, i.e. $c < c'$) and straight damping ($\Delta > 0$, i.e. $c > c'$) occurs when

$$\Delta = (cr_2/b_2)^2 - 4r_2(r_1 - cr_2/b_2) = 0,$$

which solves to give

$$c' = 2b_2[-1 + \sqrt{(1 + r_1/r_2)}]. \tag{6.36}$$

To illustrate these ideas let us now return to the numerical example used in Section 6.1.3, namely $r_1 = 1.5, r_2 = 0.25, b_1 = 0.1$ and $b_2 = 0.01$, but with the logistic prey parameter $c > 0$ included in the model to ensure deterministic stability. From (6.21) we have the equilibrium values

$$N_1^* = r_2/b_2 = 25$$

and

$$N_2^* = (1/b_1)(r_1 - cN_1^*) = 15 - 250c.$$

So for N_2^* to be positive we require $c < 15/250 = 0.06$.

The λ-values

$$\lambda_1, \lambda_2 = \tfrac{1}{2}[-25c \pm \sqrt{\{625c^2 + 25c - 1.5\}}]$$

obtained from (6.34) determine the speed of approach to equilibrium, and with

$$\Delta = 625c^2 + 25c - 1.5$$

the general solution for $n_i(t)$ ($i = 1, 2$) is given from (6.29) and (6.30) by

$$n_i(t) = \exp\{-25ct/2\}[A_i \exp\{\tfrac{1}{2}t\sqrt{\Delta}\} + B_i \exp\{-\tfrac{1}{2}t\sqrt{\Delta}\}] \tag{6.37}$$

for some constants A_i and B_i.

If $\Delta > 0$ then there is straight exponential damping since it follows from (6.34) that $\lambda_1, \lambda_2 < 0$. To study the case $\Delta < 0$ denote

$$\theta = \sqrt{(-\Delta)} = \sqrt{(1.5 - 25c - 625c^2)},$$

where from (6.36) $0 < c < c' = (0.02)(-1 + \sqrt{7}) = 0.033$. Then on noting that

$$\exp\{\pm\tfrac{1}{2}t\sqrt{\Delta}\} = \exp\{\pm\tfrac{1}{2}it\theta\}$$
$$= \cos(\tfrac{1}{2}\theta t) \pm i \sin(\tfrac{1}{2}\theta t) \quad \text{(here } i = \sqrt{(-1)}),$$

we may write the general solution (6.37) in the equivalent form

$$n_i(t) = \exp\{-25ct/2\}[C_i \cos(\tfrac{1}{2}\theta t) + D_i \sin(\tfrac{1}{2}\theta t)] \tag{6.38}$$

for some appropriate constants C_i and D_i. Thus damping occurs at rate

$\exp\{-25ct/2\}$, and the cycle period $T = 4\pi/\theta$. When $c = 0$ no damping occurs and $T = 4\pi/\sqrt{(1.5)} = 10.26$, exactly as we found before; whilst as c approaches its maximum value $c' = 0.033$, not only does the damping factor $\exp\{-0.41t\}$ increase swiftly as t increases, but also θ approaches zero which corresponds to an infinite period of oscillation. So if damping is to occur fairly slowly and the period of oscillation is to be reasonably short, then c must be substantially less than c'.

*6.2.4   The coefficient of variation

Suppose we choose c to be as small as possible consistent with the probability of extinction of either species being small over the time period of interest. Now if the equilibrium point (N_1^*, N_2^*) is far enough away from the axes relative to the likely variation in $(X_1(t), X_2(t))$, then stochastic trajectories should bypass the axes quite safely. So a sensible way of proceeding is to evaluate the coefficient of variation, namely

$$CV(X) = \sqrt{\{\text{Var}(X)\}}/N^*, \tag{6.39}$$

since if $CV(X)$ is substantially smaller than 1 then it is unlikely that excursions away from equilibrium will be large enough to result in extinction.

To determine expressions for $CV(X_1)$ and $CV(X_2)$ we shall extend the technique developed in Sections 3.9.1 and 3.9.2 for the single-species logistic process. Let δt denote a small time increment and write the deterministic equations (6.19) in the form

$$N_1(t + \delta t) - N_1(t) \simeq N_1(t)[r_1 - cN_1(t) - b_1 N_2(t)]\delta t$$
$$N_2(t + \delta t) - N_2(t) \simeq N_2(t)[-r_2 + b_2 N_1(t)]\delta t. \tag{6.40}$$

To convert (6.40) into stochastic equations by allowing for chance fluctuations in population size, we add on 'noise' components $\delta Z_1(t)$ and $\delta Z_2(t)$ to produce

$$X_1(t + \delta t) - X_1(t) \simeq X_1(t)[r_1 - cX_1(t) - b_1 X_2(t)]\delta t + \delta Z_1(t)$$
$$X_2(t + \delta t) - X_2(t) \simeq X_2(t)[-r_2 + b_2 X_1(t)]\delta t + \delta Z_2(t). \tag{6.41}$$

We shall assume that $\delta Z_1(t)$ and $\delta Z_2(t)$ are independent of $\delta Z_1(s)$ and $\delta Z_2(s)$ for all times $s \neq t$.

Suppose that the stochastic process $\{X_1(t), X_2(t)\}$ remains close to the deterministic equilibrium point (N_1^*, N_2^*), i.e. we may write

$$X_1(t) = N_1^*[1 + u_1(t)] \quad \text{and} \quad X_2(t) = N_2^*[1 + u_2(t)] \tag{6.42}$$

for small variables $u_1(t)$ and $u_2(t)$. Then on using (6.20), equations (6.41) become

$$N_1^*[u_1(t + \delta t) - u_1(t)] = N_1^*[1 + u_1(t)][-cN_1^*u_1(t)$$
$$- b_1 N_2^* u_2(t)]\delta t + \delta Z_1$$
$$N_2^*[u_2(t + \delta t) - u_2(t)] = N_2^*[1 + u_2(t)][b_2 N_1^* u_1(t)]\delta t + \delta Z_2.$$

On assuming that u_1^2, $u_1 u_2$ and u_2^2 are negligible in comparison with u_1 and u_2, these equations simplify to give the linear form

$$u_1(t + \delta t) = u_1(t) - [cN_1^* u_1(t) + b_1 N_2^* u_2(t)]\delta t + \delta Z_1/N_1^*$$
$$u_2(t + \delta t) = u_2(t) + [b_2 N_1^* u_1(t)]\delta t + \delta Z_2/N_2^*. \tag{6.43}$$

Now in the small time interval $(t, t + \delta t)$, and for (X_1, X_2) near to (N_1^*, N_2^*),

$$\Pr\{\delta Z_1 = +1\} \simeq N_1^*(r_1 - cN_1^*)\delta t \quad \text{(prey birth)}$$
$$\Pr\{\delta Z_1 = -1\} \simeq N_1^*(b_1 N_2^*)\delta t \quad \text{(prey death)}$$
$$\Pr\{\delta Z_2 = +1\} \simeq N_2^*(b_2 N_1^*)\delta t \quad \text{(predator birth)}$$
$$\Pr\{\delta Z_2 = -1\} \simeq N_2^*(r_2)\delta t \quad \text{(predator death)}. \tag{6.44}$$

So

$$E(\delta Z_1) = (+1)\Pr\{\delta Z_1 = +1\} + (-1)\Pr\{\delta Z_1 = -1\}$$
$$\simeq N_1^*(r_1 - cN_1^* - b_1 N_2^*)\delta t = 0,$$

and

$$E(\delta Z_2) \simeq N_2^*(b_2 N_1^* - r_2)\delta t = 0.$$

Thus

$$\text{Var}(\delta Z_1) \simeq (+1)^2 \Pr\{\delta Z_1 = +1\} + (-1)^2 \Pr\{\delta Z_1 = -1\}$$
$$= N_1^*(r_1 - cN_1^* + b_1 N_2^*)\delta t = 2b_1 N_1^* N_2^* \delta t \tag{6.45}$$

and

$$\text{Var}(\delta Z_2) \simeq N_2^*(r_2 + b_2 N_1^*)\delta t = 2b_2 N_1^* N_2^* \delta t; \tag{6.46}$$

whilst $\text{Cov}(\delta Z_1, \delta Z_2) \simeq E(\delta Z_1 \delta Z_2)$ is of order δt^2 and may therefore be ignored.

Next we (i) square and cross-multiply equations (6.43), (ii) take expectations on both sides of the resulting equations, and (iii) ignore terms in δt^2. On denoting $\sigma_i^2 = \text{Var}(u_i)$ and $\sigma_{12} = \text{Cov}(u_1, u_2)$, and noting that $\text{Var}[u_1(t + \delta t)] = \text{Var}[u_1(t)]$ etc., since we are essentially in an equilibrium situation, this procedure yields

$$\sigma_1^2 = \sigma_1^2 - 2(cN_1^*\sigma_1^2 + b_1 N_2^*\sigma_{12})\delta t + \text{Var}(\delta Z_1/N_1^*)$$
$$\sigma_2^2 = \sigma_2^2 + 2b_2 N_1^*\sigma_{12}\delta t + \text{Var}(\delta Z_2/N_2^*)$$
$$\sigma_{12} = \sigma_{12} + (b_2 N_1^*\sigma_1^2 - cN_1^*\sigma_{12} - b_1 N_2^*\sigma_2^2)\delta t. \tag{6.47}$$

On using results (6.45) and (6.46) these equations reduce to

$$cN_1^*\sigma_1^2 + b_1N_2^*\sigma_{12} = b_1N_2^*/N_1^*$$
$$b_2N_1^*\sigma_{12} = -b_2N_1^*/N_2^*$$
$$b_2N_1^*\sigma_1^2 - cN_1^*\sigma_{12} - b_1N_2^*\sigma_2^2 = 0,$$

which solve to yield the second-order moments

$$\sigma_{12} = -1/N_2^*$$
$$\sigma_1^2 = [1 + (N_2^*/N_1^*)](b_1/cN_1^*)$$
$$\sigma_2^2 = [r_2\sigma_1^2 + c(N_1^*/N_2^*)]/(b_1N_2^*). \tag{6.48}$$

Thus $\mathrm{Var}(X_i) = (N_i^*)^2\sigma_i^2$ and $\mathrm{Cov}(X_1, X_2) = N_1^*N_2^*\sigma_{12}$, whilst the coefficient of variation

$$CV(X_i) = \sqrt{[(N_i^*)^2\sigma_i^2]}/N_i^* = \sigma_i \qquad (i = 1, 2). \tag{6.49}$$

To illustrate the effect on $CV(X_i)$ of changing c, let us slightly generalize our Volterra example to

$$dN_1/dt = N_1(1.5 - cN_1 - b_1N_2)$$
$$dN_2/dt = N_2(-0.25 + 0.01N_1). \tag{6.50}$$

Here $-cN_1^2$ is the added (prey) logistic term, $N_1^* = 25$ remains unaffected by the value of c, but we vary b_1 with c to keep N_2^* constant at 15, i.e. from (6.20)

$$b_1 = (1.5 - 25c)/15. \tag{6.51}$$

With these values expressions (6.48) yield

$$\sigma_1^2 = (12 - 200c)/(1875c)$$

and

$$\sigma_2^2 = (2/1875c) + 10c/(9-150c). \tag{6.52}$$

Now at the transition value $c' = 0.033$, $\sigma_1 = 0.295$ and $\sigma_2 = 0.337$ are roughly equal. So as $\mathrm{Var}(u_1) \simeq \mathrm{Var}(u_2)$ the probabilities that $u_1(t) = -1$ and $u_2(t) = -1$ are also roughly equal, whence it follows that if a species becomes extinct then it is as likely to be a prey as a predator. Whilst as c approaches zero then both σ_1 and σ_2 become indefinitely large and extinction on the first cycle is virtually inevitable (as illustrated in Figure 6.3). In between these two extremes prey are more likely to become extinct first since $\sigma_1 \simeq \sqrt{6}\sigma_2 > \sigma_2$. Conversely, predators are more likely to become extinct first if $\sigma_2 > \sigma_1$.

Note that σ_1 and σ_2 also provide a rough indication of the probability that extinction occurs during a particular cycle. Since $X_i = N_i^*(1 + u_i)$, we require $u_i > -1$ to avoid $X_i = 0$ and hence extinction. Now if $\sigma_i \geqslant 1$ (say) then the chance that u_i drops below -1 during a given cycle is quite high, so $\sigma_1 = \sigma_2$

= 1 represents a crude upper bound for sustained stochastic cycles to have any chance of occurring. For example, from (6.52) $\sigma_1 = 1$ at $c = 0.0058$ and $\sigma_2 = 1$ at $c = 0.0011$, so here c must be at least 0.006.

6.2.5 Comparison with simulated runs

To see how this minimum c-value ties in with simulated runs, we first note from (6.51) that $c = 0.006$ corresponds to $b_1 = 0.09$. Thus the deterministic equations (6.50) give rise to the stochastic birth and death rates

$$B_1 = X_1(1.5 - 0.006X_1) \quad \text{and} \quad D_1 = 0.09X_1X_2 \quad \text{(prey)}$$
$$B_2 = 0.01X_1X_2 \quad \text{and} \quad D_2 = 0.25X_2 \quad \text{(predators)}.$$
(6.53)

Simulations with (6.53) generally resulted in around three cycles followed by prey extinction, and since both prey and predator populations reached 1 several times the value $c = 0.006$ is clearly a sensible lower bound. Increasing c to 0.01, to enable longer periods of cycling, reduces σ_1 from 0.98 to 0.73 and σ_2 from 0.43 to 0.35 and hence the chance of extinction during a given cycle. Indeed, with B_1 now replaced by $X_1(1.5 - 0.01X_1)$ and D_1 by $0.0833X_1X_2$ some simulations managed to keep going for quite a long time.

Figure 6.5 shows a typical simulation run over $t = 0, 0.5, 1, 1.5, \ldots, 100$ with these new parameter values, starting from $(8, 8)$, and although there is evidence of cyclic structure it is far less obvious than the simple sine-curves predicted by deterministic theory. Moreover, after a large population surge of prey during the first cycle, the process settles down to some kind of steady behaviour. This is in direct conflict with result (6.38) which predicts exponential damping at rate $\exp(-0.125t)$. By time $t = 90$ this factor is of the order of 10^{-5}, in strong contradiction to the obvious stochastic cycle present at that time. Since the prey cycles which peak around $t = 78$ and 92 are no smaller than the second which peaks around $t = 16$, for this simulation run no long-term damping occurs beyond the first cycle. The prey cycles are less regular in the middle part of the run, but overall maxima occur at times $t = 3$, 16.5, 26, 39, 51.5, 62.5(?), 77 and 92.5, which gives an average peak-to-peak cycle time of about 12.8. The predator population, though having cycles of smaller amplitude than the prey population, behaves in a similar manner except that it appears to lag about 2.5 time-units behind the prey.

6.2.6 Autocorrelation representation

To place these general impressions on a more rigorous framework, let $\{x_i\}$ for $i = 1, \ldots, N$ denote a stationary time-series. That is, $x_1, \ldots, x_N$ are N successive observations of a process which is assumed to contain no overall trend. Then observations a short distance apart may well be correlated with each other. For example, the sequence

1 1 1 0 0 0 0 1 1 1 1 1 0 0 0 0 1 1 1 0 ...

denoting wet (1) and dry (0) days shows an obvious positive correlation in weather pattern between successive days.

We have already met this idea in Section 4.5.4. There we considered the series

1 0 0 1 0 0 1 0 0 1 0 0

and noted that it consisted of 4 complete cycles of length 3, i.e. frequency 4 and period 3. However, instead of concentrating on the frequency representation, as before, we shall now work with the period which tells us that points 3 units apart are positively correlated with each other.

The usual measure of correlation between two time-series $\{x_i\}$ and $\{y_i\}$ is defined as

$$r = \sum_{i=1}^{N} (x_i - \bar{x})(y_i - \bar{y}) \Bigg/ \left[\sum_{i=1}^{N} (x_i - \bar{x})^2 \sum_{i=1}^{N} (y_i - \bar{y})^2 \right]^{1/2} \qquad (6.54)$$

Figure 6.5. Simulation corresponding to the deterministic Volterra process

$dN_1/dt = N_1(1.5 - 0.01N_1 - 0.0833N_2)$ (prey, ———)
$dN_2/dt = N_2(-0.25 + 0.01N_1)$ (predators, ----).

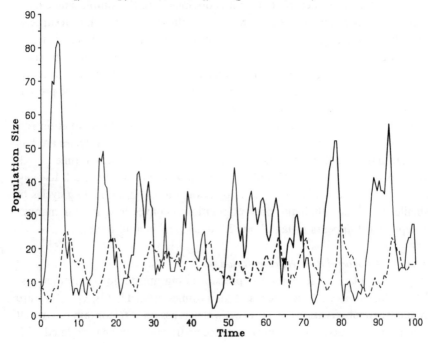

where $\bar{x} = \sum x_i/N$ and $\bar{y} = \sum y_i/N$ denote the overall means. The obvious extension to measure the correlation structure within a single series between points k time-units apart is to replace $\{y_i\}$ by $\{x_{i+k}\}$ and so define the *autocorrelation* coefficient

$$r_k = \sum_{i=1}^{N-k} (x_i - \bar{x})(x_{i+k} - \bar{x}) \bigg/ \sum_{i=1}^{N} (x_i - \bar{x})^2 \tag{6.55}$$

where the *lag* $k = 0, 1, \ldots, N - 1$. The corresponding *autocovariances* are defined to be

$$c_k = (1/N) \sum_{i=1}^{N-k} (x_i - \bar{x})(x_{i+k} - \bar{x}), \tag{6.56}$$

so

$$r_k = c_k/c_0 \qquad (k = 0, 1, \ldots, N - 1). \tag{6.57}$$

Note that some authors suggest using

$$c_k = [1/(N-k)] \sum_{i=1}^{N-k} (x_i - \bar{x})(x_{i+k} - \bar{x}) \tag{6.56a}$$

rather than (6.56), since the summation contains $(N - k)$ terms and so c_k will not be biased downwards as k increases. However, there is little difference between these two definitions if k is small relative to N.

As k increases, the summations in (6.56) and (6.56a) contain fewer and fewer terms, so it is advisable to evaluate $\{r_k\}$ up to at most lag $k = N/3$ in view of decreasing accuracy (and increasing bias in the case of (6.56)). Since the application and interpretation of autocorrelation analysis is explained in many introductory texts on time-series analysis (e.g. Chatfield, 1980), we shall just provide a brief example to illustrate the basic method.

Consider the deterministic sequence

$$1\ 0\ 0\ 1\ 0\ 0\ \ldots\ 1\ 0\ 0 \tag{6.58}$$

over $N = 24$ observations. There are 8 complete cycles of length 3, and the series repeats (i.e. correlates with itself) at times $3, 6, 9, \ldots$. As N is small the $\{r_k\}$ are easily evaluated on a pocket calculator; otherwise using a standard package such as MINITAB (command ACF) is very simple and has the visual benefit of automatically giving a line plot. Table 6.1(i) shows $\{r_k\}$ for $k = 0, \ldots, 8$, and large positive peaks are clearly evident at $k = 3$ and 6 (using (6.56a) and (6.57) gives the unbiased values $r_3 = r_6 = \cdots = 1$).

Suppose that a random sequence of 0s and 1s is now added to series (6.58), with $\Pr\{0\} = \Pr\{1\} = 0.5$. One such sequence is

$$2\ 0\ 1\ 1\ 0\ 1\ 2\ 0\ 1\ 2\ 1\ 0\ 1\ 1\ 0\ 1\ 1\ 1\ 2\ 0\ 0\ 1\ 1\ 0, \tag{6.59}$$

Table 6.1. Examples of autocorrelation structure $\{r_k\}$

k	0	1	2	3	4	5	6	7	8
(i)	1.00	−0.46	−0.48	0.88	−0.40	−0.42	0.75	−0.33	−0.35
(ii)	1.00	−0.24	−0.40	0.39	−0.04	−0.37	0.24	−0.05	−0.12

and although the underlying periodic structure is now hidden from view, the autocorrelation structure (Table 6.1(ii)) still highlights the general cyclic behaviour at lags 3, 6, ...

The graph of r_k against k is called a correlogram. To determine what constitutes a significantly large value of r_k (either positive or negative), we note that if a time-series is completely random (called white-noise) then, for large N and small $k \neq 0$, r_k is approximately distributed as a Normal random variable with mean 0 and variance $1/N$. With $N = 24$ the approximate 95% confidence interval $(-2/\sqrt{N}, 2/\sqrt{N})$ is $(-0.41, 0.41)$, so in Table 6.1(ii) $r_3 = 0.39$ is significant at about the 5% level. Though this specific result for r_3 is not in itself particularly conclusive, the overall picture painted by the whole correlogram most certainly is, showing a strong cyclic structure of period 3.

Figure 6.6 shows the autocorrelations (for lags $k = 0, ..., 60$) corresponding to the simulation run shown in Figure 6.5, and in spite of the rather jagged appearance of the simulated data both prey and predator correlograms exhibit smooth damped cycles. Moreover, they are surprisingly similar in appearance, given that the two simulated series look quite different. (With k near 50, which considerably exceeds the recommended cut-off value of $N/3 = 33$, use of (6.56a) instead of (6.56) roughly doubles the r_k values though the overall impression of matched damped cycles remains.) The first two peaks are $r_{23} = r_{24} = 0.28$ and $r_{47} = 0.19$ for prey, and $r_{22} = r_{23} = 0.29$ and $r_{47} = 0.14$ for predators. As observations are made every 0.5 time-units, the estimated cycle length is therefore $(47/2) \times 0.5 = 11.75$ time-units. Note that for both species r_{23} lies well outside the 95% confidence interval of ± 0.14.

We have already seen from (6.35) and (6.38) that the deterministic cycle period T is $4\pi/\theta$ where

$$\theta = \sqrt{(-\Delta)} = \sqrt{\{-(cN_1^*)^2 + 4r_2 b_1 N_2^*\}}.$$

This gives $\theta = 1.089$, and hence $T = 11.53$, which is close to the above simulated value of 11.75 computed from one fairly short simulation run. Moreover, the simulated prey and predator population means of 24.04 and 14.53 are in surprisingly close agreement with $N_1^* = 25$ and $N_2^* = 15$ considering the erratic nature of the simulated data. The simulated standard deviations are 15.27 (prey) and 4.88 (predators), whilst the theoretical values given from (6.48) are $N_1^*\sigma_1 = 18.26$ and $N_2^*\sigma_2 = 5.20$. Thus here simulated and theoretical values are also in good agreement, even though the latter are evaluated under the assumption that the u_i in (6.42) are a lot less than 1. Once again the local approximation seems to be remarkably robust.

6.2.7   *Mean time to extinction*

In order to derive an approximation to the mean time to extinction (T_E), let us suppose that X_1 and X_2 may be treated separately from each other. Then recourse to the single-species result (3.72) gives

$$T_E \simeq [D(1)\pi_1^{(Q)}]^{-1} \tag{6.60}$$

where, in the example of Figure 6.5,

$$D(1) \simeq 0.0833X_1N_2^* \quad \text{with} \quad X_1 = 1 \quad \text{(prey)}$$
$$D(1) \simeq 0.25X_2 \quad\quad\; \text{with} \quad X_2 = 1 \quad \text{(predators)}.$$

The quasi-equilibrium probability $\{\pi_1^{(Q)}\}$ may be evaluated by assuming that X_i is approximately distributed as a Normal random variable with

Figure 6.6. Autocorrelations of prey (——) and predators (– – –) together with their cross-correlations (— — —) for the simulated Volterra process of Figure 6.5.

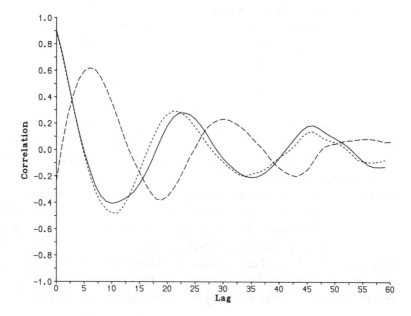

mean N_i^* and variance $(N_i^* \sigma_i)^2$, whence

$$\pi_1^{(Q)} \simeq \Pr\{0.5 \leqslant X_i \leqslant 1.5\}$$
$$\simeq \Phi[(1.5 - N_i^*)/(N_i^* \sigma_i)] - \Phi[(0.5 - N_i^*)/(N_i^* \sigma_i)]. \tag{6.61}$$

Thus for prey

$$\pi_1^{(Q)} \simeq \Phi[(1.5 - 25)/18.26] - \Phi[(0.5 - 25)/18.26] = 0.0093,$$

whilst for predators

$$\pi_1^{(Q)} \simeq \Phi[(1.5 - 15)/5.20] - \Phi[(0.5 - 15)/5.20] = 0.002\,06.$$

So on using (6.60)

$$T_E \simeq [0.0833 \times 15 \times 0.0093]^{-1} = 86 \quad \text{(prey)}$$

and

$$T_E \simeq [0.25 \times 0.002\,06]^{-1} = 1942 \quad \text{(predators)}.$$

With a mean cycle time of about 12, this gives an average of 7 cycles before prey extinction occurs and 162 cycles before predator extinction occurs. Since so many suppositions have been made en route these values should not be taken too literally, but they do at least provide a behavioural picture which is in broad agreement with our experience of simulating *this* particular model. Further research is clearly necessary to see if the approach always gives rise to worthwhile results.

6.2.8   *Cross-correlation representation*

To examine the interaction *between* prey and predators we may extend the autocorrelation definition (6.55) by defining the *cross-correlation* at lag k between two series $\{x_i\}$ and $\{y_i\}$ to be

$$s_k = \sum_{i=1}^{N-k} (x_i - \bar{x})(y_{i+k} - \bar{y}) \Big/ \left[\sum_{i=1}^{N} (x_i - \bar{x})^2 \sum_{i=1}^{N} (y_i - \bar{y})^2 \right]^{1/2}. \tag{6.62}$$

If $\{x_i\}$ and $\{y_i\}$ are uncorrelated, then for large N the s_k are approximately Normally distributed with mean 0 and variance $1/N$. Included in Figure 6.6 are the cross-correlations (evaluated by using the MINITAB command CCF) between predators (x_i) and prey (y_i) for our simulation example, and apart from a shift of phase they behave just like the autocorrelations. There is a clear maximum at lag 6 ($s_6 = 0.62$) which corresponds to 3 time-units. As the simulated cycle period $T = 11.75$, the predators therefore lag the prey by almost exactly a quarter-cycle which is in excellent agreement with the deterministic Lotka–Volterra prediction. Moreover, the correlation $s_0 = -0.238$ between the simulated predator and prey populations compares well with the theoretical value $\sigma_{12}/\sigma_1\sigma_2 = -0.264$ determined from (6.48).

6.2.9   *Some stochastic thoughts*

We have seen that the basic deterministic Lotka–Volterra model correctly predicts the general behaviour of the stochastic Volterra model, namely undamped population cycles with prey leading predators by a quarter-cycle. Moreover, the Lotka–Volterra ratio of predator to prey amplitude, $(b_1/b_2)\sqrt{(r_2/r_1)} = 3.4$, is surprisingly close to the (simulated) Volterra ratio of the two standard deviations, namely $N_1^*\sigma_1/N_2^*\sigma_2 = 3.5$.

There is clearly an analogy here between the trajectory of the two interacting populations and the trajectory of a small ball being whirled around in an elliptical orbit on the end of an elastic string. The larger the value of the parameter c, the more powerful is the elastic and the ball orbits nearer to the centre; the larger the effect of natural variation, the greater is the mass of the ball which then orbits further away from the centre. Maintaining cyclic behaviour therefore represents a balance between the mass of the ball (stochastic variation) and the strength of the elastic (c).

6.3   The Leslie and Gower model

In addition to the basic Lotka and Volterra models for predator–prey interaction, several alternative models have been proposed, each with its own mathematical peculiarities. Most of these are deterministic, and have been constructed without recourse to specific data sets. Their development has therefore not involved the usual concept of modelling, in which *biological phenomena* are interpreted with the help of solutions to simple mathematical equations, but has used instead the 'what happens if' approach. That is, if a *mathematical system* of equations is given a particular characteristic, what is the outcome?

Leslie and Gower (1960) retained the Volterra prey equation

$$dN_1/dt = N_1[r_1 - cN_1 - b_1N_2], \tag{6.63}$$

but changed the character of the predator equation to

$$dN_2/dt = N_2[r_2 - b_2(N_2/N_1)]. \tag{6.64}$$

The net predator growth rate now takes account of the relative sizes of the two populations. The larger N_2/N_1 becomes, the smaller are the number of prey available to each predator and consequently the resource available for predator growth declines.

Equations (6.63) and (6.64) have an equilibrium point at

$$N_1^* = r_1b_2/(cb_2 + b_1r_2) \quad \text{and} \quad N_2^* = r_2N_1^*/b_2, \tag{6.65}$$

which may be shown to be stable by using the same local approximation technique as for the Volterra model. With $N_i = N_i^*(1 + n_i)$ and n_i small, these equations become

$$dn_1/dt = -cN_1^* n_1 - b_1 N_2^* n_2$$
$$dn_2/dt = r_2(n_1 - n_2). \tag{6.66}$$

Hence the parameters λ_1 and λ_2 in the general form (6.28) are given by

$$\lambda_1, \lambda_2 = \tfrac{1}{2}[-(cN_1^* + r_2) \pm \sqrt{\{(cN_1^* + r_2)^2 - 4r_1 r_2\}}], \tag{6.67}$$

and we see that the real parts of λ_1 and λ_2 are both negative. Comparison with the results immediately following (6.32), in which $\Delta = (cN_1^* + r_2)^2 - 4r_1 r_2$, therefore shows that $N_i(t)$ converges exponentially towards N_i^* either as a series of damped cycles if $\Delta < 0$, i.e. if

$$(cN_1^* + r_2)^2 < 4r_1 r_2, \tag{6.68}$$

or as straight exponential damping if not.

Having obtained the local linear equations (6.66), an analogous stochastic exercise to that developed for the Volterra model (Section 6.2.4) yields the second-order moments σ_1^2, σ_2^2 and σ_{12} for n_1 and n_2, whence

$$\text{Var}(X_i) = (N_i^* \sigma_i)^2 \quad \text{and} \quad \text{Cov}(X_1, X_2) = N_1^* N_2^* \sigma_{12}.$$

Leslie and Gower suggest that the equilibrium probability distribution of the population sizes (X_1, X_2) may be approximated by a bivariate Normal distribution having these moments. Thus if this assertion is true, and if extinction is unlikely to occur in the short term, then contour plots showing lines of equal probability may be constructed from

$$\left(\frac{X_1 - N_1^*}{\sigma_1}\right)^2 - 2\rho\left(\frac{X_1 - N_1^*}{\sigma_1}\right)\left(\frac{X_2 - N_2^*}{\sigma_2}\right) + \left(\frac{X_2 - N_2^*}{\sigma_2}\right)^2$$
$$= \text{constant} \quad (6.69)$$

where $\rho = \sigma_{12}/\sigma_1 \sigma_2$. This general result is clearly applicable to any process which has a stable local equilibrium (N_1^*, N_2^*).

6.4   The Holling–Tanner model

We have seen that the basic deterministic Lotka–Volterra model gives rise to unstable cycles; whilst both introducing a logistic term into the prey equation (Volterra) and then allowing predator deaths to depend upon relative population size (Leslie and Gower) induces deterministic convergence towards a stable equilibrium point, either with or without damped cycles. A slight extension to the Leslie and Gower model now produces the interesting phenomenon of *stable limit cycles*, so enabling sustained cyclic behaviour to be constructed from a deterministic and a stochastic viewpoint simultaneously.

Several studies have shown that individual predation rate (x) rises with increasing prey population size (N_1) in the manner of Figure 6.7. In particular, Holling (1965) shows that this type of functional response is characteristic of invertebrate predators; that for vertebrate predators differs because they can learn to search for a prey species that has become more abundant. The three key features of feeding behaviour are associative learning, information channelling, and forgetting. These permit great flexibility, since they enable the predator to concentrate on a few stimuli and yet still retain the ability to exploit changes in the environment.

Two appropriate functions for x are

$$x = w(1 - \exp\{-aN_1/w\}) \tag{6.70}$$

and

$$x = wN_1/(D + N_1). \tag{6.71}$$

In both cases w represents the maximum predation rate *per predator*; there is an upper limit to how many prey a single predator may kill no matter how abundant they are. Thus x ranges from $w(1 - \exp\{-a/w\})$ and $w/(D + 1)$, respectively, when $N_1 = 1$, to the maximum value w when N_1 becomes large. The size of D and $1/a$ is directly related to the prey's ability to evade attack: the more elusive the prey, the greater D and $1/a$ become. Functions (6.70) and (6.71) produce similar theoretical results, and since (6.71) is the easier of the two to manipulate we shall retain it as the predation rate.

Figure 6.7. Prey kill rate (x) per predator, as a function of prey population size (N_1).

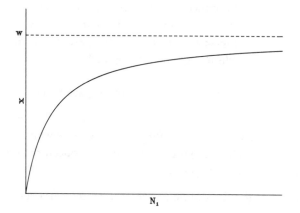

6.4.1   *Stability analysis*

Replacing b_1N_2 in the Volterra prey equation (6.63) by (6.71) gives

$$dN_1/dt = N_1[r_1 - cN_1 - \{wN_2/(D + N_1)\}];\qquad(6.72)$$

the Leslie and Gower predator equation (6.64) is left unchanged at

$$dN_2/dt = N_2[r_2 - (b_2N_2/N_1)].\qquad(6.73)$$

The associated equilibrium point (N_1^*, N_2^*) is therefore the solution of the two isocline equations

$$(r_1 - cN_1)(D + N_1) = wN_2 \quad \text{(when } dN_1/dt = 0)\qquad(6.74)$$

and

$$r_2N_1 = b_2N_2 \quad \text{(when } dN_2/dt = 0),\qquad(6.75)$$

i.e.

$$N_1^* = (1/2c)[r_1 - cD - (wr_2/b_2)$$
$$+ \sqrt{\{[r_1 - cD - (wr_2/b_2)]^2 + 4cr_1D\}}]\qquad(6.76)$$
$$N_2^* = (r_2/b_2)N_1^*.$$

The prey isocline (6.74) is a parabola, whilst the predator isocline (6.75) is a straight line through the origin, so the phase-plane diagram has the form of Figure 6.8. It follows from (6.74) that $N_2 = 0$ when $N_1 = r_1/c$ or $-D$, and $N_1 = 0$ when $N_2 = r_1D/w$. To determine the position of the parabola maximum (N_1', N_2'), we differentiate (6.74) to form

$$dN_2/dN_1 = (1/w)(r_1 - cD - 2cN_1).$$

This equals zero at $N_1' = (r_1 - cD)/(2c)$, whence $N_2' = (r_1 + cD)^2/(4cw)$.

The corresponding stability analysis is discussed in detail both by May (1974b) and Tanner (1975), so here we shall just describe the salient features. The trajectory shape is determined by the relative positions of the two isoclines featured in Figure 6.8, and four distinct situations arise.

(1) If the line and curve intersect to the right of the parabola peak, i.e. $N_1^* > N_1'$, then (N_1^*, N_2^*) is a stable equilibrium point for all values of r_2/r_1 (Figure 6.9a).

(2) If their point of intersection lies to the left of the parabola peak, i.e. $N_1^* < N_1'$, and if

(i) r_2/r_1 is large, then (N_1^*, N_2^*) is stable (Figure 6.9b); whilst if

(ii) r_2/r_1 is small then (N_1^*, N_2^*) is unstable and the trajectories converge to a stable limit cycle (Figure 6.9c).

(3) The amplitude of the limit cycle increases with the carrying capacity

r_1/c, so as c approaches zero the limit cycles approach the axes and extinction occurs (Figure 6.9d).

Summarizing Tanner (1975), the model possesses a stable (deterministic) equilibrium provided either of the conditions

$$r_2/r_1 > [N_1^*(r_1 - cD) - 2c]/[N_1^*(1 + N_1^*D)] \qquad (6.77)$$

or

$$2 > (r_1 - cD)/c \qquad (6.78)$$

are met; otherwise it has a stable limit cycle. As r_1/c increases, the rather complicated condition (6.77) reduces to

$$r_2/r_1 > r_1 b_2/r_2 w,$$

i.e.

$$r_2/r_1 > \sqrt{(b_2/w)}.$$

Thus under this model, a stable equilibrium point exists provided that either the prey population is strongly self-limiting, so that its carrying capacity is small, or the intrinsic prey growth rate (r_1) is small relative to the intrinsic predator growth rate (r_2). Study of several pairs of interacting

Figure 6.8. Phase-plane diagram for the Holling–Tanner predator–prey model (6.72) and (6.73).

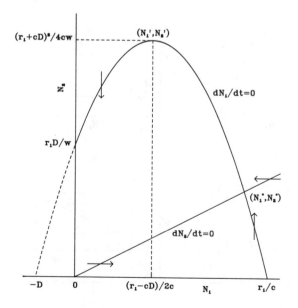

species, ranging from house sparrows and European sparrow hawk to mule deer and mountain lion (Tanner, 1975), shows that such theoretical predictions based on estimated parameter values are broadly in line with practical reality.

It must be remembered, however, that specific model structures provide little more than an indicator of possible behaviour, and they should not be credited with misplaced importance. For example, not only is the Holling–Tanner model one of many plausible mathematical forms, but when used to explain cyclic fluctuations it has the weakness that the amplitude of the cycle is sensitive to variations in certain parameter values. A seemingly small change in D may increase the amplitude sufficiently to bring the cycle close to an axis, whereupon superimposed stochastic fluctuations could soon cause extinction of one of the species.

To illustrate how the deterministic Holling–Tanner trajectories approach the stable limit cycle, Figure 6.10 shows a trajectory corresponding to the

Figure 6.9. Trajectories for the Holling–Tanner model: (a) prey carrying capacity r_1/c is small and $N_1^* > N_1'$ – stable equilibrium; (b) r_1/c is large, $N_1^* < N_1'$ and r_2/r_1 is greater than the boundary value (6.77/78) – stable equilibrium; (c) r_1/c is large, $N_1^* < N_1'$ and r_2/r_1 is less than the boundary value – stable limit cycles; (d) $c = 0$, r_2/r_1 is less than $\sqrt{(b_2/w)}$ – divergent oscillations (after Tanner, 1975).

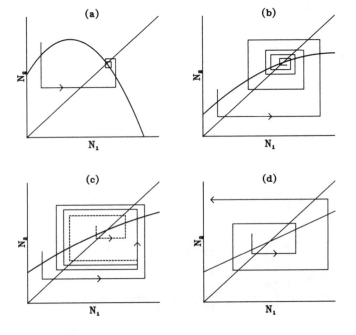

parameter values $r_1 = 1.0$, $r_2 = 0.5$, $w = 1.8$, $b_2 = 0.5$, $D = 15$ and $c = 0.001$, starting from $N_1(0) = N_2(0) = 15$. In the absence of predators the prey have carrying capacity $r_1/c = 1000$, whilst from (6.76) the equilibrium values are $N_1^* = N_2^* = 18$. Conditions (6.77) and (6.78) show that for this point to be stable we require either $0.5 > 0.518$ or $18.01 > 492.5$. Since neither of these inequalities is true, (N_1^*, N_2^*) cannot be a stable equilibrium point and so trajectories converge to a limit cycle. The trajectory shown in Figure 6.10 initially unwinds quite slowly about the equilibrium value, with $N_1(t)$ reaching a maximum of only 56 after eight full cycles have occurred. Rapid growth then takes place over the next four cycles (maximum $N_1(t) = 478$), and the limit cycle is effectively reached four cycles later (maximum $N_1(t) = 499$).

6.4.2   *Simulation of the stochastic model*
In this example, not only is the range of population size considerable,

Figure 6.10. Approach to the limit cycle for the deterministic Holling–Tanner process

$$dN_1/dt = N_1[1 - 0.001N_1 - (1.8N_2)/(15 + N_1)]$$
$$dN_2/dt = N_2[0.5 - (0.5N_2)/N_1]$$

with $N_1(0) = N_2(0) = 15$.

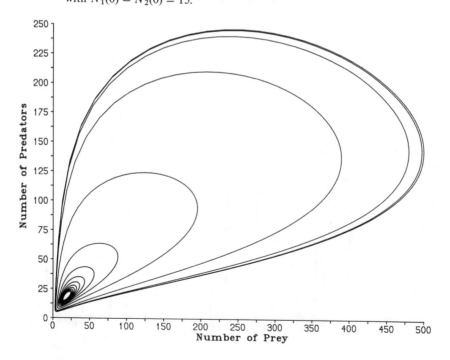

with $2.8 \leqslant N_1(t) \leqslant 499$ and $5.1 \leqslant N_2(t) \leqslant 246$, but we see that when N_1 and N_2 are both small all 12 cycles lie close to one another. Hence any stochastic perturbation in this region may result in a severe change of behaviour.

To demonstrate this effect, Figure 6.11 shows a simulation run of the associated stochastic model with birth and death rates

$$B_1 = X_1(1 - 0.001X_1) \quad \text{and} \quad D_1 = 1.8X_1X_2/(15 + X_1) \quad \text{(prey)}$$
$$B_2 = 0.5X_2 \quad\qquad \text{and} \quad D_2 = 0.5X_2^2/X_1 \quad \text{(predators)}.$$

Nine cycles occur before prey become extinct, though other simulation runs finished sooner. $X_1(t)$ and $X_2(t)$ often dropped to 1, from which position extinction follows all too readily. The process exhibits smooth cyclic behaviour, with the predator maxima being roughly coincident with the prey population size half-way through its decay period. Thus the prey lead the predators by approximately one quarter-cycle (in line with the deterministic Lotka–Volterra model!). The main feature of this simulation is the wide variation in prey peaks from 105 to 979. The latter is only just short of the full carrying capacity, and is twice the deterministic limit cycle maximum.

Figure 6.11. Simulation corresponding to the deterministic Holling–Tanner process of Figure 6.10 for prey (———) and predators (– – –).

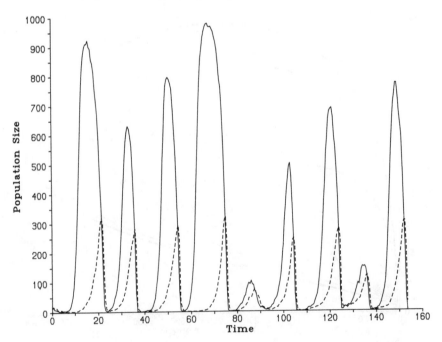

6.5   A model for prey-cover

A considerable number of theoretical deterministic models, of varying degrees of complexity, can be proposed, each of which exploits particular aspects of biological development (for examples see May, 1974b). However, when stochastic effects are introduced, many of the characteristic differences between different deterministic trajectories may become swamped by natural variation. That is, variation between several simulation runs of the same process may be just as great as variation between single runs of different processes. In practice the majority of deterministic models are therefore redundant, and for descriptive purposes the wisest policy is to select only those models which are simplest mathematically and are consistent with producing the required features of interest (e.g. stable limit cycles). The Volterra equations (6.19) and the Holling–Tanner equations (6.72) and (6.73) are an ideal pair in that between them they provide for convergence towards both a stable equilibrium point and a limit cycle. There is, however, one further model that is worth considering here, since it involves just a slight extension to the basic Lotka–Volterra model (6.3) yet guarantees that the prey cannot become extinct in stochastic simulations.

Suppose that a constant number (say k) of prey can take some form of cover which makes them inaccessible to predators. Then the N_2 predators can attack only $N_1 - k$ of the prey, so the simple Lotka–Volterra representation (6.3) becomes

$$dN_1/dt = r_1 N_1 - b_1 N_2 (N_1 - k) \qquad \text{(prey)}$$
$$dN_2/dt = -r_2 N_2 + b_2 N_2 (N_1 - k) \qquad \text{(predators)} \qquad (6.79)$$

(Maynard Smith, 1974). As

$$dN_1/dt = 0 \quad \text{at} \quad r_1 N_1 / b_1 = N_2 (N_1 - k)$$
$$dN_2/dt = 0 \quad \text{at} \quad r_2 N_2 / b_2 = N_2 (N_1 - k), \qquad (6.80)$$

there is an equilibrium point at

$$N_1^* = (r_2/b_2) + k \quad \text{and} \quad N_2^* = (r_1/b_1) + k(r_1 b_2 / r_2 b_1). \qquad (6.81)$$

Thus N_1^* and N_2^* increase linearly with the cover parameter k.

6.5.1   Stability analysis

We see from (6.80) that the two isoclines are

$$N_1 = k + (r_2/b_2) \quad \text{and} \quad N_2 = (r_1/b_1)N_1/(N_1 - k), \qquad (6.82)$$

which gives rise to the phase-plane diagram shown in Figure 6.12. Note from (6.82) that as N_1 increases, N_2 decays to r_1/b_1; whilst as N_1 approaches k, N_2 becomes indefinitely large. Superimposing trajectories then shows that the provision of cover has a stabilizing effect, since it transforms a closed Lotka–Volterra cycle into a converging spiral.

To examine the nature of this convergence write $N_i = N_i^*(1 + n_i)$, whence for small n_i equations (6.79) become

$$dn_1/dt = -k(r_1 b_2/r_2)n_1 - r_1 n_2$$
$$dn_2/dt = b_2 N_1^* n_1 \tag{6.83}$$

which are identical to the general linear equations (6.24) and (6.25) with

$$k_{11} = -kr_1 b_2/r_2, \quad k_{12} = -r_1$$
$$k_{21} = b_2 N_1^*, \qquad k_{22} = 0. \tag{6.84}$$

Substituting these values into (6.32) yields

$$\Delta = (kr_1 b_2/r_2)^2 - 4b_2 r_1 N_1^*. \tag{6.85}$$

So provided k is reasonably small, $\Delta < 0$, which results in damped cycles around the equilibrium point. The λ_i-values (6.28) become

$$\lambda_1, \lambda_2 = \tfrac{1}{2}[-(kr_1 b_2/r_2) \pm \sqrt{\Delta}], \tag{6.86}$$

Figure 6.12. Phase-plane diagram for the Lotka–Volterra process with prey-cover (6.79).

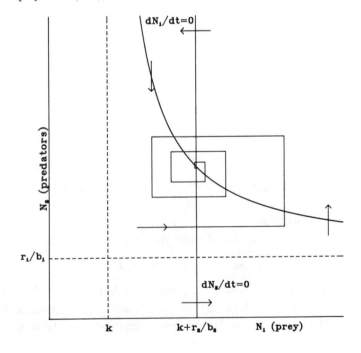

and so the general solution (6.29) and (6.30) for the $n_i(t)$ may now be written down directly.

The period of oscillation $T = 4\pi/\sqrt{(-\Delta)}$ decreases as $-\Delta$ increases. Now on writing (6.85) as

$$\Delta = (r_1 b_2/r_2)^2 [k - 2r_2^2/r_1 b_2]^2 - 4r_2(r_1 + r_2), \qquad (6.87)$$

we see that Δ reaches the minimum value $\Delta_{\min} = -4r_2(r_1 + r_2)$ when $k = 2r_2^2/r_1 b_2$, which corresponds to a minimum period of oscillation of $T_{\min} = 2\pi/\sqrt{\{r_2(r_1 + r_2)\}}$. Whilst at $k = 0$ (i.e. no cover) $\Delta = -4r_1 r_2$ so $T = T_0 = 2\pi/\sqrt{(r_1 r_2)}$. Thus increasing k first reduces T from T_0 to $T_{\min}$, and then increases T indefinitely until $\Delta = 0$, at which point oscillations cease and straight exponential damping takes over.

Finally, we note from (6.86) that the $n_i(t)$ converge to zero at a speed determined by the damping term

$$\exp\{-kr_1 b_2 t/2r_2\}, \qquad (6.88)$$

and that for fixed t this term decreases swiftly with increasing k.

6.5.2   *Deterministic versus stochastic behaviour*

As an example, consider $r_1 = 2.5$, $r_2 = 1.0$, $b_1 = 0.05$ and $b_2 = 0.01$. Then from (6.81)

$$N_1^* = 100 + k \quad \text{and} \quad N_2^* = 50 + (k/2).$$

Suppose that 10 prey are given cover, i.e. $k = 10$. Then $N_1^* = 110$ and $N_2^* = 55$, whilst (6.87) yields $\Delta = -10.94$ which gives the deterministic period of oscillation $T = 4\pi/\sqrt{(-\Delta)} = 3.80$. The damping term (6.88) becomes $\exp(-0.125t)$, so over a complete cycle the amplitude is reduced by $\exp(-0.125 \times 3.80) = 0.62$. The zero-cover period $T_0 = 3.97$, whilst $T_{\min} = 3.36$ is reached only when k takes the relatively high value of 80.

If required, $N_1(t) = 110[1 + n_1(t)]$ and $N_2(t) = 55[1 + n_2(t)]$ may be derived directly from the local solution (6.29) and (6.30). From (6.84), $k_{11} = -0.25$, $k_{12} = -1.0$, $k_{21} = 1.1$ and $k_{22} = 0$, whilst from (6.86), $\lambda_1, \lambda_2 = -0.125 \pm 1.6536i$ where i denotes $\sqrt{(-1)}$. So as $\exp(i\theta) = \cos(\theta) + i\sin(\theta)$ we have

$$n_1(t) = \exp(-0.125t)[C\cos(1.6536t) + D\sin(1.6536t)] \qquad (6.89)$$

where $C = A_1 + A_2$, $D = A_1 - A_2$ and angles are in radians (2π radians $= 360°$).

These local approximation results work well even when $N_i(t)$ and N_i^* are far apart. To demonstrate this, exact values for $N_1(t)$ and $N_2(t)$ were calculated numerically by writing equations (6.79) with $t = jh$ in the form

$$N_1[(j + 1)h] = N_1(jh) + h[r_1 N_1(jh) - b_1 N_2(jh)\{N_1(jh) - k\}]$$
$$N_2[(j + 1)h] = N_2(jh) + h[-r_2 N_2(jh) + b_2 N_2(jh)\{N_1(jh) - k\}],$$
$$\tag{6.90}$$

and then successively evaluating $N_1(jh)$ and $N_2(jh)$ with $h = 0.01$ over $j = 1, 2, \ldots, 4000$. Values for $k = 10$ and $N_1(0) = N_2(0) = 50$ are plotted in Figure 6.13 at time intervals of 0.1, and show 10 full prey cycles occurring between times $t = 1.90$ and 39.80. The corresponding cycle period, $T = 3.79$, is virtually identical to the local value 3.80 evaluated above. Moreover, after the first full cycle has occurred successive amplitudes are reduced by factors of 0.62–0.65, very close to the above local value of 0.62.

Figure 6.13 also shows a simulation of the corresponding stochastic

Figure 6.13. Stochastic and deterministic simulation corresponding to the Lotka–Volterra process with prey-cover

$$dN_1/dt = 2.5N_1 - 0.05N_2(N_1 - 10) \qquad \text{(prey: upper ---)}$$
$$dN_2/dt = -N_2 + 0.01N_2(N_1 - 10) \qquad \text{(predators: lower ---)}$$

for stochastic prey (———) and predators (— — —).

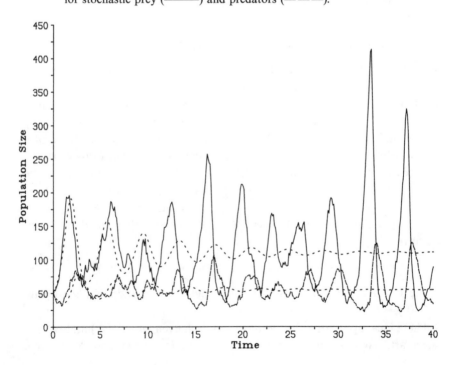

process, with birth and death rates

$$B_1 = r_1 X_1 \qquad \text{and} \quad D_1 = b_1 X_2 (X_1 - k) \qquad \text{(prey)}$$
$$B_2 = b_2 X_2 (X_1 - k) \quad \text{and} \quad D_2 = r_2 X_2 \qquad \text{(predators).} \qquad (6.91)$$

There is clearly a striking behavioural difference between the stochastic and deterministic realizations. Whereas the deterministic cycles are damped quite strongly, the stochastic cycles most certainly are not. Moreover, the large increase in prey around $t = 33$ cannot be explained by the deterministic equations, once again illustrating that damping implied by a deterministic model can be totally misleading. Indeed, if these simulated data were in fact real, then the underlying Lotka–Volterra model with prey-cover might well be excluded from consideration specifically because deterministically predicted damping did not occur.

6.6   Final comments

Arguments between the stochastic and deterministic camps have unfortunately polarized to such an extent that many researchers will use one approach to the total exclusion of the other. Clearly, both approaches should always be considered if we are to learn the most about a system. In the example above the presence/absence of sustained cycles is indeed a major difference, though in other respects there are considerable similarities. (i) The autocorrelation coefficients $\{r_j\}$ given by (6.55) peak at $j = 38$ time-units for both prey and predators. As each time-unit is of length 0.1 the cycle period is therefore 3.8 which is in exact agreement with that predicted by the local deterministic approximation $T = 4\pi/\sqrt{(-\Delta)}$. (ii) The cross-correlations $\{s_j\}$ given by (6.62) peak at $j = 8$ and 9, so the prey lead the predators by $(8.5/38) \times 360° = 81°$ which is close to the deterministic value of $85°$ obtained by comparing $N_1(t)$ and $N_2(t)$ in Figure 6.13. (iii) This figure shows initial broad agreement between stochastic and deterministic values.

In summary, deterministic analyses are useful for obtaining theoretical results for period length, lag times, etc., whilst stochastic simulation enables us to gain a behavioural overview of a process and to assess the validity of deterministic predictions.

Finally, we note that Kolmogorov (1936) developed a set of conditions which necessarily lead to the general deterministic equations

$$dN_1/dt = N_1 F(N_1, N_2)$$
$$dN_2/dt = N_2 G(N_1, N_2), \qquad (6.92)$$

where F and G are arbitrary functions of N_1 and N_2, having either a stable equilibrium point or a stable limit cycle. May (1974b) not only presents a clear summary of this work, explaining Kolmogorov's ten conditions in both mathematical and biological terms, but he goes on to claim that

> This rather robust theorem strongly suggests that those natural ecosystems which seem to exhibit a persistent pattern of reasonably regular oscillations are in fact stable limit cycles. This is altogether different from the widespread explanation of such phenomena which associates them with the oscillations in the pathological neutrally stable Lotka–Volterra system, where the amplitude depends wholly on the initial conditions.

Whilst this is indeed a reasonable assertion from a purely deterministic viewpoint, stochastic considerations lead us to conclude otherwise. We have already seen simulation experiments produce stochastic cycles under the three models:

(a) Volterra (Figure 6.5) – stable deterministic equilibrium point;
(b) Holling–Tanner (Figure 6.11) – stable deterministic limit cycle;
(c) Lotka–Volterra with cover (Figure 6.13) – as (a).

Yet only (b) produces deterministic limit cycles; natural variation ensures that both (a) and (c) oscillate stochastically *without damping*. Thus whilst May is right to discard the basic Lotka–Volterra model (with its divergent stochastic behaviour), the assertion that natural oscillations must be stable limit cycles is clearly unfounded.

7

Spatial predator–prey systems

Gause's conclusion that a predator–prey system is inherently self-annihilating without some outside interference such as immigration (Section 6.1.2) was questioned by Huffaker (1958). He claimed that Gause had used too simple a microcosm, and so set out to learn whether an adequately large and complex experiment could be constructed in which the predator–prey relation would not be self-exterminating. We therefore now ask, 'What will be the effect, if any, of accepting that individuals rarely mix homogeneously over the whole site but that they develop instead within separate sub-regions?'.

This question is an old one, for as early as 1927 A. J. Nicholson asked Bailey (1931) to investigate mathematically the abundance of two species which interact in the following manner. Members of the host species lay eggs and then die. These eggs are then searched for by members of the parasite species who traverse at random a specific area during their lifetime. Host eggs which survive this search develop into adults; those that are found are attacked and a parasite egg is deposited on them. New generations then repeat this process indefinitely.

7.1   Huffaker's experiments

Huffaker selected the six-spotted mite, *Eotetranychus sexmaculatus*, as the prey species and the predatory mite, *Typhlodromus occidentalis*, as the predator species because earlier observations had revealed this *Typhlodromus* as being a voracious enemy of the six-spotted mite. It was known to develop in great numbers on oranges infested with the prey species, to destroy essentially the entire infestation, and then to die *en masse*.

The whole site consisted of a spatial array of oranges, whilst sub-regions corresponded to individual oranges with both prey and predators being allowed to migrate between them. The oranges were partly covered with paper and paraffin to limit the available feeding area, and were sometimes intermingled with rubber balls that were similar in size. This arrangement made it possible to change either or both the available orange surface and the degree of dispersion of the oranges. Thus a simple environment in which all the prey food was concentrated in just one area of the system could be compared with ones in which the oranges were dispersed amongst the rubber balls.

To ensure that his experiment was sufficiently complex, Huffaker placed 40 oranges/balls in a 4 × 10 rectangular array on each of a number of adjacent trays (see Figure 1.2). Migration of predators across trays was restricted by inserting vaseline barriers; whilst migration of prey over these barriers was achieved by providing each tray with wooden posts from which the prey could launch themselves on a silken thread, aided by currents from an electric fan.

Theoreticians often think of biological modelling purely in terms of manipulating mathematical equations (if you cannot solve them then modify them until you can!), and completely forget the all-important biological aspects. Huffaker's series of experiments is surely a classic, involving considerable amounts of both practical expertise to construct the orange medium, and ingenuity to effect a sufficiently complex spatial scheme which would enable predator–prey cycles to be maintained. His paper should be declared mandatory reading for all spatial modellers.

In all Huffaker's seven simple one-tray systems, ranging from just 4

Figure 7.1. Densities per orange-area of the prey (*Eotetranychus sexmaculatus* – ○) and its predator (*Typhlodromus occidentalis* – △), with 20 small areas of food for the prey alternating with 20 foodless positions (after Huffaker, 1958).

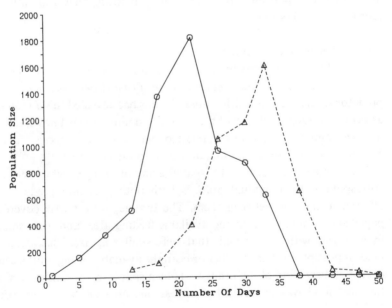

adjacent oranges to 20 oranges mixed with 20 balls, the prey and predator populations became extinct after only one cycle. Figure 7.1 shows the results from a system of 20 oranges alternating with 20 balls; initially ten female prey were introduced onto each of two of the oranges, and two female predators were introduced onto one of these two oranges 11 days later. Only one-tenth of each orange was exposed, so that the low feeding surface would encourage prey to keep on the move. The idea was that before predators could suffer a localized population crash, some prey would have already moved to other sites thereby enabling repopulation on a broader spatial basis.

The desired oscillations were, however, achieved in a three-tray system in which all 120 positions were occupied by oranges (Figure 7.2). Initially one mite was placed on each orange, and a female predator was introduced to each of 27 oranges five days later. In individual sections of the trays the mites died out, just as happened in Gause's experiments. However, by utilizing the large and more complex environment in a hop-skip-jump fashion, the prey were now able to survive by keeping one step ahead of the predators. This

Figure 7.2. Densities per orange-area of the prey (*Eotetranychus sexmaculatus* – ○) and its predator (*Typhlodromus occidentalis* – △), with 120 small areas of food for the prey (after Huffaker, 1958).

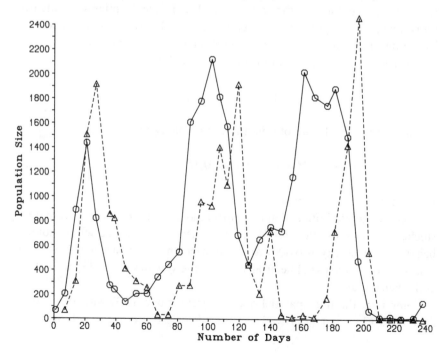

resulted in three full predator–prey oscillations before extinction finally occurred. An even more complex 252-orange system (Huffaker, Skea and Herman, 1963) produced four such oscillations, though all this really illustrates is the extreme difficulty in sustaining oscillations even with the most heavily contrived laboratory situations of spatial heterogeneity.

7.2   Simulation of the spatial Lotka–Volterra model

To simulate the effect of introducing spatial heterogeneity into a system, in particular its effect on times to extinction, we shall assume that both prey and predators are distributed in space more or less discontinuously in clumps, so forming numerous colonies, and that individuals of either type may move between adjacent colonies. This generates the so-called 'stepping-stone' model of population spread, first proposed by Kimura (1953) in a genetic context. Within any colony the species are assumed to undergo an ordinary non-spatial predator–prey process; though prey and predators may now migrate between colonies. For simplicity it is assumed that the colonies are sufficiently close together for migration to be effectively immediate; that is, the time taken to migrate is negligible compared to the life cycle.

Suppose that there are n such colonies, labelled by $i = 1, 2, \ldots, n$, with $M_i(t)$ and $N_i(t)$ denoting the (deterministic) number of prey and predators, respectively, in colony i at time t. Let u_{ij} and v_{ij} denote the migration rate per individual prey and predator, respectively, from colony i to colony j. Then the total rate of migration of prey (for example) from colony i to colony j is $u_{ij}M_i(t)$, so the overall migration rate out of colony i is

$$(u_{i1} + u_{i2} + \cdots + u_{in})M_i(t) \qquad \text{(for prey)}. \tag{7.1}$$

Similarly, the overall rate of migration into colony i is

$$u_{1i}M_1(t) + u_{2i}M_2(t) + \cdots + u_{ni}M_n(t). \tag{7.2}$$

For convenience we define $u_{ii} = 0$.

Since the simple Lotka–Volterra model has great difficulty in maintaining stochastic cycles (Section 6.1.3), we shall use it to describe within-colony behaviour. It clearly provides an ideal benchmark against which we can assess the effect of spatial heterogeneity on the system's capability to sustain cyclic behaviour.

Combining the non-spatial Lotka–Volterra model (6.3) with the spatial flow rates (7.1) and (7.2) leads to the set of $2n$ non-linear deterministic equations $(i = 1, 2, \ldots, n)$

$$
\begin{aligned}
dM_i/dt &= M_i(r_1 - b_1 N_i) - M_i(u_{i1} + \cdots + u_{in}) \\
&\quad + (M_1 u_{1i} + \cdots + M_n u_{ni}) \\
dN_i/dt &= N_i(-r_2 + b_2 M_i) - N_i(v_{i1} + \cdots + v_{in}) \\
&\quad + (N_1 v_{1i} + \cdots + N_n v_{ni})
\end{aligned}
\tag{7.3}
$$

for prey and predators, respectively. Here we have assumed that the parameters r_1, r_2, b_1 and b_2 are the same for each colony, though this is obviously not a necessity. However, as such equations are intrinsically very difficult to solve mathematically we shall go directly to a simulation analysis.

7.2.1 *Spatial simulation approach*

Denote $X_i(t)$ and $Y_i(t)$ to be the stochastic number of prey and predators in colony i at time t. Then for a small time increment h, and $i, j = 1, \ldots, n$, the non-spatial transition probabilities (6.16) extend to

(i) $\Pr[X_i(t + h) = X_i(t) + 1] = r_1 X_i(t)h$ (prey birth)
(ii) $\Pr[X_i(t + h) = X_i(t) - 1] = b_1 X_i(t)Y_i(t)h$ (prey death)
(iii) $\Pr[Y_i(t + h) = Y_i(t) + 1] = b_2 X_i(t)Y_i(t)h$ (predator birth)
(iv) $\Pr[Y_i(t + h) = Y_i(t) - 1] = r_2 Y_i(t)h$ (predator death)
(v) $\Pr[X_i(t + h) = X_i(t) - 1 \text{ and } X_j(t + h) = X_j(t) + 1]$
$\quad = u_{ij} X_i(t)h$ (prey moves from colony i to j)
(vi) $\Pr[Y_i(t + h) = Y_i(t) - 1 \text{ and } Y_j(t + h) = Y_j(t) + 1]$
$\quad = v_{ij} Y_i(t)h$ (predator moves from colony i to j). $\quad$ (7.4)

Construction of a simulation program based on these transition probabilities involves a straightforward extension to our previous non-spatial routines. First, determine the overall transition rate $R(t)$ by summing over (i) to (vi) above, viz:

$$
\begin{aligned}
R(t) = \sum_{i=1}^{n} \Bigg[X_i(t) \bigg\{ r_1 + b_1 Y_i(t) &+ \sum_{j=1}^{n} u_{ij} \bigg\} \\
+ Y_i(t) \bigg\{ r_2 + b_2 X_i(t) &+ \sum_{j=1}^{n} v_{ij} \bigg\} \Bigg].
\end{aligned}
\tag{7.5}
$$

Second, generate a uniform pseudo-random number $0 \leqslant Z \leqslant 1$. Then accumulate the computed rates $r_1 X_1(t)$, etc., in (i) to (vi) over $i, j = 1, \ldots, n$ in say SUM, and compare SUM with $Z \times R(t)$ at each step. The appropriate event type is determined when SUM $\geqslant Z \times R(t)$ for the first time. The inter-event time (s) is calculated as in Section 3.4.2, by first selecting a new random number Z and then computing

$$
s = -[\log_e(Z)]/R(t).
$$

To demonstrate this spatial approach consider migration between just three colonies (Figure 7.3). In model A all six migration directions are permitted, with equal migration rates $u_{ij} = u$ and $v_{ij} = v$ $(i \neq j)$. In model B migration is purely clockwise, so only $u_{12} = u_{23} = u_{31} = u$ and $v_{12} = v_{23} = v_{31} = v$. To enable comparison with our previous non-spatial results we shall retain the parameters used in example (6.6) ($r_1 = 1.50$, $r_2 = 0.25$, $b_1 = 0.1$ and $b_2 = 0.01$). Three simulation runs were made for each of various values of u and v, and population sizes were listed at time intervals of 0.2. The process was always started from $X_1(0) = 10$ and $Y_1(0) = 5$, the other colonies being initially empty.

7.2.2   *Results for model A*

(i) $u = v = 0$
The process is confined solely to colony 1, and so is completely non-spatial. The simulation runs were similar to the outer one shown in Figure 6.3, with prey extinction occurring before the end of the first cycle. The three times to extinction (T_{ext}) were in the range 5.2–5.7.

(ii) $u = v = 0.01$
Colony 1 develops much as before, and as prey soon outnumber predators the first migration into colony 2 or 3 is far more likely to be a prey than a predator. This means that the time gap between the arrival of the first prey and the first predator was usually large enough to allow the prey population to explode (to over 15 000 prey in two of the runs) before the predator population could start to control it. This vast food supply then caused the predator population to explode, resulting in a sudden dramatic crash in the prey population as they were all quickly eaten. In one run colony 1 did cycle

Figure 7.3. A three-colony model with (A) migration between all colonies, (B) clockwise migration only.

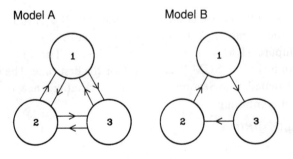

Model A   Model B

twice, but otherwise this migration configuration did not lead to sustained cycles. Times to extinction were raised slightly to 8.2–10.2, since it took a little time for colonies 2 and 3 to get going.

(iii) $u = v = 0.1$
Though increasing both migration rates did reduce the maximum prey size, colonies 2 and 3 just developed more quickly resulting in a reduction of extinction times to 5.4–6.6.

(iv) $u = 0.01, v = 0.1$
These results suggest that we should keep the prey migration rate down, whilst making the predators sufficiently mobile so that they can hunt and kill the relatively inert prey. Using $v = 10u = 0.1$ immediately extended the extinction times to 21.2–61.2, and in the 'best' run each colony successfully executed five cycles before prey extinction finally occurred. Increasing the rates to $v = 10u = 1$, however, caused the colonies to become synchronized, with each one containing roughly the same number of individuals. Spatial structure is thereby effectively removed, and extinction occurs before the end of the first cycle.

(v) $u = 0.001, v = 0.01$ (10 colonies)
The three-colony process (iv) can maintain a limited number of cycles because if two colonies are devoid of prey then there is a reasonable chance that prey in the third colony will migrate to them before total extinction occurs. Increasing the number of colonies to say 10 should therefore increase the chance of at least one or two colonies containing prey at any given time, and thereby increase the mean time to extinction. Before demonstrating this, we must note from (7.1) that the individual total migration rate out of a colony is $(n - 1)u$ for prey and $(n - 1)v$ for predators. With $u = 0.01, v = 0.1$ and $n = 10$ this gives values of 0.09 and 0.9, respectively, which are high enough to cause some degree of synchronization between the colonies. To effect a non-synchronous spatial structure we therefore reduced u and v to $v = 10u = 0.01$. Extinction times of 304 and 696 were obtained for the first two simulations, whilst in the third the process was still proceeding quite happily at time 1435 when the process was terminated because a limit of 7000 seconds c.p.u. time had been reached! The average number of colonies containing prey was 6, whilst for 85% of the time this number lay between 4 and 8. The lowest such value was generally 2. Taking the non-spatial cycle period of $T = 10.26$ (Section 6.1.1), the smallest time to extinction of 304 corresponds to about 30 cycles. Thus even with as few as 10 colonies in the system, extinction has to be viewed as a fairly unlikely event over any reasonable time scale.

These trial results show that sustained cycles can be achieved if:

(a) there are around 10 or more colonies;
(b) predators are more mobile than prey;
(c) v is high enough to prevent a prey population explosion in a colony before predators can become established there;
(d) u and v are not so large that they synchronize the spatial structure out of the system.

Thus even with a small degree of spatial complexity (10 colonies is by no means a large system), the primary determinant of persistence is the relative dispersal abilities of predator and prey. An earlier simulation experiment by Hilborn (1975) produced exactly the same conclusion.

Figure 7.4. Simulation of a three-colony predator–prey process with clockwise migration, and parameter values $r_1 = 1.50$, $b_1 = 0.1$, $r_2 = 0.25$, $b_2 = 0.01$, $u_{12} = u_{23} = u_{31} = 0.01$ and $v_{12} = v_{23} = v_{31} = 0.1$ showing (a) prey and (b) predator population sizes for colonies 1 (———),
2 (— — —) and 3 (---).

(a)

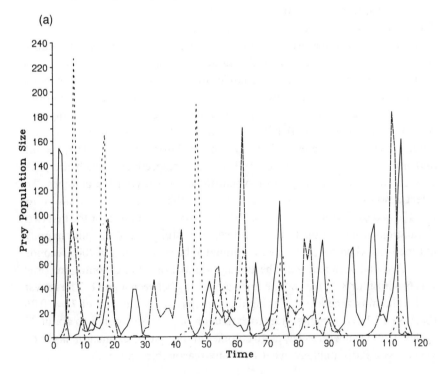

7.2.3   *Results for model B*

A major computing problem is that run-time swiftly increases as the number of colonies (n) increases. Under model A each colony has four types of birth/death and $2(n-1)$ migration events associated with it, giving a total of $2n(n+1)$ events in all. The number of 'IF' statements to be computed therefore increases with n^2, quickly rendering this simulation procedure impractical.

Switching to model B (Figure 7.3) substantially reduces the number of possible event types, and hence the associated computing cost, since the number of migration routes is reduced from $n(n-1)$ to n. Survival under B relies purely on the ability of a developing colony to send enough prey and predators ahead of it so that the next colony-cycle can start before the current one collapses.

(vi) $u = 0.01$, $v = 0.1$ (3 colonies)

In three simulations starting with just 10 prey and 5 predators in colony 1, the prey became extinct at times 30, 69 and 117 which are broadly in line with those from simulation (iv) (model A). Figure 7.4 shows the population

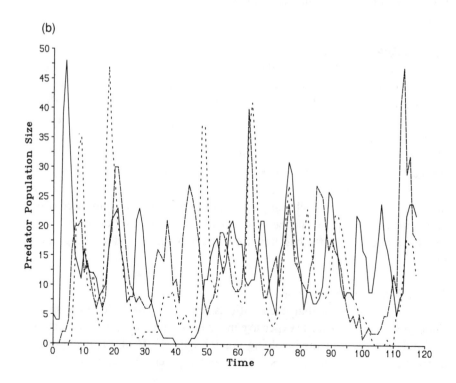

development for the third run, and it can be seen that in spite of each colony being devoid of prey at various times an influx of prey from the neighbouring colony starts the cycle off again. Extinction occurred when the total prey population dropped to 1 and was unable to survive against the far more numerous predators (52). The prey size dropped very low several times during the run (e.g. 3 prey and 59 predators at time 22), and long-term survival with this particular process is clearly unlikely. Even so, model B's ability to sustain cycles is vastly superior to the related non-spatial process. Within colonies, the predator population generally peaks two time units after the prey peak, the three cross-correlations at lag 2 being 0.75, 0.73 and 0.75. However, analysis of the autocorrelations provides little, if any, evidence for regular cycling within individual colonies.

(vii) $u = 0.01$, $v = 0.1$ (10 colonies)
With just 10 migration routes, far less than the 90 under model A, the three extinction times of 29, 91 and 574 are substantially down on those of (v). Population development is concentrated in one, or sometimes two, clusters of 3 to 4 neighbouring colonies travelling clockwise around the circle. Since increasing n will just increase the number of clusters, this in itself will not lead to a dramatic rise in the mean time to extinction as happens under model A.

7.2.4   *Follow-up remarks*

These simulation experiments, restricted though they are, suggest that Huffaker's conclusion that cycles may be sustained only through a complex arrangement of many colonies may be false. Even with just three colonies cycling can be achieved (experiments (iv) and (vi)) *provided that the migration rates are suitably chosen*. Herein lies the clue to Huffaker's problem. If the migration rates are not conducive to sustained cycles in say a simple 10-colony array of oranges, then no amount of complicated habitat extension with artificial aids and barriers to movement (using 252 oranges in his most extreme case) will really compensate for this. Thus an interesting follow-up experiment would be to use just a small number of orange-sites with the aim of finding the best balance between predator and prey migration rates.

7.3   Matrix representation

We have seen that even with as few as $n = 10$ colonies a single simulation run can be fairly expensive in terms of c.p.u. time, so construction of simulation experiments involving much larger values of n is likely to be impractical unless we have access to a supercomputer. The problem is that a

tremendous amount of information on event types and inter-event times is
developed within a simulation run, yet virtually all of it is ignored if all we
require is a qualitative assessment of how the process is developing.

7.3.1   *Four-state representation*

In the simplest description a particular colony can be in any one of
four states:

> E   empty
> H   contains only prey
> P   contains only predators
> M   contains both prey and predators (mixed).   (7.6)

Development of the process at 'times' $t = 0, 1, 2, \ldots$ may therefore be
represented (albeit very crudely) by the transition probability matrix

$$
\begin{array}{cccc}
\text{time } t + 1 \\
E \quad\; H \quad\; P \quad\; M \\
\left[\begin{array}{cccc}
p_{EE} & p_{EH} & p_{EP} & p_{EM} \\
p_{HE} & p_{HH} & p_{HP} & p_{HM} \\
p_{PE} & p_{PH} & p_{PP} & p_{PM} \\
p_{ME} & p_{MH} & p_{MP} & p_{MM}
\end{array}\right]
\begin{array}{c} E \\ H \\ P \\ M \end{array}
\end{array}
\qquad \text{time } t. \qquad (7.7)
$$

This representation is easily extended to cover multi-species situations by
expanding the number of states; Crowley (1979), for example, uses a 9×9
matrix to analyze a spatial system in which two prey-species interact with a
single predator-species.

To illustrate how the matrix formulation (7.7) may be used to study times
to extinction, consider run 3 of experiment (iv) (Section 7.2.2) in which colony
1 experiences the following transitions between times $t = 0, 1, \ldots, 60$

$$
\begin{array}{llll}
P \to P & 11; & P \to M & 4; \\
M \to P & 5; & M \to M & 40.
\end{array}
$$

Since states E and H do not feature here, the transition probabilities (7.7)
comprise four non-zero values estimated by

$$
\begin{array}{ll}
p_{PP} = 11/15; & p_{PM} = 4/15; \\
p_{MP} = 5/45; & p_{MM} = 40/45.
\end{array}
\qquad (7.8)
$$

Let $p_P(t)$ and $p_M(t)$ denote the probabilities of a given colony being in states
P and M, respectively, at time t. Then

$$
p_P(t + 1) = \text{Pr(in state P at time } t \text{ and remains there)}
$$
$$
+ \text{Pr(in state M at time } t \text{ and moves to P)}
$$

i.e.

$$p_P(t+1) = p_P(t)p_{PP} + p_M(t)p_{MP}. \tag{7.9}$$

Similarly,

$$p_M(t+1) = p_P(t)p_{PM} + p_M(t)p_{MM}.$$

Now as t increases, the probabilities $p_P(t)$ and $p_M(t)$ will approach steady-state values π_P and π_M, independent of t. Equations (7.9) then become

$$\begin{aligned}\pi_P &= \pi_P p_{PP} + \pi_M p_{MP} \\ \pi_M &= \pi_P p_{PM} + \pi_M p_{MM},\end{aligned} \tag{7.10}$$

with the solution

$$\pi_P = p_{MP}/(p_{PP} + p_{MM}) \quad \text{and} \quad \pi_M = p_{PM}/(p_{PP} + p_{MM}). \tag{7.11}$$

Hence (7.8) gives rise to the estimates $\pi_P = 5/17$ and $\pi_M = 12/17$ for the probabilities of being in states P and M, respectively.

Although this analysis relates to just a single colony, a rough assessment of the whole system can be achieved by assuming that the three colonies develop independently from each other (even though they plainly do not). The equilibrium probability that the process is devoid of prey at a particular time is then given by $\pi_P^3 = (5/17)^3 = 0.0254$. If we make the further assumption that the probability of prey extinction occurring exactly at time t is given by

$$\begin{aligned}&\text{Pr(each of the first } t-1 \text{ times has prey somewhere)} \\ &\quad \times \text{Pr(no prey anywhere at time } t) \\ &= (1 - 0.0254)^{t-1} \times 0.0254,\end{aligned} \tag{7.12}$$

then the time to prey extinction has a geometric distribution with parameter $g = 0.0254$. Thus a ball-park estimate for the mean time to prey extinction is $T_E = (1 - g)/g = 38$. Note that this is close to the mean value of 31 from the three simulation runs.

The derivation of result (7.12) clearly involves grossly oversimplifying assumptions, and before we can give this approach general credibility further work is needed to discover just how critical the underlying assumptions are. Simulation provides a natural first attack; mathematically oriented results, including estimates of the intensity of the fluctuations in the numbers of occupied patches, may be found in Gurney and Nisbet (1978).

7.3.2   *Eight-state representation*

The four-state representation (7.6) is very restricted, and Maynard Smith (1974) suggests expanding it to cover the following eight situations:

E   empty
HA   few prey
HB   increasing prey
HC   many prey
MA   many prey, few predators
MB   many prey, increasing predators
MC   many prey, many predators
MD   few prey, many predators.

Thus

H denotes prey but no predators,
M denotes both prey and predators.

This list is obviously not exhaustive, but to enlarge it further would spoil the whole point of having a simple representation. We shall assume that the transitions MA → MB, MB → MC and MC → MD must occur, that the predator death rate is sufficiently high for MD → E, whilst a migrant predator arriving at an E cell soon starves. Moreover, a migrant predator converts an HA, HB or HC cell into an MA cell, whilst a migrant prey converts an E cell into an HA cell. There is clearly nothing sacred about these rules, but they do allow easy interpretation and very fast simulation.

In this model the stochastic structure relates to the mobility of prey and predators, so it is ideal for studying the effect of changing the migration strategy. For a given colony (i) let x denote the number of neighbouring colonies with many prey (i.e. HC, MA, MB, MC) that are connected to it. Then if α denotes the probability that in one time unit prey will migrate from a connected colony (j) to i, we have

$$\Pr(\text{no prey migrate from } j \text{ to } i) = (1 - \alpha),$$

so

$$\Pr(\text{no prey migrate from any connecting colony to } i) = (1 - \alpha)^x$$

since colonies send out migrants independently from each other. Thus $\Pr(E \to HA)$ is given by

$$a = \Pr(\text{at least 1 prey reaches } i \text{ from any connected colony})$$
$$= 1 - (1 - \alpha)^x. \tag{7.13}$$

Similarly, if y denotes the number of colonies connected to i which contain migratory predators, and β denotes the predator migration rate, then

$$b = \Pr(\text{at least 1 predator reaches } i \text{ from any connected colony})$$
$$= 1 - (1 - \beta)^y. \tag{7.14}$$

Table 7.1. Transition probabilities for an eight-state predator/prey representation

E	HA	HB	HC	MA	MB	MC	MD	
$1-a$	a	0	0	0	0	0	0	E
0	0	$1-b$	0	b	0	0	0	HA
0	0	0	$1-b$	b	0	0	0	HB
0	0	0	$1-b$	b	0	0	0	HC
0	0	0	0	0	1	0	0	MA
0	0	0	0	0	0	1	0	MB
0	0	0	0	0	0	0	1	MC
1	0	0	0	0	0	0	0	MD

If predators migrate only when they are hungry, then y is the number of connected MD colonies. Whilst if any predator may migrate then y is the number of connected MB, MC and MD colonies (we exclude MA – few predators). Table 7.1 shows the full set of transition probabilities for this latter situation with

$$\text{Pr}(\text{HA} \rightarrow \text{MA}) = \text{Pr}(\text{HB} \rightarrow \text{MA}) = \text{Pr}(\text{HC} \rightarrow \text{MA}) = b$$

and

$$\text{Pr}(\text{HA} \rightarrow \text{HB}) = \text{Pr}(\text{HB} \rightarrow \text{HC}) = \text{Pr}(\text{HC} \rightarrow \text{HC}) = 1 - b.$$

Note that any degree of spatial complexity can be incorporated into this model, even Huffaker's elaborate system with vaseline barriers and air currents! All we need are individual cell values for the probabilities α and β, and knowledge of the connectivity between cells so that x and y can be determined for each colony.

Let us now consider the nature of the quasi-equilibrium distribution (i.e. both prey and predators are assumed to be present somewhere). With the transition probabilities given in Table 7.1, the equilibrium equations become

$$\begin{aligned}
\pi_E &= (1-a)\pi_E + \pi_{MD} & \pi_{MA} &= b(\pi_{HA} + \pi_{HB} + \pi_{HC}) \\
\pi_{HA} &= a\pi_E & \pi_{MB} &= \pi_{MA} \\
\pi_{HB} &= (1-b)\pi_{HA} & \pi_{MC} &= \pi_{MB} \\
\pi_{HC} &= (1-b)(\pi_{HB} + \pi_{HC}) & \pi_{MD} &= \pi_{MC}.
\end{aligned} \tag{7.15}$$

If we are to make mathematical headway then some simplification is necessary since the coefficients a and b in (7.15) are not constant but depend on the variables x and y. Let us therefore replace x and y by their 'average' values $\bar{x}$ and $\bar{y}$. Then on writing

$$\pi_H = \pi_{HA} + \pi_{HB} + \pi_{HC} \quad \text{(prey, but no predators)}$$

and

$$\pi_M = \pi_{MA} + \pi_{MB} + \pi_{MC} + \pi_{MD} \qquad \text{(prey and predators)},$$

equations (7.15) reduce to

$$\pi_{HA} = a\pi_E = \pi_{MD} = \pi_{MC} = \pi_{MB} = \pi_{MA} = b\pi_H = \pi_M/4$$
$$\pi_{HB} = (1 - b)\pi_{HA}$$
$$b\pi_{HC} = (1 - b)\pi_{HB}. \tag{7.16}$$

But the process must be in one of the states E, H or M, i.e.

$$1 = \pi_E + \pi_H + \pi_M = \pi_M[(1/4a) + (1/4b) + 1].$$

Thus

$$\pi_M = 4ab/(a + b + 4ab)$$
$$\pi_H = a/(a + b + 4ab)$$

and

$$\pi_E = b/(a + b + 4ab). \tag{7.17}$$

It now follows from (7.16) that

$$\pi_E = b/D, \qquad \pi_{HA} = ab/D, \qquad \pi_{HB} = ab(1 - b)/D,$$
$$\pi_{HC} = a(1 - b)^2/D \quad \text{and} \quad \pi_{MA} = \pi_{MB} = \pi_{MC} = \pi_{MD} = ab/D \tag{7.18}$$

where $D = a + b + 4ab$.

On examining expressions (7.17) we see that for the special cases

	$\pi_E =$	$\pi_H =$	$\pi_M =$
(i) $a \simeq 0, b \simeq 0$:	$b/(a + b)$	$a/(a + b)$	0
(ii) $a \simeq 0, b \simeq 1$:	1	0	0
(iii) $a \simeq 1, b \simeq 0$:	0	1	0
(iv) $a \simeq 1, b \simeq 1$:	1/6	1/6	2/3

$$\tag{7.19}$$

Inspection of Table 7.1 shows that if $a \simeq 0$ then once the process reaches state E it remains there (case (ii)), whilst if $b \simeq 0$ then transfer into the M-states and thereby subsequent return to E is very unlikely (case (iii)). Moreover, on writing $\pi_M = 1/\{(1/4a) + (1/4b) + 1\}$ and noting that π_M increases as either a or b increases, it follows that π_M has its maximum size of 2/3 at $a = b = 1$ (case (iv)).

This matrix representation is ideal for studying how different types of network (i.e. the way in which colonies are connected) affect the equilibrium structure. For example, in 'island' models migration may occur between many, if not all, of the n colonies; whilst in 'stepping-stone' models migration may occur between nearest neighbours only. Clearly the average connection values $\bar{x}$ and $\bar{y}$ will vary substantially between these two situations, from up to $n - 1$ in the first to around 4 in the second. So if extinction probabilities are to be compared it seems sensible to keep a in (7.13) and b in (7.14) constant (across models) by using the migration rates

$$\alpha \simeq 1 - (1 - a)^{1/\bar{x}} \quad \text{and} \quad \beta \simeq 1 - (1 - b)^{1/\bar{y}}. \tag{7.20}$$

Under condition (7.20), simulations by Maynard Smith (1974) suggest that the island model is more persistent than the stepping-stone model, though the difference is not very great. The inference is that in nature persistence would be favoured if occasional individuals could migrate long distances.

7.3.3   Simulation of the eight-state representation

Though the above approach does take account of how colonies are connected, it suffers from the defect that the equilibrium probabilities relate to a single colony. We cannot therefore use it to develop exact probability statements, such as for the probability that the predator population is extinct by time t. Fortunately, a simulation program is easily constructed which incorporates the transition probabilities of Table 7.1 (or any other appropriate probability structure) into the development of an $m \times n$ array of colonies.

(i) Denote the states E, HA, HB, ..., MC, MD by 1, 2, ..., 8.
(ii) Let the arrays $\{X_{ij}(t)\}$ and $\{Y_{ij}(t + 1)\}$ denote the state of each colony at times t and $t + 1$, for $1 \leqslant i \leqslant m$ and $1 \leqslant j \leqslant n$.
(iii) Use Table 7.1 in conjunction with expressions (7.13) and (7.14) to determine $\{Y_{ij}(t + 1)\}$ from $\{X_{ij}(t)\}$. For example, if $X_{ij}(t) = 1$ (state E) then $Y_{ij}(t + 1)$ equals 1 and 2 (state HA) with probabilities $1 - a$ and a, respectively.
(iv) Put $\{X_{ij}(t + 1)\} = \{Y_{ij}(t + 1)\}$ and print $\{X_{ij}(t + 1)\}$.
(v) Return to (ii) with t replaced by $t + 1$.

Edge effects may be dealt with by maintaining a border of empty colonies (i.e. 1s) just outside the $m \times n$ array.

Consider, for example, simulating the stepping-stone model over a 6×6 array of colonies with migration between four nearest-neighbours. Let the migration rates α and β take values 0.1, 0.5 and 0.9, with the corresponding probabilities a and b being given from (7.13) and (7.14) by

α or β			x or y		
	0	1	2	3	4
0.1	0	0.1	0.19	0.27	0.35
0.5	0	0.5	0.75	0.86	0.94
0.9	0	0.9	0.99	1.00	1.00

Though a full simulation study would involve running this process several times for each of a wide variety of (α, β)-values, the general flavour can be determined by considering the following five single runs (Table 7.2) starting from the symmetric configuration

$$\{X_{ij}(0)\} = \begin{bmatrix} 1 & 1 & 1 & 1 & 1 & 1 \\ 1 & 5 & 1 & 1 & 5 & 1 \\ 1 & 1 & 1 & 1 & 1 & 1 \\ 1 & 1 & 1 & 1 & 1 & 1 \\ 1 & 5 & 1 & 1 & 5 & 1 \\ 1 & 1 & 1 & 1 & 1 & 1 \end{bmatrix}.$$

That is, all colonies are initially empty (E) except for four which contain many prey and few predators (MA).

(i) $\alpha = \beta = 0.1$ – both prey and predators have low mobility – not only does predator migration into a predator-free colony (H → MA) happen relatively infrequently, but any such transitions that do occur cycle through to E. Predators therefore soon become extinct (here by $t = 4$), and prey then gradually spread throughout the array.

(ii) $\alpha = 0.1$, $\beta = 0.5$ – prey low and predators moderate mobility – E still acts as a trap, though predators can persist a little longer (here until $t = 7$).

(iii) $\alpha = 0.1$, $\beta = 0.9$ – prey low and predators very high mobility – strong predator mobility quickly converts all H-states into Ms from where local extinction of both species may occur. With prey having low mobility they are unable to move to neighbouring colonies sufficiently fast to keep the process going (here prey and predators simultaneously become extinct at $t = 12$).

(iv) $\alpha = 0.5$, $\beta = 0.1$ – prey moderate and predators low mobility – with E no longer acting as a trap prey soon become highly abundant, but the predators cannot migrate fast enough to start up new cycles in neighbouring colonies before their own cycle collapses (here predators are extinct by $t = 12$).

(v) $\alpha = \beta = 0.5$ – both prey and predators moderately mobile – prey soon recolonize empty cells and predators are sufficiently mobile to maintain a fresh supply of cycles in neighbouring colonies. This process is not only persistent (still going strong after 1000 generations), but it is also cheap! On an ICL 2988 machine this 6×6 matrix requires only 20 seconds c.p.u. time per 1000 generations.

Some indication of the final outcome of the process may be gleaned from the equilibrium probabilities (7.19). For example, case (iii) $\alpha = 0.1$, $\beta = 0.9$ may be likened to $a \simeq 0$, $b \simeq 1$ with the predicted outcome E (i.e. $\pi_E = 1$); case (iv) $\alpha = 0.5$, $\beta = 0.1$ may be likened to $a \simeq 1$, $b \simeq 0$ with the predicted outcome H (i.e. $\pi_H = 1$). Case (v) $\alpha = \beta = 0.5$ may be likened (at a stretch!) to

Table 7.2. Simulated spatial predator–prey process shown at times $t = 1, 5, 10$ and 100 for: (i) $\alpha = \beta = 0.1$; (ii) $\alpha = 0.1, \beta = 0.5$; (iii) $\alpha = 0.1, \beta = 0.9$; (iv) $\alpha = 0.5$, $\beta = 0.1$; and (v) $\alpha = \beta = 0.5$.

	$t = 1$	$t = 5$	$t = 10$	$t = 100$
(i)	1 1 1 1 1 1	1 1 1 1 4 1	1 1 1 4 4 4	4 4 4 4 4 4
	1 6 1 1 6 1	1 1 1 1 1 1	1 1 1 1 1 2	4 4 4 4 4 4
	1 1 1 1 2 1	1 1 1 1 4 1	1 1 1 1 4 4	4 4 4 4 4 4
	1 1 1 1 1 1	1 1 1 1 1 1	1 1 1 4 1 4	4 4 4 4 4 4
	1 6 1 1 6 1	1 1 2 4 1 4	1 1 4 4 4 4	4 4 4 4 4 4
	1 1 1 1 1 1	1 1 1 1 1 1	1 1 1 3 1 4	4 4 4 4 4 4
(ii)	1 1 1 1 1 1	1 1 1 1 6 1	1 1 1 1 1 1	4 4 4 4 4 4
	1 6 1 1 6 1	1 1 1 1 1 1	1 1 1 1 1 1	4 4 4 4 4 4
	1 1 1 1 1 1	1 1 1 1 1 1	1 1 1 1 1 1	4 4 4 4 4 4
	1 1 1 1 2 1	1 1 1 2 8 1	1 1 1 4 1 1	4 4 4 4 4 4
	1 6 1 1 6 1	1 1 1 1 1 1	1 1 1 1 1 1	4 4 4 4 4 4
	1 1 1 1 1 1	1 1 1 1 1 1	1 1 1 1 1 1	4 4 4 4 4 4
(iii)	1 1 1 1 1 1	1 1 1 1 1 1	1 1 1 1 1 1	1 1 1 1 1 1
	1 6 1 1 6 1	1 1 6 1 1 1	1 1 8 1 1 1	1 1 1 1 1 1
	1 1 1 1 1 1	1 1 1 1 1 1	1 1 1 1 1 1	1 1 1 1 1 1
	1 2 1 1 1 1	5 8 1 1 1 1	1 1 1 1 1 1	1 1 1 1 1 1
	1 6 1 1 6 2	7 2 7 1 1 8	1 1 1 1 1 1	1 1 1 1 1 1
	1 1 1 1 1 1	5 6 1 1 1 1	1 1 7 1 1 1	1 1 1 1 1 1
(iv)	1 1 1 1 2 1	2 4 2 1 4 3	4 4 4 4 4 4	4 4 4 4 4 4
	1 6 1 1 6 1	4 2 4 2 2 4	4 4 4 4 4 4	4 4 4 4 4 4
	1 2 1 1 1 1	3 6 2 1 1 2	4 3 7 4 4 4	4 4 4 4 4 4
	1 1 1 1 1 1	2 4 1 3 1 1	4 7 4 4 4 4	4 4 4 4 4 4
	2 6 1 2 6 1	4 2 4 4 2 4	4 4 4 4 4 4	4 4 4 4 4 4
	1 1 1 1 1 1	3 4 1 2 4 2	4 4 4 4 4 4	4 4 4 4 4 4
(v)	1 1 1 1 2 1	2 6 2 2 7 6	1 5 1 1 6 1	7 3 8 1 6 2
	2 6 1 2 6 2	7 2 6 7 2 8	4 1 5 6 1 7	1 5 7 5 8 8
	1 1 1 1 2 1	1 1 2 4 8 5	5 8 8 1 6 2	6 4 1 7 1 6
	1 1 1 1 2 1	1 6 4 3 4 3	4 5 1 7 1 8	8 1 1 7 7 2
	1 6 2 1 6 1	1 2 8 7 2 4	5 1 7 6 8 5	2 7 8 6 1 1
	1 2 1 1 1 1	1 8 6 5 1 1	2 7 3 2 6 4	7 3 6 1 7 1

$a \simeq 1$, $b \simeq 1$ with $\pi_E = \pi_H = 1/6$ and $\pi_M = 2/3$; counting the 180 states generated through the five times $t = 10, 15, 20, 25$ and 30 gave 36 Es, 37 Hs and 107 Ms which is in broad agreement with the values $180\pi_E = 30$, $180\pi_H = 30$ and $180\pi_M = 120$ predicted from the equilibrium probabilities.

8

Fluctuating environments

The fascination of natural communities of plants and animals lies in their endless variety. Not only do no two places share identical histories, climates or topography, but also climate and other environmental factors are constantly fluctuating. Such systems will therefore not exhibit the crisp determinacy which characterizes so much of the physical sciences (May, 1974a). Now in the preceding chapters we have implicitly assumed that the environment is unvarying; birth and death rates, carrying capacities, etc., have all been held constant through time and space. Thus our stochastic models have involved variation in the sense that random events occur with probabilities which depend only on population size.

However, the most striking features of life on this planet are directly attributable to the diurnal rotation of the Earth and its annual journey around the Sun. Indeed, the behaviour and reproductive cycles of living organisms are closely adapted to the regular alternation of summer and winter, or of wet season and dry season (Skellam, 1967). So as real environments are themselves uncertain, all parameters which characterize populations must exhibit random or periodic fluctuations to at least some degree. Thus even deterministic equilibrium is not an absolute fixed state, but is instead a 'fuzzy' value around which the biological system fluctuates.

We have already seen in Section 4.8 that static environment blowfly models (for example) do not produce sufficient variability, and so by admitting the reality of *environmental* variation we have a second, powerful source of variability at our disposal. Thus not only are four types of model now available, corresponding to the combinations of

> events: deterministic or stochastic
> environment: static   or fluctuating,

but the fluctuating environment component can itself be either deterministic or stochastic.

8.1   Deterministic variability

To use the general birth–death process of Chapter 3 in a dynamic environment, we shall assume that some external influence affects the birth and death parameters to make them time-dependent. This means that equation (3.9) for the probabilities $\{p_N(t)\}$ takes the even more general form

$$dp_N(t)/dt = D(N+1, t)p_{N+1}(t) - [B(N, t) + D(N, t)]p_N(t) \\ + B(N-1, t)p_{N-1}(t) \tag{8.1}$$

for $N = 0, 1, 2, \ldots$.

Not only are special-case solutions to equations (8.1) too mathematically demanding to be of interest here, but also a general equilibrium solution cannot be constructed since the probabilities $p_N(t)$ continually fluctuate with changing $B(N, t)$ and $D(N, t)$. We can, however, generate numerical solutions over some finite range $0 \leqslant N \leqslant N'$, since computer routines for such first-order linear systems are readily available. Ideally N' should be chosen large enough for $p_{N'}(t)$ to remain negligibly small as t increases, yet N' chosen small enough to ensure computational feasibility. To prevent probability 'escaping' during the numerical integration we place $B(N', t) = 0$ for all t, since this prohibits births occurring when $N = N'$.

The extinction probability $p_0(t)$ is computed directly, whilst the mean and variance of population size are constructed from

$$m(t) = \sum_{N=0}^{N'} N p_N(t) \tag{8.2}$$

and

$$\sigma^2(t) = \sum_{N=0}^{N'} N^2 p_N(t) - m^2(t), \tag{8.3}$$

respectively. In the static environment case we were especially interested in quasi-equilibrium probabilities $\{\pi_N^{(Q)}\}$, defined in terms of non-extinct realizations (e.g. Sections 3.3 and 3.4.1). The analogous time-dependent situation is easily developed by defining

$$p_N^{(Q)}(t) = p_N(t)/[1 - p_0(t)] \qquad (N = 1, 2, \ldots) \tag{8.4}$$

as the probability that the population is of size N at time t conditional on N being positive. It immediately follows that the corresponding conditional mean and variance are determined from

$$m^{(Q)}(t) = m(t)/[1 - p_0(t)]$$

and

$$[\sigma^{(Q)}(t)]^2 + [m^{(Q)}(t)]^2 = [\sigma^2(t) + m^2(t)]/[1 - p_0(t)], \tag{8.5}$$

respectively. Plots can therefore be made of the extinction probability $p_0(t)$, together with confidence bands such as $m(t) \pm 2\sigma(t)$ (for all realizations) and $m^{(Q)}(t) \pm 2\sigma^{(Q)}(t)$ (for all non-extinct realizations) within which the population is expected to lie roughly 95% of the time.

8.1.1 *Examples*

Consider a logistic process having the simple time-dependent structure

$$B(N, t) = rN \quad \text{and} \quad D(N, t) = (rN^2/K)(1 + a\cos(\omega t)). \tag{8.6}$$

The birth rate is assumed to be a function of the individual, and thereby static, whilst the death rate is assumed to be affected by the environment in that it oscillates as a cosine wave around the value rN^2/K.

This is, of course, by no means the only way of driving the process. We could just as easily force the birth rate and keep the death rate static by putting

$$B(N, t) = rN(1 + b\cos(\omega t)) \quad \text{and} \quad D(N, t) = rN^2/K; \tag{8.7}$$

or else we could let the carrying capacity fluctuate so that

$$B(N, t) = rN \quad \text{and} \quad D(N, t) = rN^2/K(t)$$

where

$$K(t) = K_0(1 + c\cos(\omega t)). \tag{8.8}$$

If required, features (8.6)–(8.8) can be combined to give more complicated models.

For demonstration purposes let us study model (8.6) with birth rate $r = 0.2$, a large death rate amplitude $a = 0.7$, and an extremely low carrying capacity $K = 5$. We shall calculate the probabilities $\{p_N(t)\}$ over the range $0 \leqslant t \leqslant 100$, taking the period of oscillation (T) of the death rate to be 10 time units. The *frequency* (f) of the oscillation is defined to be $1/T$, and the *angular frequency* (ω) is $2\pi f = 2\pi/T$ (radians per unit time) where 2π radians corresponds to 360°. In our example $T = 10$, so $\omega = 2\pi/10$.

Equations (8.1) were integrated numerically by sequentially evaluating

$$p_N(t + h) = p_N(t) + h\{D(N + 1, t)p_{N+1}(t)$$
$$- [B(N, t) + D(N, t)]p_N(t) + B(N-1, t)p_{N-1}(t)\} \tag{8.9}$$

for $N = 0, 1, \ldots, N'$ over $t = 0, h, 2h, \ldots, (100/h) - 1$ with $h = 0.002$. If necessary more precise numerical integration procedures can be used (as contained, for example, in the NAG-routines D02), but for our purposes scheme (8.9) is accurate enough. A suitable value for the upper bound N' can be quickly determined by trial and error; here $p_N(100)$ is less than 10^{-6} for $N \geqslant 16$, so we took $N' = 20$.

Figure 8.1a shows the computed extinction probability $p_0(t)$ correspond-ing to an initial population size $X(0) = 1$. As new individuals are not allowed to enter the population from outside the system, extinction is irreversible, and so $p_0(t)$ can only increase in time. However, it does so in a series of steps, rises occurring when the population size drops to relatively low values.

Figure 8.1b contrasts the gradual decline in the mean population size $m(t)$ with the sustained oscillations of the conditional mean $m^{(Q)}(t)$. Note that $m^{(Q)}(t)$ eventually oscillates in the range 3.3 to 5.1, so for most of the time it lies below the carrying capacity $K = 5$. Even when $a = 0$ and the model reduces to an ordinary logistic, $m^{(Q)}(t)$ approaches 4.0 which is still less than K. Though such differences between mean and deterministic behaviour will often be minor, we shall later see (in Chapter 10) that this is not always the case. Note also the two-unit time-lag between the death rate $D(N, t)$ and mean $m^{(Q)}(t)$; the former have minima at $t = 5, 15, 25, \ldots, 95$, and the latter at $8, 17, 27, \ldots, 97$. We shall make a detailed study of this phase difference in Section 8.1.3.

Figure 8.1. Development of the logistic time-dependent model (8.6) with $r = 0.2$, $a = 0.7$, $K = 5$ and $\omega = 2\pi/10$: (a) extinction probability $p_0(t)$; (b) conditional mean $m^{(Q)}(t)$ (———), unconditional mean $m(t)$ (– – –) and conditional confidence band $m^{(Q)}(t) \pm \sigma^{(Q)}(t)$ (- - -).

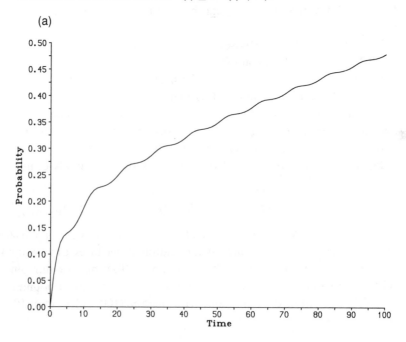

(a)

8.1.2   *A simulation run*

Since here environmental fluctuations are purely deterministic, randomness can only occur through stochastic events. Simulation runs are therefore easy to generate by replacing the birth and death rates $B(N)$ and $D(N)$ in the program GENONE (Appendix to Chapter 3) by the time-dependent rates (8.6). Figure 8.2 shows such a simulation over $0 \leqslant t \leqslant 100$ starting from $X(0) = 10$ with $r = 0.2$, $a = 0.5$, $K = 50$ and time period $T = 25$, i.e. the death rate completes exactly 4 cycles. Also shown is the mean $m^{(Q)}(t)$ and confidence band $m^{(Q)}(t) \pm 2\sigma^{(Q)}(t)$; the latter provides a useful guide to the likely region within which the process will wander. For example, this run generally lies within the band though it closely follows the upper limit on the second and third cycles and the lower limit on the fourth. Note that since $\sigma^{(Q)}(t)$ ranges from 10.5 (near the peaks) down to 5.7 (near the troughs) the scope for stochastic wandering increases and decreases with $m^{(Q)}(t)$.

Even at its lowest point (near $t = 80$) $\{X(t)\}$ is still over 4 standard deviations away from zero, and so early extinction is highly improbable ($p_N(100) < 10^{-6}$ for $N \leqslant 15$). Hence as $p_0(t)$ is effectively zero, $m(t) = m^{(Q)}(t)$ and $\sigma(t) = \sigma^{(Q)}(t)$.

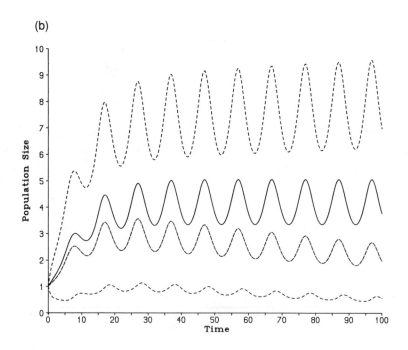

(b)

*8.1.3 *Local solutions*

Suppose that the birth and death rates B and D depend on time through some particular function $G(t)$, i.e.

$$dN(t)/dt = B(N, G(t)) - D(N, G(t)). \tag{8.10}$$

Let the average value of G over a long time period be G^*. Then if G were to remain constant with the value G^*, equation (8.10) would reduce to

$$dN(t)/dt = B(N, G^*) - D(N, G^*) \tag{8.11}$$

which is identical to the general form (3.2) already studied in Chapter 3.

Let us therefore suppose that equation (8.11) possesses a stable state N^*, and that small fluctuations in G about G^* cause small fluctuations in the population size N about N^*. Then on writing

$$N = N^* + n \quad \text{and} \quad G = G^* + g, \tag{8.12}$$

and making the usual assumption that n and g are small enough for terms in n^2, g^2, etc. to be ignored, equation (8.10) may be expanded as a Taylor series

Figure 8.2. Development of the logistic time-dependent model (8.6) with $r = 0.2$, $a = 0.5$, $K = 50$ and $\omega = 2\pi/25$, showing $m^{(Q)}(t)$ ($---$), $m^{(Q)}(t) \pm 2\sigma^{(Q)}(t)$ ($-----$), and a stochastic realization (———).

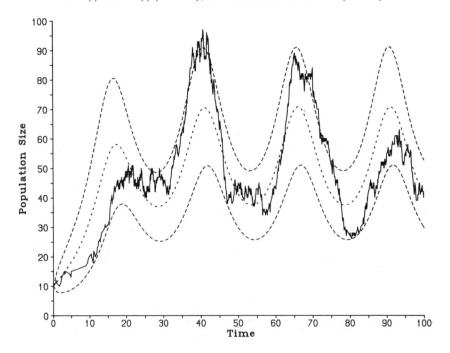

about $N*$ and $G*$ to yield the first-order approximation

$$\mathrm{d}n/\mathrm{d}t = \alpha n + \beta g \tag{8.13}$$

where

$$\alpha = \left[\frac{\partial B}{\partial N} - \frac{\partial D}{\partial N}\right]_{\substack{N=N* \\ G=G*}} \quad \text{and} \quad \beta = \left[\frac{\partial B}{\partial G} - \frac{\partial D}{\partial G}\right]_{\substack{N=N* \\ G=G*}}. \tag{8.14}$$

Equation (8.13) is a standard first-order differential equation (see, for example, Piaggio, 1962), and may be solved as follows.

(i) Set $g = 0$ and then integrate; this yields the *complementary function*

$$n(t) = A \exp(\alpha t) \quad \text{(for some constant } A).$$

(ii) Find (generally by guessing!) any solution of (8.13), say $v(t)$. This is called a *particular integral*. Then the general solution to (8.13) is given by

$$n(t) = A \exp(\alpha t) + v(t). \tag{8.15}$$

At time $t = 0$ we have $n(0) = A + v(0)$, so the final solution is

$$n(t) = [n(0) - v(0)] \exp(\alpha t) + v(t). \tag{8.16}$$

Provided $\alpha < 0$ we see that, for large t, $n(t) \simeq v(t)$. So if $\alpha < 0$ the first term in (8.16) is a *transient* part which just affects the *initial* development of the process; whilst the second term, $v(t)$, is the *persistent* part which describes its *long-term* behaviour.

To illustrate this technique, consider the logistic process with time-dependent rates (8.6), namely $B = rN$ and $D = (rN^2/K)G$ where $G(t) = 1 + a \cos(\omega t)$. Since the long-term average $G* = 1$, equating B and D gives $N* = K$. So from (8.14)

$$\alpha = -r \quad \text{and} \quad \beta = -rK, \tag{8.17}$$

whence (8.13) becomes

$$\mathrm{d}n/\mathrm{d}t = -rn - rKg \tag{8.18}$$

where $g = G - G* = a \cos(\omega t)$.

As the driving fluctuation $G(t)$ varies sinusoidally with frequency ω, a sensible guess is that $n(t)$ will also vary sinusoidally with the same frequency, though possibly out of phase with $G(t)$. This makes biological sense, since populations normally take at least a little time to respond to changes in the surrounding environment. Let us therefore try the particular integral

$$v(t) = B \cos(\omega t - \phi), \tag{8.19}$$

and see whether values of the constant amplitude B and phase ϕ can be found which make $v(t)$ satisfy equation (8.18). That is, we require

$$-B\omega \sin(\omega t - \phi) = -rB \cos(\omega t - \phi) - rKa \cos(\omega t),$$

the sine and cosine terms of which expand to give

$$B\omega[\sin(\omega t) \cos(\phi) - \cos(\omega t) \sin(\phi)]$$
$$= rB[\cos(\omega t) \cos(\phi) + \sin(\omega t) \sin(\phi)] + rKa \cos(\omega t).$$

For this expression to be valid at all times t the coefficients of $\sin(\omega t)$ and $\cos(\omega t)$ must vanish, i.e.

$$B\omega \cos(\phi) = rB \sin(\phi)$$
$$-B\omega \sin(\phi) = rB \cos(\phi) + rKa.$$

From the first of these equations we have

$$\tan(\phi) = \omega/r. \tag{8.20}$$

So

$$\sin(\phi) = \omega/\sqrt{(r^2 + \omega^2)} \quad \text{and} \quad \cos(\phi) = r/\sqrt{(r^2 + \omega^2)},$$

whence the second equation leads to

$$B = -rKa/\sqrt{(r^2 + \omega^2)}. \tag{8.21}$$

Thus

$$v(0) = B \cos(\phi) = -r^2 Ka/(r^2 + \omega^2).$$

The complete solution (8.16) is therefore given by

$$n(t) = [n(0) + \{r^2 Ka/(r^2 + \omega^2)\}] \exp(-rt)$$
$$- [rKa/\sqrt{(r^2 + \omega^2)}] \cos(\omega t - \phi) \tag{8.22}$$

where ϕ is determined from (8.20).

Though here we have allowed the death rate to fluctuate, similar solutions can clearly be found for processes with other fluctuating features. For example, if the carrying capacity K varies, then under model (8.8)

$$B = rN \quad \text{and} \quad D = rN^2/(K_0 G) \quad \text{where } G(t) = 1 + c \cos(\omega t). \tag{8.23}$$

As before, $N^* = K_0$ and $G^* = 1$, so from (8.14)

$$\alpha = \left[r - \frac{2rN}{K_0 G} \right]_{\substack{N=K_0 \\ G=1}} = -r \quad \text{and} \quad \beta = \left[0 + \frac{rN^2}{K_0 G^2} \right]_{\substack{N=K_0 \\ G=1}} = rK_0. \tag{8.24}$$

Apart from K changing into $-K_0$, the parameter values (8.17) and (8.24) are identical. Hence modifying solution (8.22) immediately leads to

$$n(t) = [n(0) - \{r^2 K_0 c/(r^2 + \omega^2)\}] \exp(-rt)$$
$$+ [rK_0 c/\sqrt{(r^2 + \omega^2)}] \cos(\omega t - \phi). \tag{8.25}$$

Thus as far as this local approximation approach is concerned, fluctuating death rates and carrying capacities have the same effect, except that the two solutions are exactly π out of phase. This phase difference is to be expected, since large population size is associated with high carrying capacity but low death rate.

8.1.4   *Conclusions*

Solutions (8.22) and (8.25) exhibit three particular features.

(i) The transient part ceases to have any effect once $\exp(-rt)$ is sufficiently small. For example, $\exp(-rt) < 0.01$ once $t > -\log_e(0.01)/r = 4.6/r$. After this time, behaviour is effectively described by the persistent component $v(t)$.

(ii) If the angular frequency ω of the driving oscillation is small compared to the growth rate r, then it follows from (8.20) that $\tan(\phi) \simeq 0$ whence $\phi \simeq 0$. Hence as

$$v(t) \simeq -Ka \cos(\omega t) = Ka \cos(\omega t - \pi)$$

and

$$D(N, t) = (rN^2/K)[1 + a \cos(\omega t)],$$

$v(t)$ and $D(t)$ oscillate half a period out of phase with each other. So if the environment changes slowly enough then the population can keep almost perfect track of it, with maximum population size occurring almost simultaneously with minimum death rate.

(iii) Conversely, if $\omega/r = \tan(\phi)$ is large, i.e. the environment changes rapidly, then the population cannot respond fast enough. In the limiting case when $\phi \simeq 90°$, expressions (8.19) and (8.21) yield

$$v(t) \simeq -(Kar/\omega) \cos(\omega t - \tfrac{1}{2}\pi) = (Kar/\omega) \cos(\omega t - 3\pi/2).$$

Thus the amplitude $B = Kar/\omega$ is virtually zero, whilst $v(t)$ and the death rate $D(N, t)$ are out of phase by nearly three-quarters of a period. The inability of $v(t)$ to keep up effectively causes the environmental fluctuations to be ironed out.

On rewriting (8.21) as $B = -Ka/\sqrt{[1 + (\omega/r)^2]}$, and noting from (8.20) that $\tan(\phi) = (\omega/r)$, we see that the position of any process relative to the two extremes (ii) and (iii) is solely determined by ω/r. For example, in Figure 8.2 $r = 0.2$ and $\omega = 2\pi/25$. Thus $\omega/r = 1.26$, giving $\phi = 51°$, and so the process

232   *Fluctuating environments*

lies midway between (ii) and (iii). Since the amplitude B takes its maximum value $B(\text{max}) = Ka$ when $\omega = 0$,

$$B/B(\text{max}) = 1/\sqrt{[1 + (\omega/r)^2]} = 0.62,$$

an amplitude loss of about one-third. Note that the $51°$ phase lag corresponds to 3.6 time-units between maximum population size and minimum death rate, which is in good agreement with the value $3\frac{1}{2}$ taken from the numerically exact $m^{(Q)}(t)$-plot shown in Figure 8.2. We also see that the process settles down to purely persistent behaviour at about time $t = 20$, in accord with the value predicted in (i) of around $4.6/r = 23$.

In contrast, Figure 8.3 shows behaviour nearer to that of the two extremes (ii) and (iii). The parameter values $K = 50$ and $a = 0.5$ are retained, r is reduced to 0.1, and $X(0) = 30$. To illustrate (ii) we extend the time period T

Figure 8.3. Development of the logistic time-dependent model (8.6) with $r = 0.1$, $a = 0.5$ and $K = 50$, showing a stochastic realization (———, rough), mean $m(t)$ (———, smooth upper), local deterministic approximation $N(t) = K + n(t)$ (---), and death rate $D(N, t)$ (———, smooth lower) for: (a) low frequency case, $T = 200$; (b) high frequency case, $T = 5$.

(a)

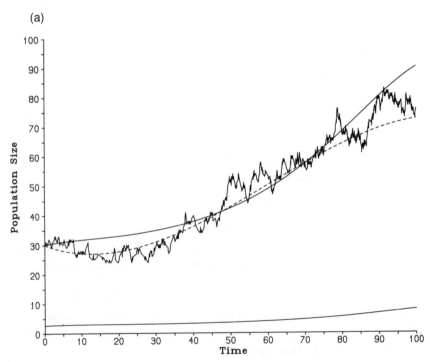

to 200, so that $\cos(\omega t)$ only goes through a half-cycle during the time-span $0 \leqslant t \leqslant 100$. This long wavelength corresponds to a low angular frequency of $\omega = 2\pi/200$, whence $\tan(\phi) = \omega/r = 2\pi/(200 \times 0.1) = 0.3142$ which gives a phase-lag of $\phi = 17°$. Note the disparity between the local deterministic approximation $N(t) = K + n(t)$ and the stochastic mean $m(t)$.

To illustrate the other extreme (iii), we lower T to 5 so that $\cos(\omega t)$ now goes through 20 full cycles. Thus $\omega = 2\pi/5$ and so $\phi = 85°$. $\{X(t)\}$ shows strong cyclic fluctuations (Figure 8.3b), which though more wild than the deterministic ones still have the same period of length 5. The local solution $N(t) = K + n(t)$ runs just below $m(t)$ once the initial transient behaviour has died away.

8.2   Jillson's flour beetle experiment

Lest it be thought that such studies exist solely in the minds of mathematical modellers, we should note that the above approach has been directly paralleled in the laboratory. Jillson (1980) investigated the response of populations of the flour beetle *Tribolium castaneum* cultured in a series of regularly fluctuating environments. These environmental fluctuations were created by alternating the amount of culture medium between 32 g and 8 g at

(b)

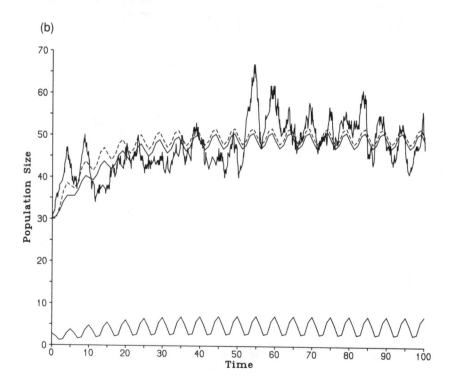

regular intervals. Since *Tribolium* population size is known to be positively correlated with the amount of culture medium present, this experiment represents fluctuations in carrying capacity. Five particular regimes were used: a constant 20 g environment (as control), and fluctuating environments with 4, 8, 12 and 16 week periods. Thirty such populations were cultured in 20 g medium for the initial 18 weeks of the experiment to allow the populations to reach demographic equilibrium. At week 18, six populations were randomly assigned to each of the five treatments, and the environment was then allowed to fluctuate for the next 48 weeks.

Regular, though fortunately small, fluctuations occurred in the control (20 g) environment, due in part to cannibalistic interactions – if necessary, our simple model could be extended to incorporate cannibalism of vulnerable pupae. The proportion of adult beetles in the control and 4-week (i.e. highest frequency) cycle was relatively constant; the latter did not show any strong correlation with environmental change which is consistent with the large-ω results in (iii) above. In contrast, the population changes in the 8, 12 and 16 week treatments did follow the environmental fluctuations. Figure 8.4 shows

Figure 8.4. Mean population size (———) of six replicates of a *Tribolium castaneum* experiment with a square wave (– – –) food supply, period 12 weeks (redrawn from Jillson, 1980).

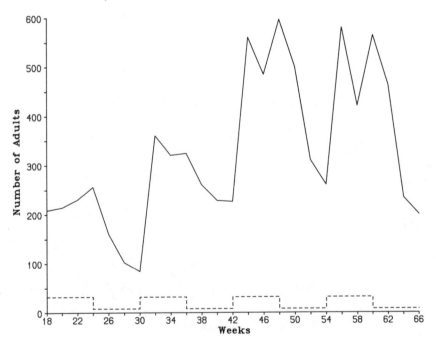

the average of the six replicates of the 12-week treatment, together with the oscillating 'square-wave' food supply of 32 g and 8 g. The population clearly rises and falls directly in line with the provision of high and low amounts of culture medium (though note the unexplained double peaks).

Constructing a logistic model with a fluctuating square-wave carrying capacity $K(t)$ presents no really new problem, since $K(t)$ simply flip-flops between two constant values K_1 and K_2. In between times the process behaves as a simple logistic process exactly as described in Chapter 3. To generate a deterministic solution we therefore modify the basic logistic equation (3.20) to

$$dN(t)/dt = rN(t)[1 - \{N(t)/K(t)\}] \tag{8.26}$$

where $K(t)$ equals K_1 for $0 \leqslant t \leqslant T/2$, K_2 for $T/2 < t \leqslant T$, K_1 for $T < t \leqslant 3T/2$, etc. It then follows immediately from (3.22) that the solution to (8.26) is given by

$$N(t) = K_1/[1 + \{(K_1 - N(0))/N(0)\} \exp(-rt)] \qquad (0 \leqslant t \leqslant T/2),$$
$$N(t) = K_2/[1 + \{(K_2 - N(\tfrac{1}{2}T))/N(\tfrac{1}{2}T)\} \exp\{-r(t - \tfrac{1}{2}T)\}]$$
$$(T/2 < t \leqslant T),$$
$$N(t) = K_1/[1 + \{(K_1 - N(T))/N(T)\} \exp\{-r(t - T)\}]$$
$$(T < t \leqslant 3T/2), \tag{8.27}$$

etc.

Stochastic realizations of this process are easy to construct from the general program GENONE (Appendix to Chapter 3) by modifying the death rate D as follows:

> evaluate $G = \mathrm{mod}(t, T)$ (remainder on dividing t by T, e.g. $\mathrm{mod}(63, 25) = 13$ since $63 = 2 \times 25 + 13$);
> if $G \leqslant T/2$ then put $K = K_1$ (first half of square wave);
> if $G > T/2$ then put $K = K_2$ (second half of square wave);
> put $D = rN^2/K$.

Figure 8.5 shows such a realization for $K_1 = 600$, $K_2 = 150$ and $T = 12$, these being values which roughly correspond to Jillson's regime (Figure 8.4). The chosen r-value 0.4 ensures that $\{X(t)\}$ can get near to the target carrying capacities. There is a crude similarity between Figures 8.4 (averaged data) and 8.5 (single simulation), though our simple model clearly does not fit in any statistical sense. The important result is a qualitative one, namely that in both theory and practice an oscillating food supply forces a corresponding oscillatory population change.

8.3   Stochastic behaviour with deterministic variability

Equations (8.1) for the probabilities $\{p_N(t)\}$ are particularly nasty because of the general time-dependent parameters $B(N, t)$ and $D(N, t)$. However, some progress can be made by steering a middle course between the purely stochastic equations (8.1) and the deterministic equation

$$\mathrm{d}N(t)/\mathrm{d}t = B(N, t) - D(N, t). \tag{8.28}$$

8.3.1   *The stochastic equation*

First, replace (8.28) by the discrete-time version

$$N(t + h) = N(t) + h[B(N, t) - D(N, t)] \tag{8.29}$$

where h is small. Second, replace the deterministic population size $N(t)$ by the random variable $X(t)$, whereupon (8.29) becomes

$$X(t + h) = X(t) + h[B(X, t) - D(X, t)] + \text{a stochastic term.}$$
$$\tag{8.30}$$

This stochastic term will have mean zero, since the rest of the equation

Figure 8.5. Realization (———) of a time-dependent logistic process with $r = 0.4$ and a square-wave carrying capacity (– – –) oscillating between $K = 600$ and 150 with a 12-week period.

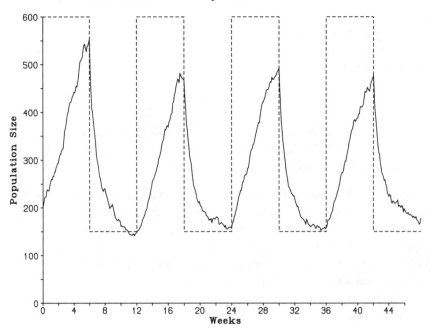

describes the mean population change between times t and $t + h$, namely

$$
\begin{aligned}
m(t, t + h) &= (+1)\,\Pr\{X(t + h) = X(t) + 1\} \\
&\quad + (0)\,\Pr\{X(t + h) = X(t)\} \\
&\quad + (-1)\,\Pr\{X(t + h) = X(t) - 1\} \\
&= h\{B(X, t) - D(X, t)\}.
\end{aligned}
\tag{8.31}
$$

The corresponding variance is

$$
\begin{aligned}
&(+1)^2\,\Pr\{X(t + h) = X(t) + 1\} + (0)^2\,\Pr\{X(t + h) = X(t)\} \\
&\quad + (-1)^2\,\Pr\{X(t + h) = X(t) - 1\} \\
&\quad - [m(t, t + h)]^2 \\
&= h[B(X, t) + D(X, t)] \\
&\quad - h^2[B(X, t) - D(X, t)] \\
&\simeq h[B(X, t) + D(X, t)]
\end{aligned}
\tag{8.32}
$$

since h^2 is negligibly small in comparison to h. Thus on denoting $Z(0)$, $Z(h)$, $Z(2h)$, ... to be a sequence of independent Normal random variables with mean zero and unit variance (often called 'white noise'), we may write (8.30) as

$$
\begin{aligned}
X(t + h) - X(t) &= h\{B(X, t) - D(X, t)\} \\
&\quad + Z(t)\sqrt{[h\{B(X, t) + D(X, t)\}]}.
\end{aligned}
\tag{8.33}
$$

To be mathematically precise we should now let $h \to 0$ and write (8.33) as the stochastic differential equation

$$
\mathrm{d}X(t) = \{B(X, t) - D(X, t)\}\,\mathrm{d}t + \{B(X, t) + D(X, t)\}^{\frac{1}{2}}\,\mathrm{d}Z(t),
\tag{8.34}
$$

where $\mathrm{d}X(t)$, $\mathrm{d}Z(t)$ and $\mathrm{d}t$ respectively denote infinitesimally small elements of population size, white noise, and time. However, the associated mathematical theory is well beyond the level of this text.

Since the local linearization approach generally provides good deterministic approximations, we shall develop a parallel technique to generate stochastic results. This involves linearizing the random variable $\{X(t)\}$ about the deterministic steady state N^* that would arise if $G(t)$ remained constant with value G^*. On setting

$$
X(t) = N^* + x(t) \quad \text{and} \quad G(t) = G^* + g(t)
\tag{8.35}
$$

for small x and g, comparison of the local deterministic equation (8.13) with the stochastic differential equation (8.34) then yields

$$
\mathrm{d}x = \alpha x\,\mathrm{d}t + \beta g(t)\,\mathrm{d}t + Q^{\frac{1}{2}}\,\mathrm{d}Z(t),
\tag{8.36}
$$

where the time-dependent $(B + D)$-term in (8.34) is replaced by the constant

$$Q = B(N^*, G^*) + D(N^*, G^*). \tag{8.37}$$

8.3.2   *Autocovariance results*

Following Nisbet and Gurney (1982), some algebraic juggling then leads to the autocovariance function $\{c_k\}$ (see Section 6.2.6) of the process $\{X(t)\}$. In particular, for the sinusoidal driving function

$$g(t) = g_0 \cos(\omega_0 t), \tag{8.38}$$

we have

$$\begin{aligned} c_k = & [g_0^2 \beta^2 / 2(\omega_0^2 + \alpha^2)] \cos(\omega_0 k) \\ & + (Q/2|\alpha|) \exp(-|\alpha|k) \quad (k \geqslant 0) \end{aligned} \tag{8.39}$$

where $|\alpha|$ denotes the magnitude of α (i.e. disregarding sign). Evaluation of the autocorrelation function $r_k = c_k/c_0$ follows directly since

$$c_0 = \mathrm{Var}\{X(t)\} = [g_0^2 \beta^2 / 2(\omega_0^2 + \alpha^2)] + [Q/2|\alpha|].$$

As k increases, c_k does not approach zero but tends instead to the cosine wave

$$c_k \simeq D \cos(\omega_0 k) \quad \text{where} \quad D = g_0^2 \beta^2 / 2(\omega_0^2 + \alpha^2). \tag{8.40}$$

For example, we see from (8.17) that for the fluctuating death rate process (8.6) $\alpha = -r$, $\beta = -rK$ and $g_0 = a$, so

$$D = \tfrac{1}{2} a^2 K^2 / [(\omega_0/r)^2 + 1]. \tag{8.41}$$

Thus when $\omega_0/r \simeq 0$ (extremely low frequency fluctuations) $D \simeq \tfrac{1}{2} a^2 K^2$, whilst as $\omega_0/r \to \infty$ (extremely high frequency fluctuations) $D \to 0$. So although the stochastic population size $\{X(t)\}$ exhibits sustained cycles with constant frequency ω_0, regardless of how large ω_0 is, the cycle amplitude diminishes to zero as ω_0 increases. The reason is that when ω_0 is very large the environment fluctuates too fast for the population to respond, and so $\{X(t)\}$ behaves like a simple logistic process with constant parameters. Note that these results are exactly in line with our previous deterministic conclusions.

The effect that ω_0 has on the cyclic component is highlighted by the two simulations of Figure 8.3; whilst the stochastic excursions about the deterministic population cycles have similar magnitude, the amplitudes of the underlying oscillations are quite different. The corresponding effect on the autocorrelations is clearly demonstrated in Figure 8.6 which shows the first 400 autocorrelations of $X(0), X(1), \ldots, X(1000)$ in each case, namely:

(a) $T = 200$ (low frequency); and (b) $T = 5$ (high frequency). Here we used the alternative estimator (6.56a) to (6.56), namely

$$\tilde{c}_k = (N + 1 - k)^{-1} \sum_{t=0}^{N-k} \{X(t) - \bar{X}\}\{X(t + k) - \bar{X}\} \tag{8.42}$$

where $\bar{X} = [X(0) + \cdots + X(N)]/(N + 1)$ and $N = 1000$. This prevents bias hiding the persistent cyclic effect predicted by result (8.40).

Substituting the parameter values $r = 0.1$, $a = 0.5$ and $K = 50$ for Figure 8.3 into (8.40) gives (a) $D = 284.43$ ($T = 200$) and (b) the much smaller $D = 1.9665$ ($T = 5$). Moreover, since $N^* = K = 50$ and $G^* = 1$, we see from (8.37) that $Q = 2rK = 10$. Hence as $\alpha = -r = -0.1$ and $\omega_0 = 2\pi/T$, (8.39) becomes

$$c_k = D \cos(2\pi k/T) + 50 \exp(-0.1k). \tag{8.43}$$

Thus the autocorrelations

$$r_k = c_k/c_0 = [D \cos(2\pi k/T) + 50 \exp(-0.1k)]/(D + 50). \tag{8.44}$$

Figure 8.6. Autocorrelations $\{r_k\}$ for the two simulations shown in Figure 8.3 evaluated over $t = 0, 1, 2, \ldots, 1000$: (a) $T = 200$ (———), (b) $T = 5$ (– – –).

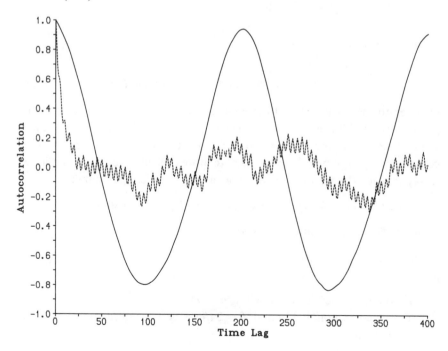

The transient part of r_k attenuates as $\exp(-0.1k)$ irrespective of the period T. So as $\exp(-0.1 \times 40) \simeq 0.02$, from about $k = 40$ onwards $\{r_k\}$ settles down to purely persistent behaviour. From (8.44) we have

(a) $r_k = 0.8505 \cos(2\pi k/200) + 0.1495 \exp(-0.1k)$ $(T = 200)$
(b) $r_k = 0.0378 \cos(2\pi k/5) + 0.9622 \exp(-0.1k)$ $(T = 5)$,

so the persistent amplitude (0.85) is near one for the low frequency case, whilst that for the high frequency case (0.04) is near zero.

Inspection of Figure 8.6 shows that these local linearization results compare well with the simulated values. For realization (a) the $\{r_k\}$ exhibit an almost pure cosine wave, with the first peak $\bar{c}_{200} = 0.94$ being reasonably close to the local theoretical value of 0.85. For realization (b) the $\{r_k\}$ meander quite considerably, though the sustained cycles of period 5 are very much in evidence. Subtracting this 'meander' from $\{r_k\}$ yields an amplitude of about 0.047, in good agreement with the local theoretical value of 0.038. Local stochastic approximation is clearly a useful technique.

8.3.3   *Mean time to extinction*

An approximate expression for the mean time to extinction, T_E, can be derived along lines similar to those developed in Section 3.8.1. From (3.70)

$$T_E = \int_0^\infty [1 - p_0(t)]\, \mathrm{d}t, \tag{8.45}$$

where (3.68) is now written as

$$\begin{aligned}
\mathrm{d}p_0(t)/\mathrm{d}t &= D(1, G(t))p_1(t) \\
&= D(1, G(t))\pi_1^{(Q)}(t)[1 - p_0(t)] \quad \text{(using (3.67))}
\end{aligned} \tag{8.46}$$

in order to take account of the time-dependent death rate. Note that the probabilities $\{\pi_i^{(Q)}\}$ are now time-dependent, yet it was their constancy in the static case which led to the simplicity of the final result (3.72). The corresponding argument here is therefore more difficult to develop. Suffice it to say that for the fluctuating carrying capacity model (8.8), namely

$$B(N, t) = rN \quad \text{and} \quad D(N, t) = rN^2/[K_0(1 + c \cos(\omega t))],$$

we have (Nisbet and Gurney, 1982)

$$\log_e(rT_E) \sim K_0(1 - c) \tag{8.47}$$

if:

(i) the frequency is very low (i.e. $\omega \ll r$);
(ii) the corresponding period $T = 2\pi/\omega$ is very small in comparison to T_E; and

(iii) the average carrying capacity K_0 is very large.

This result, though approximate and riddled with assumptions, is ideal for exposing the effect of a fluctuating environment on the mean time to extinction

$$T_E \sim (1/r) \exp\{K_0(1 - c)\} = [(1/r) \exp\{K_0\}] \exp\{-K_0 c\}.$$

When K_0 is large this value reduces dramatically as c increases, from $(1/r) \exp(K_0)$ at $c = 0$ to $1/r$ at $c = 1$. For example, if we take the fairly low value $K_0 = 100$, then for

$c =$	0	0.2	0.5	0.9
$rT_E =$	2.7×10^{43}	5.6×10^{34}	5.2×10^{21}	2.2×10^4.

Thus a 20% modulation in carrying capacity will reduce the mean extinction time by a factor of nearly 500 million, whilst a fairly wild swing of 90% will reduce T_E from being effectively infinite to a value low enough to be meaningful. The reason is that the probability of extinction near the low part of the $K(t)$-cycle rises swiftly as $\min[K(t)] = K_0(1 - c)$ decreases.

The precise nature of these rT_E-values is, of course, irrelevant, though the conclusion to be drawn from them is most certainly not. As Nisbet and Gurney (1982) state so succinctly, 'with all but the smallest populations the mean times to extinction predicted by static birth and death models are so large as to be meaningless in an ecological context, and we are forced to conclude that virtually all ecologically useful models of extinction will involve environmental variability'.

8.4   Random environments

We have so far considered an exact deterministic representation for $G(t)$, the external force which drives the birth and death rates $B(N, G(t))$ and $D(N, G(t))$. However, any environment is bound to be subject to random fluctuations of one sort or another. So the assumption that an environment involves purely deterministic fluctuations is only appropriate if this random component is relatively small. We clearly need to be able to cover situations in which it is not.

In a random environment the population process $\{X(t)\}$ is *doubly stochastic*. Not only are there an infinite number of possible realizations $\{X(t)\}$ for any *fixed* sequence $\{G(t)\}$, but also every *different* set of $\{G(t)\}$-values will have its own set of $\{X(t)\}$-realizations associated with it. Note that the concept of mean population size carries through exactly as before, except that $m(t)$ now represents an average over all possible events *and* environments.

8.4.1 *Four particular models*

The stochastic driving term $G(t)$ can be modelled in a variety of ways depending on how we envisage the environment actually fluctuates. Four possibilities are as follows.

(A) The environment may be subject to 'random jitter', in that

$$G(t) = Z(t) \tag{8.48}$$

where $\{Z(t)\}$ is a sequence of independent random variables with mean zero and common variance σ^2 (i.e. white noise).

(B) $G(t)$ may be the sum of a deterministic component and random jitter, e.g.

$$G(t) = [1 + a\cos(\omega t)] + Z(t). \tag{8.49}$$

(C) Alternatively, the stochastic element may feature in the frequency term, e.g.

$$G(t) = 1 + a\cos[\omega t + Z(t)]. \tag{8.50}$$

(D) The sequence $\{G(t)\}$ may perform a random walk in which its value at time t is closely related to its immediate past history, e.g.

$$G(t) = bG(t-1) + Z(t) \tag{8.51}$$

for some constant b.

For simplicity let us dispense with the added complication of random events and just consider stochastic environments with deterministic events. For computational reasons we shall allow $G(t)$ to change only at the discrete time points $t = 0, h, 2h, \ldots$ for appropriately small h.

The basic deterministic logistic equation

$$dN/dt = rN(1 - N/K) \tag{3.20}$$

has the exact solution

$$N(t) = K/[1 + \{(K - N(0))/N(0)\}\exp(-rt)]. \tag{3.22}$$

So the corresponding stochastic environment logistic equation with variable growth rate $r(t)$ and carrying capacity $K(t)$, namely

$$dX(t)/dt = r(t)[1 - X(t)/K(t)], \tag{8.52}$$

has the segmented solution

$$X(t) = K(0)/[1 + \{(K(0) - X(0))/X(0)\}\exp\{-r(0)t\}] \quad (0 \leqslant t \leqslant h)$$
$$X(t) = K(h)/[1 + \{(K(h) - X(h))/X(h)\}\exp\{-r(h)t\}] \quad (h < t \leqslant 2h)$$

etc. Hence in general

$$X(t) = K(nh)/[1 + \{(K(nh) - X(nh))/X(nh)\} \exp\{-r(nh)t\}]$$
$$(nh < t \leqslant (n+1)h) \quad (8.53)$$

where $n = 0, 1, 2, \ldots$. If h is small then the discrete sequence $X(0), X(h), X(2h), \ldots$ is a 'jittery' solution. If it is not, so that the driving term $G(t)$ changes infrequently (e.g. once a month when measurements are taken daily), then (8.53) generates smooth changes between widely spaced time points.

If $r(t)$ is constant then we have a similar scenario to (8.8) except that $K(t)$ now varies randomly and not deterministically. Let us therefore use solution (8.53) with $h = 1$ to obtain $X(0), X(1), X(2), \ldots$ directly from simulated $K(t)$-values.

First consider case (A) with

$$K(t) = K_0 + G(t) = K_0 + Z(t) \qquad (8.54)$$

for some constant mean carrying capacity K_0 and $\{Z(t)\}$ a set of pseudo-random numbers generated from the Normal distribution with mean zero and variance σ^2 (using, for example, the NAG-routine G05DDF). Figure 8.7a shows both $\{X(t)\}$ and $\{Z(t)\}$ for one such simulated run over $t = 0, 1, \ldots, 250$ with $r = 0.2$, $K_0 = 50$ and $\sigma = 10$. Now $X(t + 1)$ must lie partway between $X(t)$ and $K(t)$. So although the population size will try to follow the random carrying capacity, it will move with smaller fluctuations. Indeed, for this particular realization the standard deviation of $\{X(t)\}$ is 3.69, considerably less than $\sigma = 10$. Thus in its pursuit of $\{K(t)\}$, the process $\{X(t)\}$ performs a random walk about $K_0 = 50$ but with variance smaller than σ^2.

Case (B) with

$$K(t) = K_0 G(t) = K_0[1 + a \cos(\omega t) + Z(t)] \qquad (8.55)$$

is a stochastic environment version of the stochastic event model simulated in Figure 8.2 ($r = 0.2$, $a = 0.5$, $K_0 = 50$ and $\omega = 2\pi/25$). In a simulation run with $\sigma = 0.2$ (Figure 8.7b) $\{K(t)\}$ exhibits substantial jitter on top of the deterministic cosine component. Again $\{X(t)\}$ pursues $\{K(t)\}$, smoothing the very rough track that $\{K(t)\}$ takes since $\{X(t)\}$ short-cuts the peaks and troughs. Thus (8.55) leads to regular cycles with mildly varying amplitude.

Use of the random frequency process (model C)

$$K(t) = K_0 G(t) = K_0[1 + a \cos\{\omega t + Z(t)\}] \qquad (8.56)$$

with $\sigma = 0.5$ produces similar behaviour, since as far as $\{X(t)\}$ is concerned the random phases $\{Z(t)\}$ effectively act in the same way as the random jitter in (8.55). To get truly irregular cycles we must therefore let ω itself be a slowly changing function of t.

Suppose we replace the constant period $T = 25$ by

$$T(t) = 25 + v(t) \quad \text{where} \quad v(t) = bv(t-1) + Y(t). \tag{8.57}$$

Here $\{Y(t)\}$ is a new white-noise sequence with mean zero and variance σ_Y^2, and b is a constant just less than one. So now

$$K(t) = K_0[1 + a\cos\{2\pi/(25 + v(t))\} + Z(t)]. \tag{8.58}$$

Figure 8.7c shows a simulation run using the parameter values of Figure 8.7b together with $b = 0.99$ and $\sigma_Y = 0.5$, i.e.

$$v(t) \simeq v(t-1) + \text{ a small perturbation,}$$

Figure 8.7. Simulation of $\{X(t)\}$ (———) for the logistic model with $r = 0.2$ and randomly varying carrying capacity $K(t)$, where:

(a) $K(t) = 50 + Z(t) \quad (\sigma = 10; {-}{-}{-} \text{ denotes } Z(t))$;
(b) $K(t) = 50[1 + 0.5\cos(2\pi t/25) + Z(t)] \quad (\sigma = 0.2; {-}{-}{-} \text{ denotes } K(t)$;
(c) $K(t) = 50[1 + 0.5\cos\{2\pi t/(25 + v(t))\} + Z(t)]$ with $v(t) = 0.99v(t-1)$
$+ Y(t)$ and $\sigma_Y = 0.5 \quad (\sigma = 0.2; {-}{-}{-} \text{ denotes } K(t))$.

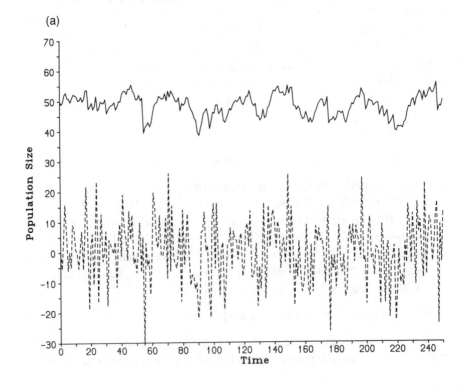

(a)

and comparison of these two figures highlights the loss of regularity in the cyclic structure under model (8.58) as ω varies around $2\pi/25$.

8.4.2   *The autoregressive model (D)*

To demonstrate complete irregularity in cyclic behaviour we now turn to model (D) with

$$K(t) = K_0 + G(t) \quad \text{where} \quad G(t) = bG(t-1) + Z(t) \tag{8.59}$$

for some constant b. In time-series language $\{G(t)\}$ is called an autoregressive process of order one, and has been used in stochastic environment models by Roughgarden (1975).

Suppose $G(0) = 0$. Then on successively applying rule (8.59) we have

$$G(1) = bG(0) + Z(1) = Z(1)$$
$$G(2) = bG(1) + Z(2) = bZ(1) + Z(2),$$

and in general

$$G(t) = b^{t-1}Z(1) + b^{t-2}Z(2) + \cdots + bZ(t-1) + Z(t). \tag{8.60}$$

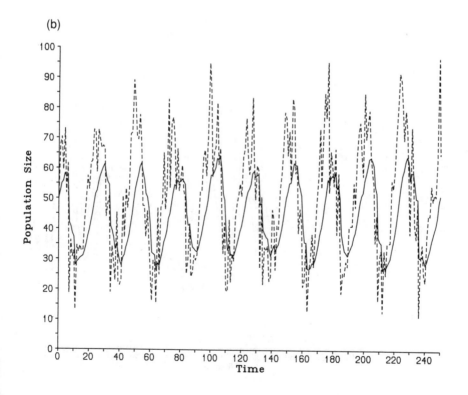

(b)

Since $\{Z(t)\}$ are independent random variables with constant variance σ^2, and since for any constant c $\mathrm{Var}(cZ) = c^2\,\mathrm{Var}(Z)$, we may use (8.60) to show that

$$\mathrm{Var}[K(t)] = \mathrm{Var}[K_0 + G(t)] = \mathrm{Var}[G(t)]$$
$$= \sigma^2[b^{2t-2} + b^{2t-4} + \cdots + b^2 + 1]. \tag{8.61}$$

This geometric series sums to give

$$\mathrm{Var}[K(t)] = \sigma^2(1 - b^{2t})/(1 - b^2). \tag{8.62}$$

Thus for $-1 < b < 1$ the variance of $K(t)$ increases from σ^2 when $t = 1$ to $\sigma^2/(1 - b^2)$ when t becomes large.

If $b = \pm 1$ then (8.61) becomes

$$\mathrm{Var}[K(t)] = \sigma^2(1 + 1 + \cdots + 1) = \sigma^2 t, \tag{8.63}$$

which steadily increases to infinity with increasing t. When $b = +1$ the model is known as the simple random walk (Section 2.3.5) and has many interesting properties (see, for example, Feller, 1966). In particular, it can wander off to $+\infty$ or $-\infty$ and then return to zero, albeit after an infinite time! If $b < -1$

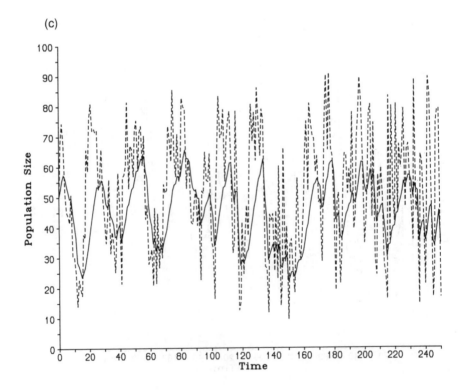

(c)

or $b > 1$ then the situation is even more extreme, since now (8.62) yields

$$\text{Var}[K(t)] \sim \sigma^2 b^{2t}/(b^2 - 1) \tag{8.64}$$

which explodes to infinity geometrically fast as t increases. To make biological sense we therefore have to restrict b to the range $-1 < b < 1$. Indeed, since $K(t)$ is unlikely to flip-flop about K_0 as happens when b is negative, we can be even more restrictive and place $0 < b < 1$.

The nearer b is to $+1$ the greater is the variance of $K(t)$, and so the more the carrying capacity fluctuates about its mean value K_0. For example, with $\sigma^2 = 1$ and t large we have

b	0	0.1	0.5	0.9	0.99	0.999	1
$\text{Var}\{K(t)\}$	1	1.01	1.33	5.26	50.25	500.25	∞

Figure 8.8. (a) Simulation of the autoregressive process $G(t) = bG(t-1) + Z(t)$ with $\sigma^2 = 1$ and $b = 0.5$ (———), 0.9 (— — —) and 1.0 (- - -).
(b) Simulation of $\{X(t)\}$ (———) with $K(t) = 50 + G(t)$ (— — —) and $G(t) = 0.95G(t-1) + Z(t)$ for $\sigma^2 = 1$.

(a)

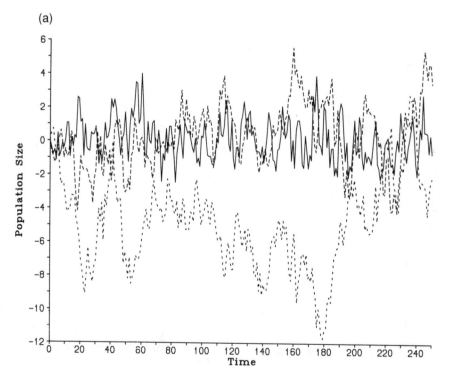

Figure 8.8a shows $\{G(t)\}$ for three simulations with $b = 0.5, 0.9$ and 1.0. The first two realizations remain fairly close to zero, though the variance associated with $b = 0.9$ is clearly greater than that with $b = 0.5$. However, by time $t = 180$ the simple random walk $(b = 1)$ has already shown signs of wandering off, and the longer the simulation persists the larger such departures will be.

Figure 8.8b shows a simulation of $\{X(t)\}$ when b is near 1, generated (via (8.53) with $h = 1$) for

$$K(t) = 50 + G(t) \quad \text{where} \quad G(t) = 0.95G(t-1) + Z(t) \qquad (8.65)$$

and $\sigma^2 = 1$. Since successive values of $K(t)$ are highly correlated, this produces a jittery meandering $K(t)$-path centred around $K_0 = 50$. $\{X(t)\}$ follows the same track, though rather more smoothly.

8.4.3   *Comparison of the autocovariances*
To study the simulation runs of Figures 8.7 and 8.8 in more detail, let

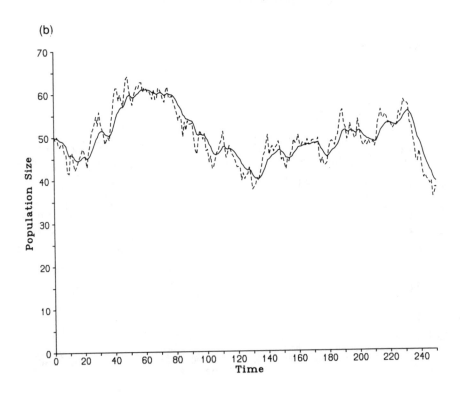

(b)

us use the (almost) unbiased autocovariance estimator (8.42), namely

$$\tilde{c}_k = (N + 1 - k)^{-1} \sum_{t=0}^{N-k} \{X(t) - \bar{X}\}\{X(t+k) - \bar{X}\} \qquad (8.66)$$

with $N = 250$, to determine the autocorrelations

$$r_k = \tilde{c}_k/\tilde{c}_0. \qquad (8.67)$$

Though a purely random $\{K(t)\}$-sequence (Figure 8.7a) has independent jumps of variable size, since $\{X(t)\}$ pursues $\{K(t)\}$ the associated $\{X(t)\}$-sequence might take a few steps to recover from the larger $K(t)$-excursions. Thus whilst neighbouring $X(t)$-values are related, more widely separated values are virtually independent (called short-term memory). A plot of the first 125 autocorrelations (Figure 8.9a) verifies this; only the first five or so autocorrelations are substantially larger than the rest with r_k wandering haphazardly about zero thereafter. In contrast, the simulated autoregressive structure shown in Figure 8.8b has a much longer term memory, highlighted by r_k now being positive over the first 36 time-lags.

Whereas these two processes are non-cyclic, Figure 8.7b shows the other

Figure 8.9. Autocorrelations $\{r_k\}$ for simulated $\{X(t)\}$ shown in: (a) Figures 8.7a (————) and 8.8b (– – –); (b) Figures 8.7b (————) and 8.7c (– – –).

(a)

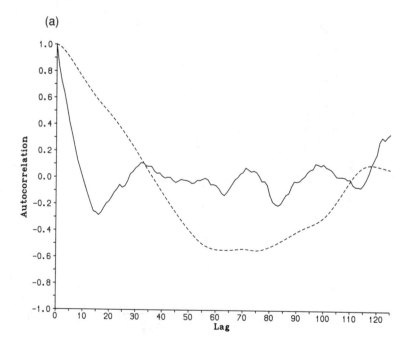

extreme of regular oscillations (with added noise). The corresponding autocorrelations (Figure 8.9b) not only pick out the underlying deterministic cosine wave, but they also do not attenuate to zero. No matter how large t is, the population size $\{X(t)\}$ will always peak around time multiples of 25. We therefore say that this $\{X(t)\}$-process has an infinitely long memory.

Modelling the period T through the autoregressive process (8.57) (see Figure 8.7c) provides a compromise between short and infinitely long memory. Although small-lag autocorrelations (Figure 8.9b) do reflect the initial period 25, the nature of the r_k clearly attenuates towards that of a short-term memory process (see Figure 8.9a) as k increases.

8.4.4   *Coherence time*

Many biological variables resemble a random walk in their behaviour over fairly short time intervals, yet over longer time spans they appear to be controlled about some particular value. Examination of $\{X(t)\}$ in Figure 8.8b and its autocorrelation function in Figure 8.9a suggests that a way of incorporating both types of behaviour into a single model is to consider the autoregressive process (8.59), namely

$$K(t) = K_0 + G(t) \quad \text{for} \quad G(t) = bG(t-1) + Z(t),$$

where b is only just less than 1.

(b)

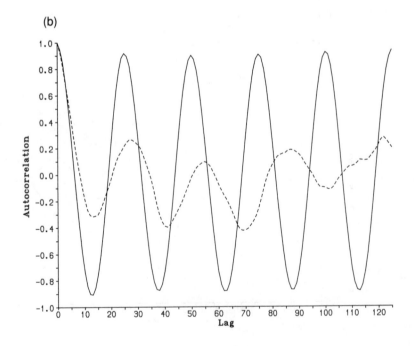

Write

$$b^2 = 1 - 1/t_c \tag{8.68}$$

for some large constant t_c. This constant can be interpreted as the 'coherence time', i.e. the fluctuations appear like a random walk over times much less than t_c and appear regulated on time scales greatly exceeding t_c. Then the variance (8.62) becomes

$$\text{Var}[K(t)] = \sigma^2 t_c [1 - (1 - 1/t_c)^t]$$
$$= \sigma^2 t_c [1 - \{(1 - 1/(t_c)^{t_c}\}^{t/t_c}].$$

But by definition of e^x,

$$(1 - 1/t_c)^{t_c} \to e^{-1} \quad \text{as} \quad t_c \to \infty. \tag{8.69}$$

So for large t_c

$$\text{Var}[K(t)] \simeq \sigma^2 t_c (1 - e^{-t/t_c}). \tag{8.70}$$

If t is much less than t_c, then on expanding $\exp(-t/t_c)$ we have

$$\text{Var}[K(t)] \simeq \sigma^2 t_c \left[1 - \left\{ 1 + \frac{(-t/t_c)}{1!} + \frac{(-t/t_c)^2}{2!} + \cdots \right\} \right] \simeq \sigma^2 t$$

which is precisely result (8.63) for the variance of a simple random walk. Whilst if t is much greater than t_c, then $\exp(-t/t_c) \simeq 0$ and so

$$\text{Var}[K(t)] \simeq \sigma^2 t_c$$

which, being a constant, corresponds to the random jitter model (8.54). Hence both situations are covered.

Since $\{X(t)\}$ pursues $\{K(t)\}$, evaluating the theoretical autocorrelation function of $\{K(t)\}$ provides some idea of how the population process itself will behave. From (8.60)

$$G(t) = b^{t-1} Z(1) + b^{t-2} Z(2) + \cdots + b Z(t-1) + Z(t)$$

whence

$$G(t + k) = b^{t+k-1} Z(1) + b^{t+k-2} Z(2) + \cdots + Z(t + k).$$

As $\{Z(t)\}$ is a white-noise process with

$$\text{Cov}[Z(t), Z(s)] = \begin{cases} \sigma^2 & \text{if } s = t \\ 0 & \text{otherwise,} \end{cases}$$

the autocovariances

$$c_k(t) \equiv \text{Cov}[K(t), K(t + k)] = \text{Cov}[G(t), G(t + k)]$$
$$= \sigma^2[b^{2t+k-2} + b^{2t+k-4} + \cdots + b^k]$$
$$= \sigma^2 b^k(1 - b^{2t})/(1 - b^2). \tag{8.71}$$

Replacing b^2 by $1 - 1/t_c$ then yields

$$c_k(t) = \sigma^2 t_c(1 - 1/t_c)^{k/2}[1 - (1 - 1/t_c)^t]$$
$$= \sigma^2 t_c\{(1 - 1/t_c)^{t_c}\}^{k/2t_c}[1 - \{(1 - 1/t_c)^{t_c}\}^{t/t_c}],$$

which from (8.69) gives (for large t_c)

$$c_k(t) \simeq \sigma^2 t_c \exp\{-k/2t_c\}(1 - \exp\{-t/t_c\})$$
$$\simeq \text{Var}[K(t)] \exp\{-k/2t_c\} \quad \text{(from (8.70))}. \tag{8.72}$$

Hence the autocorrelations

$$r_k(t) = \exp\{-k/2t_c\} \tag{8.73}$$

are independent of t, and so have a rate of decay determined solely by the coherence time t_c.

Result (8.73) shows that if

(i) $k \ll t_c$ then $r_k \simeq 1$
(ii) $k \simeq t_c$ then $r_k \simeq \exp(-\frac{1}{2}) \simeq 0.6$
(iii) $k \gg t_c$ then $r_k \simeq 0$.

So interpretation of a set of biological observations may be greatly affected by the length of the observed time-span (n) in relation to the coherence time (t_c), since three totally different conclusions may be drawn.

(i) $n \ll t_c$: Because r_k remains close to 1 all values of $\{K(t)\}$ remain highly correlated with each other and so the environment appears to be 'static'.
(ii) $n \simeq t_c$: The exponential decline in r_k indicates that the process ultimately approaches a random walk as n increases, colourfully called a 'pink-noise' environment.
(iii) $n \gg t_c$: Virtually all the $r_k \simeq 0$, corresponding to a 'white-noise' environment.

In practice, most biological situations will be pink to some extent, with the static and white-noise cases being useful approximations near the two extremes of pinkness.

8.4.5   *Period remembering or period forgetting?*
When fitting a model to a particular data set the selection of a static,

pink or white environment is clearly a qualitative one involving the identification of time scales. For example, we may wish to choose between either (a) a period-remembering process – e.g. model (8.55) (Figure 8.7b), or (b) a period-forgetting process – e.g. model (8.58) (Figure 8.7c). In (a) the autocorrelations $\{r_k\}$ show a slight initial decline followed by cycles of essentially constant amplitude, whilst in (b) the $\{r_k\}$ comprise damped oscillations (see Figure 8.9b).

A simple way of describing these two situations is to write

(a') $r_k \simeq A \exp(-ck) + B \cos(\omega k)$

and   (8.74)

(b') $r_k \simeq D \exp(-ck) \cos(\omega k)$

for $k \geqslant 0$ and A, B, D, c appropriate constants. To identify the periodic part of (a'), k must be sufficiently large to ensure that the transient term $A \exp(-ck)$ is much smaller than B, i.e.

$$k \gg (1/c) \log_e(A/B). \tag{8.75}$$

Whilst to detect cycles in case (b'), the cycle period (T) must be shorter than the memory of the process. In crude terms, this means that when $k = 1/c$, $\cos(\omega k)$ must have traversed at least a full cycle so that we know oscillations exist. Thus we require

$$2\pi < \omega k = \omega/c, \quad \text{i.e. } \omega > 2\pi c,$$

which is equivalent to

$$T = 2\pi/\omega < c^{-1}. \tag{8.76}$$

This suggests that c^{-1} may be interpreted as being the coherence time t_c of the process.

8.5   The Canadian lynx data

In spite of such difficulties of model identification, great delight is often taken in concentrating on one particular set of data and subjecting it to the 'death of a thousand models'! Irrespective of what the (unknown) *biological* or *physical* generating process might be that actually led to the data in the first place, there is an apparent compulsion to develop a myriad of mathematical models with the sole aim of obtaining the best *statistical* fit.

One such data set is the well-known time-series comprising the annual records of the numbers of Canadian lynx trapped in the Mackenzie River district of North-west Canada for the period 1821–1934 (Figure 8.10). These data appear in Elton and Nicholson (1942); a paper which not only contains a large amount of interesting descriptive information relating to lynx trapping records in various regions of Canada over some 200 years, but which also provides an excellent account of the practical difficulties involved in accurately constructing such data sets from widely varying sources of archive material.

The most striking feature of these data is the presence of persistent oscillations which exhibit a very regular period of about ten years and substantial changes in amplitude which show no systematic trend. This feature is also prominent in trapping records of lynx and other animals over the whole of Canada. Indeed, such population cycles have been reported in many parts of the world and have been familiar to biologists for a long time (see, for example, references in Bulmer, 1974; Elton and Nicholson, 1942). Three well-known examples in natural populations are the four-year cycle in the Arctic, the ten-year cycle in the coniferous forest zone, and the eight-year cycle of the larch bud moth in the Swiss Alps.

Moran (1953a) provided the first detailed statistical analysis of the lynx data, and in view of their extremely asymmetric nature he worked with the logarithmic transformation $Y(t) = \log_{10}\{X(t)\}$. He could have used a simple

Figure 8.10. Number of lynx trapped in the Mackenzie River district from 1821 to 1934.

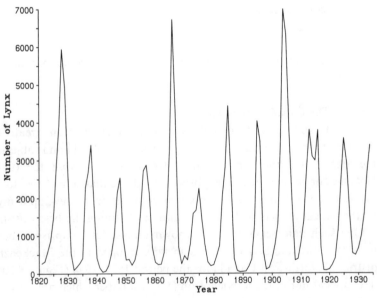

fluctuating environment process such as

$$Y(t) = K(t) + Z(t) \quad \text{for} \quad K(t) = \mu + A\cos(\omega t + \phi), \qquad (8.77)$$

where μ and ϕ denote mean log-population size and phase, respectively. However, he discarded this idea in favour of the *stochastic event* process

$$Y(t) = \mu + G(t)$$

where

$$G(t) + \theta_1 G(t-1) + \cdots + \theta_p G(t-p) = Z(t) \qquad (8.78)$$

for appropriate parameters $\theta_1, \ldots, \theta_p$. Moran's reasoning lay more along meteorological lines (Moran, 1953b) than biological ones, and since weather contains no intrinsic cycle he did not wish to invoke one artificially.

Using the autoregressive approach (8.78) Moran obtained the model

$$Y(t) = 1.0549 + 1.4101 Y(t-1) - 0.7734 Y(t-2) + Z(t) \qquad (8.79)$$

where $\{Z(t)\}$ has estimated standard deviation 0.2143. This process generates a theoretical autocorrelation function which is close to the empirical one (Figure 8.11), so in this sense (8.79) encapsulates the regular period of oscillation over the observed time span even though the $\{Y(t)\}$ are not driven by a regular fluctuating term (as is the case in (8.77)).

Figure 8.11. Autocorrelations $\{r_k\}$ of $\log_{10}$ (lynx data) for $k = 0, 1, \ldots, 40$.

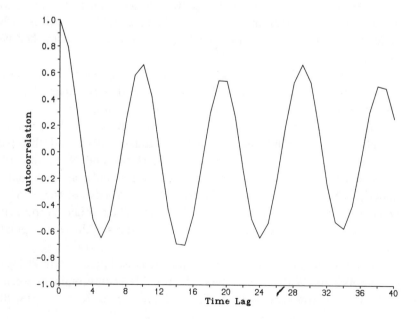

However, there is a biological reason for favouring a fluctuating environment model, since the Canadian lynx's food is known to consist almost entirely of the snowshoe rabbit (*Lepus americanus*). Indeed, as Seton (1912, chapter 14) remarks, 'It lives on Rabbits, follows the Rabbits, thinks Rabbits, tastes like Rabbits, increases with them, and on their failure dies of starvation in the unrabbited woods'. So we may interpret the cosine part of (8.77) as a driving term which reflects the cyclic fluctuation in the rabbit population. Moreover, since the autocorrelations $\{r_k\}$ (calculated via (8.66) and (8.67)) peak at

k	0	10	19	29	38
r_k	1	0.66	0.55	0.67	0.51

there is little, if any, evidence of damping beyond $k = 10$. So $\{r_k\}$ may be represented by (8.74a') with $\exp(-10c) < 0.1$ (say), i.e. $c > 0.23$.

An alternative approach is to consider the rabbits and lynx as a two-species population process, with lynx controlling rabbits as well as rabbits controlling lynx. This scenario gives rise to a predator–prey situation (Chapter 6) in which stochastic population cycles can be maintained virtually indefinitely. However, since the snowshoe rabbit is a dominant item in the food of four cyclic predators in northern North America, namely lynx, coyote, red fox and fisher, an interaction model with just two species is really too naive to be useful. A possible compromise is to assume that the rabbit cycle drives the lynx cycle, but that it itself is due to some other factor, such as a plant–herbivore interaction (Bulmer, 1976).

Campbell and Walker (1977) provide a comprehensive account of subsequent statistical analyses of these data, and it is apparent that the biological story has been submerged in a wave of mathematical enthusiasm. For as well as the dependence of lynx on rabbits, in the late 1800s a series of terrible smallpox pandemics affected the Indian tribes and this partly destroyed the basis of the interior fur trade (Elton and Nicholson, 1942). Thus since the proposed statistical models generally assume that the underlying mean and variance of the process remain constant, they cannot be taken too seriously as descriptions of reality. Their prime purpose is to provide a convincing description of the main features of the lynx population data.

From a purely mathematical viewpoint the most interesting model is due to Campbell and Walker themselves, who provide a compromise between the fluctuating environment model (8.77) and the autoregressive scheme (8.78) by developing a hybrid of the two, namely

$$Y(t) = 1.0302\,Y(t-1) - 0.3228\,Y(t-2) + 0.8496$$
$$- 0.2531\,\cos[(2\pi/9.5)(t-2.02)] + Z(t) \qquad (8.80)$$

where $Z(t)$ has estimated standard deviation 0.2098. Bulmer (1974) provides a similar analysis, though he uses the slightly longer period of $9.63 = 183/19$ since there are 19 complete cycles in the full Canadian data between the peak years of 1752 and 1935. There are strong indications of the constancy of the cycle period (though not of amplitude or phase) across the whole of Canada.

There is clearly no simple answer as to which model is 'best', for choice depends on the data, the situation, and of course the purpose for which the model will be used. Do we wish to understand what is going on, or do we just want to produce forecasts? If the former, then we should consider a wide variety of model types and make our selection on both biological and statistical grounds. In this situation the temptation of working in a mathematical vacuum must clearly be avoided. The latter usually involves finding a purely mathematical description which produces the best statistical fit, and so an autoregressive scheme or even a hybrid such as (8.80) might be most appropriate.

Note that we have by no means exhausted all possible model types. As well as

  (i) fluctuating environments,
 (ii) autoregressive schemes, and
(iii) predator–prey processes (Chapter 6),

there are also

 (iv) non-linear models (Chapter 4), and
  (v) models in which (for example) expression (8.77) is replaced by

$$Y(t) = \mu + A_t \cos(\omega t + \phi_t) + Z(t) \qquad (8.81)$$

where the amplitude A_t and phase ϕ_t change slowly with time.

Opportunities for imaginative model construction are clearly rife.

9

Spatial population dynamics

The geographic distribution of a species over its range of habitats, and the associated dynamics of population growth, are inseparably related, a fact which no complete study of population development can afford to ignore (see Levin, 1974). Thus whilst the assumption that populations develop at a single location is ideal for mathematical purposes, in real life we must accept that individuals seldom mix homogeneously over the whole region available to them but develop instead within separate sub-regions. Indeed, this is precisely the reason why we extended the non-spatial predator–prey process (Chapter 6) to allow individuals of either species to migrate between separate sites (Chapter 7). Having shown that spatial and non-spatial predator–prey behaviour can be very different, we clearly need to extend our spatial framework to cover more general population processes.

Most species attempt to migrate for a variety of both individual and population reasons, including: search for food; territorial extension for increasing population needs; widening the available gene pool; and minimizing the probability of extinction. Migration can range from being purely local, e.g. aquatic life in a small pond, to extensive migration patterns covering a fair part of the Earth's surface, e.g. birds, locusts, salmon, caribou, and viruses. Moreover, migration can occur either between distinct sites, such as neighbouring valleys or islands in an archipaelego, or else it can occur within continuous media such as the air or sea. The type and extent of the spatial environment will clearly have a very important effect on the way that specific populations develop (e.g. Huffaker's orange and mite experiment (Section 7.1)), and the consequence of adding a spatial dimension merits serious attention if we are to further our understanding of population growth. Spatial model enthusiasts should note that Andrewartha and Birch (1954, chapter 5) present a rich array of field dispersal studies for natural populations, all of which are ripe for further investigation. Hengeveld's (1989) account of documented invasion scenarios contains many varied examples. These include holocene tree invasions, red deer in New Zealand, cholera and the European starling in North America, the gallant and shaggy soldier plant in Britain, spread of muskrat centred on Prague 1905, immigration of

Neolithic farmers into Europe, the spread of stripe rust in wheat, the European invasion of the collared dove, expansion of cattle egret in North and South America, and the progression of rabies in Central Europe. Since we are concentrating on modelling, rather than the description of spatial phenomena, our own suite of examples is by no means exhaustive. Hengeveld's text therefore complements this chapter extremely well, a feature which readers of both books will find to be of considerable use.

In line with earlier chapters mathematical detail will be kept to a minimum, but it would be lamentable if attention were not drawn to the stimulating and entertaining theoretical world of spatial population processes. Until recently, virtually all ecological literature ignored the role of spatial heterogeneity, so this approach is relatively new. Exceptions have occurred, such as the classic work of Skellam (1951), but as Levin (1986) remarks, 'despite much earlier developments in the related areas of epidemics (Brownlee, 1911) and genetics (Fisher, 1937; Haldane, 1948), it has only been in the last few years that theoretical studies in ecology have begun to take account of the fundamental importance of geographic distribution'. He cites both recent work and other reviews, and considers a variety of ecological models which parallel island colonization studies (e.g. Simberloff, 1976) and the associated theory of island biogeography (MacArthur and Wilson, 1967). It is important to realize that even ostensibly non-insular environments, such as forests, farms and rocky shores, may often be viewed as mosaics of islands with different characteristics (Levin and Paine, 1974, 1975).

9.1   The simple random walk

Before we attempt to combine birth, death and migration, let us first consider migration by itself. The simplest way is to assume that an individual moves along a linear habitat, such as a river or shore line, that has been suitably zoned into neighbouring sites labelled by the integers $i = \ldots,$ $-1, 0, 1, 2, \ldots.$ Suppose that successive moves between sites are independent of each other, with

$$\Pr(\text{move from site } i \text{ to } i + 1) = p$$

and

$$\Pr(\text{move from site } i \text{ to } i - 1) = q = 1 - p. \tag{9.1}$$

These assumptions generate a simple random walk process (Figure 9.1) which has already been introduced, albeit in a different context, in Section 2.3.5. There the position X_n after n moves denotes population size, whereas here it refers to the geographic location of a single individual.

9.1.1   *Position after n steps*

Suppose this individual moves $n - i$ steps upwards (i.e. in a positive direction) and i steps downwards, starting at $X_0 = 0$. Then $X_n = n - 2i$, with

$$\Pr(X_n = n - 2i) = \overbrace{(p \times p \times \cdots \times p)}^{(n-i\,\text{times})} \times \overbrace{(q \times q \times \cdots \times q)}^{(i\,\text{times})}$$

$$\times \text{ (no. ways of choosing } i \text{ steps from } n)$$

$$= \binom{n}{i} p^{n-i} q^i \quad \left(\text{where } \binom{n}{i} \text{ denotes } \frac{n!}{i!(n-i)!} \right) \tag{9.2}$$

which is the well-known binomial distribution. Hence on writing $j = n - 2i$, we have

$$\Pr(X_n = j) = \binom{n}{\frac{1}{2}(n-j)} p^{(n+j)/2} q^{(n-j)/2} \tag{9.3}$$

where j takes the values $j = -n, -n+2, \ldots, n$.

When $p = q = \frac{1}{2}$ this expression becomes

$$\Pr(X_n = j) = \binom{n}{\frac{1}{2}(n-j)} (\tfrac{1}{2})^n, \tag{9.4}$$

and so the probability of returning to the starting point after $2n$ steps (this is impossible after an odd number of steps) is given by

$$\Pr(X_{2n} = 0) = \binom{2n}{n} (\tfrac{1}{2})^{2n}. \tag{9.5}$$

Now an excellent approximation to $n!$, which works well even when n is as low as 6, is

$$n! \sim \sqrt{(2\pi)} n^{n+\frac{1}{2}} e^{-n} \quad \text{(Stirling's formula)}. \tag{9.6}$$

So (9.5) reduces to the attractively simple result

$$\Pr(X_{2n} = 0) \sim 1/\sqrt{(\pi n)}. \tag{9.7}$$

Note that although result (9.3) looks rather cumbersome to evaluate, the computer package MINITAB (for example) will compute the binomial

Figure 9.1. The unrestricted simple random walk.

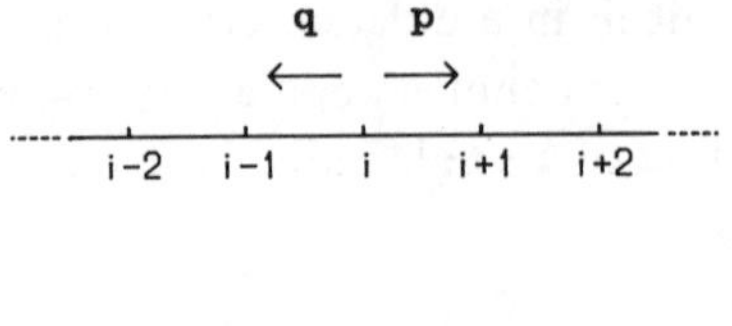

probabilities (9.2) together with their cumulative sum for $i = 0, 1, ..., n$. It can also be used to simulate i-values for given n and p, and hence the occupation positions $j = n - 2i$.

9.1.2 *Use of the Normal approximation*

Alternatively, we can use the Normal approximation. Denote the ith step by Z_i, and take the initial position to be 0. Then X_n is the sum of the first n steps, i.e.

$$X_n = Z_1 + Z_2 + \cdots + Z_n. \tag{9.8}$$

Thus if the Z_i are independently distributed with mean μ and variance σ^2, then

$$\text{mean}(X_n) = n\mu \quad \text{and} \quad \text{Var}(X_n) = n\sigma^2, \tag{9.9}$$

whence

$$\Pr(a \leqslant X_n \leqslant b) \simeq \int_{a-\frac{1}{2}}^{b+\frac{1}{2}} (2\pi n\sigma^2)^{-\frac{1}{2}} \exp\{-(x - n\mu)^2/2n\sigma^2\}\, dx$$
$$= \Phi[(b + \tfrac{1}{2} - n\mu)/\sqrt{(n\sigma^2)}] - \Phi[(a - \tfrac{1}{2} - n\mu)/\sqrt{(n\sigma^2)}]. \tag{9.10}$$

Here $\Phi(.)$ denotes the tabulated standard Normal distribution with mean 0 and variance 1. For example, the simple random walk with $p = 0.6$ has $\mu = p - q = 0.2$ and $\sigma^2 = 4pq = 0.96$. So after say 40 steps

$$\Pr(X_{40} \leqslant 10) \simeq \Phi[(10 + \tfrac{1}{2} - 8)/\sqrt{(38.4)}] = \Phi(0.4034) = 0.657,$$

which compares quite well with the exact value 0.683 calculated from (9.3).

This Normal result is particularly useful in telling us within what range X_n is likely to lie. Since the probability that a Normal random variable lies further than 4 standard deviations from its mean is less than 0.0001, we can write

$$\Pr(-4\sigma\sqrt{n} \leqslant X_n - n\mu \leqslant 4\sigma\sqrt{n}) < 0.0001. \tag{9.11}$$

Hence X_n has 'drift' $n\mu$, whilst the deviation from $n\mu$ increases only with $\sqrt{n}$ and not with n. For example, the symmetric random walk $p = q = \frac{1}{2}$ has mean $\mu = 0$ and variance $\sigma^2 = 4pq = 1$, whence (9.11) becomes $\Pr(-4\sqrt{n} \leqslant X_n \leqslant 4\sqrt{n}) < 0.0001$. Thus even after a million steps X_n will almost certainly lie within a mere 4000 units of its starting point.

9.1.3 *Absorbing barriers*

In many situations an individual will not be able to roam without bounds, but will be restricted in the range available to it; for example, a fish feeding in a particular stretch of river. Let us therefore insert two barriers into our simple random walk model, say at 0 and a. Within the range of mobility $(0 < i < a)$ the model will hold as before, but the behaviour at the boundaries $i = 0$ and $i = a$ will depend on the type of situation we have in mind.

Suppose there are traps at the boundaries, so that an individual will be captured, or 'absorbed', there. Now this absorbing barrier situation (Figure 9.2 with $r = 1 - p - q = 0$) has already been considered in Section 2.3.5. So on using result (2.60) with

$$q_k = \text{Pr(eventual absorption at 0 given } X_0 = k)$$

and

$$p_k = \text{Pr(eventual absorption at } a \text{ given } X_0 = k),$$

we have

$$q_k = \frac{(q/p)^a - (q/p)^k}{(q/p)^a - 1} \quad \text{and} \quad p_k = \frac{(q/p)^k - 1}{(q/p)^a - 1}. \tag{9.12}$$

In particular, if say $p > q$ so that there is a drift to the right, and if a is large enough for $(q/p)^a$ to be ignored, then

$$q_k \simeq (q/p)^k \quad \text{and} \quad p_k \simeq 1 - (q/p)^k. \tag{9.13}$$

Thus the further k is away from zero the less chance there is of a downwards excursion reaching the trap there before the upward drift carries the individual towards the trap at a.

The corresponding number of steps (d_k) taken before either trap 0 or a is reached, starting from position k, is given by expressions (2.65) and (2.66). In particular, when $p = q = \frac{1}{2}$ we have $d_k = k(a - k)$ which increases from ka near $k = 0$ to $a^2/4$ near $k = a/2$. This latter value rises fast with increasing a.

Figure 9.2. The simple random walk with two absorbing barriers.

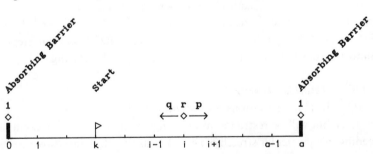

For example, with

$$a = 10 \quad : \quad d_1 = 9 \quad \text{and} \quad d_5 = 25$$
$$a = 1000: \quad d_1 = 999 \quad \text{and} \quad d_{500} = 250\,000.$$

9.1.4 *Reflecting barriers*

Often barriers are merely a restriction to movement. Consider, for example, a fish swimming in a long thin tank notionally zoned into one-metre sections from 0 to a. Suppose that in each time unit the fish moves from zone i to zone $i + 1$, i or $i - 1$ with probabilities $p, r = 1 - p - q$ and q, respectively. At the boundary $i = 0$ the fish is unable to move left, so we assume that it stays there with probability $q + r = 1 - p$; similarly, the fish stays at the upper boundary $i = a$ with probability $r + p = 1 - q$. Thus we now have a 'reflecting barrier' situation (Figure 9.3).

After the fish has spent some time swimming up and down the tank the effect of its initial position will have worn off, i.e. the fish's position will have settled down to a condition of statistical equilibrium. So let

$$\pi_i = \Pr(\text{fish is at position } i).$$

Then considering the fish's previous move gives for $0 < i < a$

$$\pi_i = \Pr(\text{fish at } i - 1 \text{ and moves to } i)$$
$$+ \Pr(\text{fish at } i \text{ and stays there})$$
$$+ \Pr(\text{fish at } i + 1 \text{ and moves to } i),$$

i.e.

$$\pi_i = p\pi_{i-1} + (1 - p - q)\pi_i + q\pi_{i+1} \qquad (0 < i < a); \tag{9.14}$$

whilst at the boundaries

$$\pi_0 = (1 - p)\pi_0 + q\pi_1 \qquad (i = 0) \tag{9.15}$$

Figure 9.3. The simple random walk with two reflecting barriers.

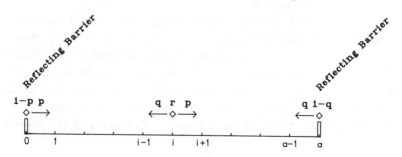

and

$$\pi_a = p\pi_{a-1} + (1-q)\pi_a \qquad (i=a). \tag{9.16}$$

One way of solving these equations is to write (9.14) as

$$q\pi_{i+1} - p\pi_i = q\pi_i - p\pi_{i-1},$$

and then note that as the two sides of this equation are identical in structure they must be independent of i. Thus

$$q\pi_i - p\pi_{i-1} = \text{constant}.$$

When $i = 1$

$$q\pi_1 - p\pi_0 = \text{constant} = 0 \qquad \text{(from (9.15))},$$

and so

$$\pi_i = (p/q)\pi_{i-1} = (p/q)^2\pi_{i-2} = \cdots = (p/q)^i\pi_0. \tag{9.17}$$

Since this value automatically satisfies equation (9.16), all that remains is to determine π_0 from the probability sum

$$1 = \sum_{i=0}^{a} \pi_i = \pi_0[1 + (p/q) + \cdots + (p/q)^a]$$
$$= \pi_0[1 - (p/q)^{a+1}]/[1 - (p/q)].$$

Thus the equilibrium probability

$$\pi_i = (p/q)^i[1 - (p/q)]/[1 - (p/q)^{a+1}] \qquad (i=0, \ldots, a). \tag{9.18}$$

We see that if $p > q$ ($p < q$) then π_i decreases geometrically fast away from the upper (lower) reflecting barrier; whilst if $p = q$ then it follows from (9.17) that

$$\pi_i = 1/(a+1) \qquad (i = 0, \ldots, a),$$

and so all positions are equally likely.

9.2   Brownian motion

Whilst the notional zoning of a tank into one-metre sections is quite appropriate when describing the motion of a fairly large individual such as a trout, it is obviously inadequate when dealing with individuals which change their direction of motion almost continuously, e.g. molecules, plankton and small insects. In such situations a one-metre zone is ridiculously large, so let us replace it by δx (say) where δx can be taken to be as small as we wish. Moreover, since changes of position may now occur rapidly, let δt denote the (small) unit of time. Each step Z_i of this scaled random walk then has

$$\text{mean } m = p(+\delta x) + q(-\delta x) = (p - q)\delta x$$

and

$$\text{variance } \sigma^2 = [p(+\delta x)^2 + q(-\delta x)^2] - m^2$$
$$= 4pq(\delta x)^2 \quad \text{(on taking } p + q = 1).$$

As $t/\delta t$ steps occur in a time period of length t, the individual's position $\{X(t)\}$ therefore has

$$\text{mean } m(t) = (p - q)t \, \delta x/\delta t$$

and

$$\text{variance } \sigma^2(t) = 4pqt(\delta x)^2/\delta t. \tag{9.19}$$

If we now allow $\delta x \to 0$ and $\delta t \to 0$ then this mean and variance must remain finite if the process is to make practical sense. In particular, suppose we require the limiting process to have mean a and variance D^2 in unit time. Then δx and δt must tend to zero in such a way that

$$m(1) = (p-q)\,\delta x/\delta t \to a$$
$$\sigma^2(1) = 4pq(\delta x)^2/\delta t \to D^2, \tag{9.20}$$

and these conditions are satisfied by taking

$$\delta x = D\sqrt{(\delta t)}, \quad p = \tfrac{1}{2}\{1 + [a\sqrt{(\delta t)}/D]\}$$
$$\text{and} \quad q = \tfrac{1}{2}\{1 - [a\sqrt{(\delta t)}/D]\}. \tag{9.21}$$

The limiting process $\{X(t)\}$ then has a Normal distribution with mean at and variance $D^2 t$, viz:

$$f(x; t)\,\delta x = \Pr(x \leqslant X(t) \leqslant x + \delta x)$$
$$= (2\pi D^2 t)^{-\frac{1}{2}} \exp\{-(x-at)^2/2D^2 t\}\,\delta x, \tag{9.22}$$

and is known as Brownian motion with drift a and variance parameter D^2.

Conditions (9.21) imply that p and q must lie close to $\tfrac{1}{2}$ for the process to give sensible results, and also that δx must be of order $\sqrt{(\delta t)}$. Thus the individual's velocity, being

$$\delta x/\delta t = D\sqrt{(\delta t)}/\delta t = D/\sqrt{(\delta t)},$$

becomes infinite in the limit as $\delta t \to 0$. In spite of this biological absurdity the Brownian motion process has proved to be an excellent representation of random movement in continuous media, and it forms an important basis for understanding the dynamics of population spread. Readers interested in this wide-ranging field of 'diffusion processes' should consult the entertaining text by Okubo (1980).

9.3   Application of diffusion processes

Skellam (1951) used this diffusion approach in a pioneering study of animal and plant dispersal. Suppose we are interested in the two-dimensional spread $\{X(t), Y(t)\}$ of a large number of individuals from a central point. Then solution (9.22) has first to be replaced by its two-dimensional counterpart, namely the bivariate Normal distribution. If there is no preferred direction of motion, i.e. there is zero drift, then this distribution has the symmetric form

$$f(x, y; t) = (2\pi D^2 t)^{-1} \exp\{-(x^2 + y^2)/2D^2 t\} \tag{9.23}$$

where $\text{Var}[X(1)] = \text{Var}[Y(1)] = D^2$.

To exploit this symmetry let us make the transformation

$$x = r\cos(\theta), \quad y = r\sin(\theta) \quad \text{and} \quad dx\,dy = r\,d\theta\,dr, \tag{9.24}$$

and denote $b^2 = 2D^2$ as the mean square deviation in unit time. Then $f(x, y; t)\,dx\,dy$ has the polar form

$$\phi(r, \theta; t)\,d\theta\,dr = (\pi b^2 t)^{-1} \exp\{-r^2/(b^2 t)\}r\,d\theta\,dr. \tag{9.25}$$

If we are concerned purely with the distance $R(t)$ of an individual from the origin at time t, then we can integrate (9.25) over $\theta = 0$ to 2π to obtain

$$\Pr(r \leqslant R(t) \leqslant r + dr) = \phi(r; t)\,dr = (2r/b^2 t)\exp\{-r^2/b^2 t\}\,dr. \tag{9.26}$$

We can now determine the rate at which the population disperses through the region. Since there is no drift the probability density (9.25) will spread out in ever-expanding circles. Suppose we define the 'wavefront' $R(t)$ of the process at time t to be that value for which we expect to find just one individual at a distance r further than $R(t)$. Then given a population of size N this means that

$$1 = N \int_{R(t)}^{\infty} \phi(r; t)\,dr = N \int_{R(t)}^{\infty} (2r/b^2 t)\exp\{-r^2/b^2 t\}\,dr.$$

On putting $w = r^2$, so that $dw = 2r\,dr$, this equation becomes

$$1 = N \int_{R^2(t)}^{\infty} (1/b^2 t)\exp\{-w/b^2 t\}\,dw = N\exp\{-R^2(t)/b^2 t\}.$$

Hence

$$\log_e(1/N) = -R^2(t)/b^2 t,$$

giving

$$R^2(t)/t = b^2 \log_e(N). \tag{9.27}$$

Thus the area of the circle containing the diffusing population is directly proportional to the elapsed time t. Equivalently, we can write (9.27) as

$$\text{radial velocity} = R(t)/t = [b\sqrt{\{\log_e(N)\}}]/\sqrt{t} \tag{9.28}$$

which decreases as $t^{-\frac{1}{2}}$.

In many situations the population size will not remain constant but will increase. Suppose we assume deterministic exponential growth at rate λ, so that a single individual will develop into $N(t) = \exp(\lambda t)$ individuals by time t. Then we may argue that with N in (9.27) replaced by $\exp(\lambda t)$,

$$R^2(t)/t = b^2 \lambda t$$

i.e.

$$R(t) = bt\sqrt{\lambda}, \tag{9.29}$$

which implies that the velocity of propagation of the wavefront remains constant at $b\sqrt{\lambda}$. Note that the combination of population growth and diffusion is clearly essential if spatial expansion is not to fade out as local density becomes too low. After reproduction, offspring can invade neighbouring areas and so prevent the wave from exhausting itself.

An added complication is that there is ample evidence that some migrating species avoid crowding. For example, Carl (1971) has observed that arctic ground squirrels migrate from densely populated areas into sparsely populated areas, even when the latter provide a less favourable habitat. Gurtin and MacCamy (1977) develop the associated diffusion argument, and show that contrary to the birth–migration result (9.29) only a finite region of space may ultimately be colonized.

9.3.1   *Skellam's examples*

Skellam (1951) applied the above results to the following three situations.

(A) *The dispersal of oaks.*   Oaks do not produce acorns until they are 60 or 70 years old, after which time they produce them for several hundred years. So the generation time can be taken as 60 years at an absolute minimum. Moreover, the final recession of the ice sheet started at about 18 000 BC, yet oak forests were apparently well-established throughout Britain in Roman times (Tansley, 1939). Skellam therefore took the number of generations to be $t = 18\,000/60 = 300$, with $N(1) = \exp(\lambda) = 9\,000\,000$ as an upper bound to the number of mature daughter oaks produced by a single parent oak. With these values, expression (9.29) gives

$$R(t)/b = t\sqrt{\lambda} \leqslant 300\sqrt{\{\log_e(9 \times 10^6)\}} = 1200.$$

Now in the original stated form of the problem Reid (1899) takes R to be about 600 miles. So if we assume that the spread of oaks may be described by our basic diffusion process, then it follows that b (the root mean square distance, r.m.s.d., of daughter oaks about their parent) is at least $600/1200$, i.e. a half-mile. Thus under these assumptions Reid's conclusion that animals such as rooks must have played a major role in the dispersal of oaks appears to be conclusive.

(B) *The dispersal of small animals.* In his second example Skellam supposes a mean square dispersion per minute of a wingless ground beetle wandering at random to be 1 yard^2. So after a season of say 6 months, without rest, the r.m.s.d. is $\sqrt{(6 \times 30 \times 24 \times 60)} \simeq 500$ yards. Without external help, a million seasons would therefore raise the r.m.s.d. to only $500\sqrt{(10^6)}$ yards, i.e. just under 300 miles. These admittedly basic calculations show that if global distribution is to be achieved then accidental displacements due to external agencies such as gales, floating material, and the muddy feet of birds and mammals are clearly necessary.

(C) *Empirical confirmation.* The parameter values in these last two situations have simply been guessed, and conclusions are able to be drawn only because theory predicts results orders of magnitude different from reality. Information is well documented, however, on the spread of the muskrat (*Ondatra zibethica* L.) in Central Europe since its introduction in 1905 (Ulbrich, 1930). Skellam takes the reported wavefronts (Figure 9.4a) for certain years, and uses the area enclosed by them to provide an estimate of $\pi R^2(t)$. His plot of $\sqrt{(\text{area})}$ against time (Figure 9.4b) shows an excellent linear fit, thereby providing empirical confirmation of result (9.29).

These three examples provide an excellent demonstration of the *qualitative* use of mathematical modelling. Since even though we cannot expect such simplified arguments to produce results which agree exactly with reality, they do permit convincing conclusions to be drawn about rates of spread that are possible in different circumstances.

9.3.2   *Broadbent and Kendall's example*

As well as extending the diffusion result (9.27) by allowing for change in the population size N, we can also vary the length of time t over which each particular individual moves. For example, the larvae of the helminth

Trichostrongylus retortaeformis hatch from eggs in the excreta of sheep and rabbits, and wander apparently at random until they climb and remain on blades of grass. Later they are eaten by another animal, so beginning a new cycle.

Let $H(t)$ denote the probability that a particular larva climbs onto a blade of grass at or before time t. Then if we assume that the probability that this larva stops moving and starts climbing during the small time interval $(t, t + dt)$ is $\lambda\,dt$, we have

Pr(stops by time $(t + dt)$) = Pr(stops by time t)
  + Pr(moving at time t but stops during $(t, t + dt)$)

i.e.

$$H(t + dt) = H(t) + \lambda\,dt[1 - H(t)].$$

As $dt \to 0$ this yields

$$dH(t)/dt = -\lambda H(t)$$

which integrates to give

$$H(t) = 1 - \exp(-\lambda t)$$

with density function

$$h(t) = dH(t)/dt = \lambda \exp(-\lambda t). \tag{9.30}$$

Figure 9.4. (a) A map showing the spread of muskrat in Central Europe, the contours of which provide (b) a linear plot of $\sqrt{\text{(area)}}$ against time. (Reproduced from Skellam, 1951, by permission of the Biometrika Trustees.)

(a) (b)

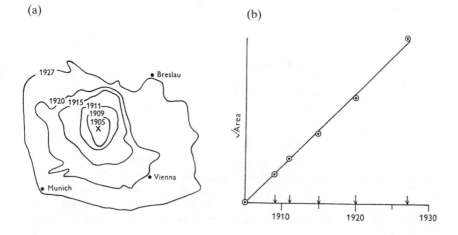

By combining (9.26) and (9.30) we see that the radial distribution of the larvae after all movement has ceased has density $g(r)$ where

$$g(r)\,\mathrm{d}r = \int_0^\infty \text{Pr(final position is in } (r, r + \mathrm{d}r) \text{ and}$$

$$\text{motion stops at time } t)\,\mathrm{d}t$$

$$= \mathrm{d}r \int_0^\infty \phi(r; t) h(t)\,\mathrm{d}t$$

$$= \mathrm{d}r \int_0^\infty (2r/b^2 t) \exp\{-r^2/b^2 t\} \lambda\,\mathrm{e}^{-\lambda t}\,\mathrm{d}t \qquad (9.31)$$

(Broadbent and Kendall, 1953).

Expression (9.31) is unlike all the other results we have so far met in that nothing much can be done to simplify it! Unfortunately such mathematical inconvenience occurs in a great many ecological modelling situations and is something we just have to live with. However, such results can often be written down in terms of *known functions*, in the sense that they have already appeared so often in the literature that most of their properties are well documented.

For example, the integral (9.31) is closely related to the function

$$K_\nu(z) = \tfrac{1}{2}(z/2)^\nu \int_0^\infty \exp\{-s - (z^2/4s)\} s^{-\nu-1}\,\mathrm{d}s \qquad (9.32)$$

called the modified Bessel function of the second kind (Watson, 1952, p. 183). Since this function is well tabulated (e.g. Watson, 1952), it does not really matter that we cannot solve the integral (9.31) in a neat form since we can always construct numerical values. Specifically, on putting $s = \lambda t$ in (9.31) we have

$$g(r) = \int_0^\infty (2\lambda r/b^2 s) \exp\{-s - (\lambda r^2/b^2 s)\}\,\mathrm{d}s \qquad (9.33)$$

which is nearly identical to (9.32) when $\nu = 0$ and $z^2/4 = \lambda r^2/b^2$. In fact

$$g(r) = (4\lambda r/b^2) K_0(2r\sqrt{\lambda}/b),$$

which is even simpler still when written in the form

$$g(r)\,\mathrm{d}r = z K_0(z)\,\mathrm{d}z. \qquad (9.34)$$

The function $K_0(z)$ declines swiftly to zero as z increases, for example

z	0	0.1	0.5	1.0	2.0	5.0
$K_0(z)$	∞	2.43	0.92	0.42	0.11	0.004

whilst $zK_0(z)$ equals zero at $z = 0$ (there can be no larvae in a circle of radius zero), rises very rapidly to a maximum at $z \simeq 0.6$, and then tails off to zero (Figure 9.5).

Note that if larvae do not necessarily stop at the first blade of grass they reach, but instead have probability p of stopping at each new blade they visit, then the same result (9.34) holds but with λ replaced by $p\lambda$.

Out of a total of N larvae the expected number finally positioned in the annulus between circles of radius z and $z + dz$ is $NzK_0(z)\, dz$. The area of this annulus is $2\pi z\, dz$, so (9.34) tells us that the density of larvae at a distance z from the centre is $(N/2\pi)K_0(z)$. Thus as $z = 2r\sqrt{\lambda/b}$ changes only with $2\sqrt{\lambda/b}$, where $\lambda = 1/($expected travel time of a larva$)$ and b^2 is the mean square displacement in unit time under continual motion, it is quite possible for two species which spread according to this model to generate identical spatial distributions if one of them diffuses slowly with a long mean travel time whilst the other diffuses rapidly but with a short mean travel time (Pielou, 1977).

This diffusion model is of general applicability, and Williams (1961) notes that it provides a satisfactory representation of the infestation of fruit by larvae. By releasing a large number of moths at one point in a fairly uniform orchard, Wildbolz and Baggiolini (1959) were able to study the effects of the random movement and egg laying of the moths by assessing the numbers of

Figure 9.5. The radial density $zK_0(z)$ of population size for organisms which diffuse from a given point and stop at random times.

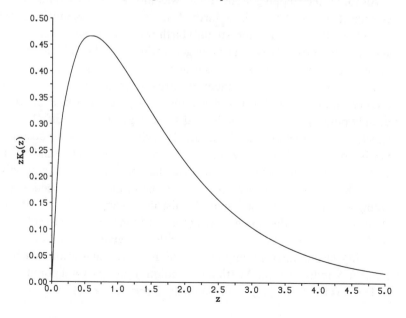

fruit infested with larvae emerging from the eggs. The observed spatial distribution of infested fruit is described quite well by a Bessel function with suitably chosen parameters.

However, Williams warns that just because a phenomenon obeys some mathematical law it does not mean that the assumptions which lead to the law are necessarily the same set which give rise to the phenomenon. Different causes can often lead to the same mathematical law, and he cites various assumptions, by no means exhaustive, which lead to the same form of Bessel distribution. Where a mathematical modelling approach scores is that it not only enables us to limit the kinds of causal systems that might be applicable, but it also guides our choice of experiments to decide between them.

9.4   Stepping-stone models (1)

In the last two sections we first developed the concept of Brownian motion through a continuous region, and then applied it to populations which were increasing in size as well as diffusing spatially. A similar extension can be applied to the simple random walk model (Section 9.1), where individuals now not only move between neighbouring sites but they also give birth and die. We have, of course, already met this 'stepping-stone' model in a predator–prey context (Chapter 7).

Note the intrinsic difference between the diffusion model, in which individuals can move anywhere in space, and the stepping-stone model, in which they have to live at specific sites.

Although the stepping-stone model was introduced by Kimura (1953), in essence it was first used in a birth–death context by Kendall (1948). He considered the basic (i.e. non-spatial) birth process and wished to change the standard exponential distribution of generation times (S). This distribution is often different from the distributions of generation times actually observed in practice, even for such elementary organisms as bacteria. Unlike the theoretical exponential distribution, observed generation time distributions usually possess a non-zero mode, and so he suggested a modified process in which S is distributed as a χ^2_{2k}-variate. He achieved this by supposing that a newly-born individual passes through k 'phases', spending a random time in each, and that only after the kth phase has been reached can subdivision occur. Kendall's multiphase process may therefore be envisaged as representing a population which is spatially distributed amongst k sites or colonies (Figure 9.6). Individuals in colony i $(1 \leqslant i < k)$ may migrate to colony $i + 1$, whilst in the event of an individual in colony k giving birth, both it and its offspring instantaneously migrate to colony 1. Kendall examined both the number of individuals in the ith phase (colony), and the total population size.

This class of stepping-stone models forms a special type of Markov population process, in which the state of the process at time t is represented by the k values $(X_1(t), \ldots, X_k(t))$ where $X_i(t)$ denotes the number of individuals in the ith colony at time t (Kingman, 1969). 'Markov' means that the next event depends only on the current population sizes and not on the previous history of the process; a necessary requirement if mathematical headway is to be made. Events may be of three types, corresponding to the arrival of a new individual, the departure of an existing one, or the transfer of an individual from one colony to another, and we may denote the transition rates out of the states $\{X_i(t) = n_i\}$ by

$$\begin{aligned}
\alpha_i(\mathbf{n}): \quad &\text{arrival at } i &&(\text{i.e. } X_i(t) \to n_i + 1) \\
\beta_i(\mathbf{n}): \quad &\text{departure from } i &&(\text{i.e. } X_i(t) \to n_i - 1) \\
\gamma_{ij}(\mathbf{n}): \quad &\text{transfer from } i \text{ to } j &&(\text{i.e. } X_i(t) \to n_i - 1 \text{ and} \\
& && X_j(t) \to n_j + 1) \qquad (9.35)
\end{aligned}$$

where $\mathbf{n}$ denotes $(n_1, \ldots, n_k)$.

Arrival, departure and transfer rates often depend only on the numbers in the colonies affected by the transition. In such circumstances we may write

$$\alpha_i(\mathbf{n}) = \alpha_i(n_i), \quad \beta_i(\mathbf{n}) = \beta_i(n_i) \quad \text{and} \quad \gamma_{ij}(\mathbf{n}) = \gamma_{ij}(n_i, n_j), \qquad (9.36)$$

and describe the process as being *simple*. Readers interested in discovering the wide range of models generated by these transition rates should consult Renshaw (1986); all we shall do here is to abstract some of the more easily obtained results.

9.4.1 *Equilibrium probabilities*

If an equilibrium population size distribution $\{\pi_i(\mathbf{n})\}$ exists, so that neither extinction nor population explosion can occur, then the equilibrium probability equations corresponding to the transition rates (9.35) may be constructed fairly easily. Unfortunately there is no hope of finding a general solution to them, so if progress is to be made then some simplification is necessary.

One way is to consider the simple system (9.36) with

$$\alpha_i(\mathbf{n}) = a_i, \quad \beta_i(\mathbf{n}) = b_i f_i(n_i) \quad \text{and} \quad \gamma_{ij}(\mathbf{n}) = g_{ij} f_i(n_i), \qquad (9.37)$$

Figure 9.6. Kendall's multiphase birth process viewed as a stepping-stone model.

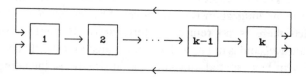

where a_i, b_i and g_{ij} are constants, and $f_i(n_i)$ are functions of the population size n_i that are common to both death and migration. Note that the migration rate from i to j depends only on the number in i and not j. Denote the multiplication symbol

$$\prod_{i=1}^{k} (x_i) = x_1 \times x_2 \times x_3 \times \cdots \times x_k.$$

Then Whittle (1968) shows that the equilibrium probability

$$\pi(\mathbf{n}) = C \prod_{i=1}^{k} \{\lambda_i^{n_i}/f_i(1) f_i(2)\ldots f_i(n_i)\}, \tag{9.38}$$

where the λ_i are determined from the linear simultaneous equations

$$\left(b_i + \sum_{j=1}^{k} g_{ij}\right)\lambda_i = a_i + \sum_{j=1}^{k} g_{ji}\lambda_j \tag{9.39}$$

and C is a normalizing constant to ensure that these probabilities sum to 1. This solution is not as complicated as it seems, since the unknown parameters $\{\lambda_i\}$ are easy to determine numerically by using a standard package such as MINITAB.

A special case of (9.37) is the simple immigration–death–migration system

$$\alpha_i(\mathbf{n}) = a_i, \quad \beta_i(\mathbf{n}) = b_i n_i \quad \text{and} \quad \gamma_{ij}(\mathbf{n}) = g_{ij} n_i, \tag{9.40}$$

for which $f_i(n_i) = n_i$. From (9.38)

$$\pi(\mathbf{n}) = \prod_{i=1}^{k} \lambda_i^{n_i} e^{-\lambda_i}/(n_i!) \tag{9.41}$$

(Bartlett, 1949), so in equilibrium the k population sizes behave as though they are from independent Poisson distributions with parameter λ_i.

Whilst Whittle's solution (9.38) holds for the general migration rate $\gamma_{ij}(\mathbf{n})$ $= g_{ij} f_i(n_i)$, the corresponding arrival rate $\alpha_i(\mathbf{n}) = a_i$ is very restrictive as it denotes pure immigration and does not allow even a simple birth process (i.e. $\alpha_i(\mathbf{n}) = a_i n_i$) to operate. However, before we go on to include birth, which itself introduces severe mathematical difficulties, it is worthwhile mentioning a specific situation generated under assumptions (9.37).

If both birth and death are absent (i.e. $a_i = b_i = 0$) but there is general migration between all k colonies, then the probability distribution of population size is known (expression (39) of Bartlett, 1949). Moreover, departure by death or emigration is easily covered by introducing a further state from which there is no return. Unfortunately, Bartlett's result is complicated to unravel, and is not as widely appreciated as it should be. It has wide applicability in both spatial and non-spatial situations, including: the approach to statistical equilibrium of non-interacting gas molecules (Siegart, 1949); the movement of bull spermatozoa (Patil, 1957); the recruitment and promotion of individuals in various grades of an organization (Bartholomew, 1967); and the illness–death model in which there are movements both between the various states of illness and the various types of death (Chiang, 1968).

9.5   The two-colony model

As soon as birth is introduced the general probability equations for the colony sizes $\{X_1(t), \ldots, X_k(t)\}$ become intractable to direct solution. Indeed, a straightforward form for the probabilities $\Pr\{X_1(t) = n_1, X_2(t) = n_2\}$ has yet to be determined even for the simple two-colony birth–death–immigration–migration process (Figure 9.7) with transition rates (9.35) given by

$$\alpha_i(\mathbf{n}) = \alpha_i + \lambda_i n_i, \quad \beta_i(\mathbf{n}) = \mu_i n_i \quad \text{and} \quad \gamma_{ij}(\mathbf{n}) = v_i n_i \tag{9.42}$$

$(i, j = 1, 2)$. So the challenge is to extract as much information as we can from the process in order to increase our understanding of how migration affects population growth, extinction, etc.

In addition to obvious geographic situations, model (9.42) has considerable importance in various areas of application where migration is a paradigm for transfer of state rather than physical movement. For example, in genetics migration is equivalent to mutation (Armitage, 1952), whilst models for phage reproduction within a bacterium also fall under this scheme (Gani, 1965).

*9.5.1   Mean values

Let $m_i(t)$ denote the mean number of individuals in colony i at time t. Then following the usual deterministic argument we have

$$dm_1(t)/dt = \underset{\text{(net growth in 1)}}{(\lambda_1 - \mu_1 - v_1)m_1(t)} + \underset{\text{(immigr.)}}{\alpha_1} + \underset{\text{(migr. 2 to 1)}}{v_2 m_2(t)}. \tag{9.43}$$

Similarly, for colony 2

$$dm_2(t)/dt = (\lambda_2 - \mu_2 - v_2)m_2(t) + \alpha_2 + v_1 m_1(t). \tag{9.44}$$

Figure 9.7. The simple birth–death–immigration–migration process.

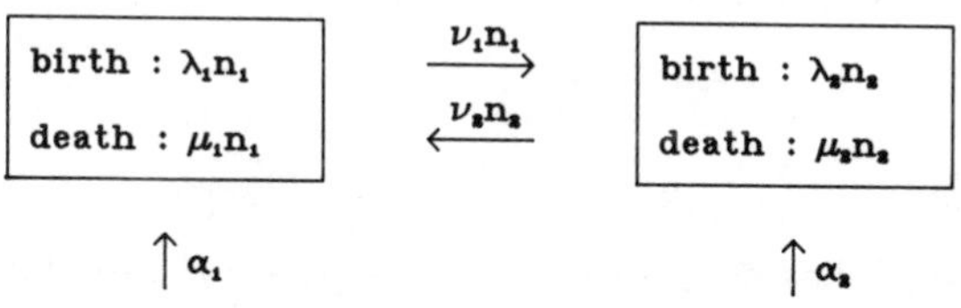

These equations are linear in $m_1(t)$ and $m_2(t)$, and have the rather messy solution (Renshaw, 1973b)

$$
\begin{aligned}
m_1(t) = (\omega_1 - \omega_2)^{-1}\{&\exp(\omega_1 t)[a_1\omega_1 + \alpha_1 - a_1\xi_2 + v_2 a_2 \\
&+ \omega_1^{-1}(\alpha_2 v_2 - \alpha_1\xi_2)] \\
-\exp(\omega_2 t)[&a_1\omega_2 + \alpha_1 - a_1\xi_2 + v_2 a_2 + \omega_2^{-1}(\alpha_2 v_2 - \alpha_1\xi_2)]\} \\
+ (\omega_1\omega_2)^{-1}&(\alpha_2 v_2 - \alpha_1\xi_2),
\end{aligned} \tag{9.45}
$$

with a similar expression for $m_2(t)$ (interchange subscripts 1 and 2, but leave ω_1 and ω_2 unaltered). Here $\xi_i = \lambda_i - \mu_i - v_i$,

$$
\omega_1, \omega_2 = \tfrac{1}{2}[(\xi_1 + \xi_2) \pm \sqrt{\{(\xi_1 - \xi_2)^2 + 4v_1 v_2\}}], \tag{9.46}
$$

and $a_i = X_i(0)$ denote the initial population sizes.

If we examine the relative proportions of $m_1(t)$ and $m_2(t)$ by defining

$$
\psi(t) = m_1(t)/(m_1(t) + m_2(t)),
$$

then it follows from (9.45) that in the absence of immigration

$$
\psi(t) \to v_2/(\lambda_2 - \mu_2 - \omega_2) \qquad (\omega_1 \neq \omega_2)
$$

as t increases. Thus migration maintains a balance between the two mean population sizes even though they have different intrinsic growth rates $\lambda_i - \mu_i$.

Since $\omega_1 > \omega_2$, and $\exp(\omega_1 t)$ appears both in $m_1(t)$ and $m_2(t)$, ω_1 dominates the behaviour of both colonies. If $\omega_1 > 0$ then $m_1(t)$ and $m_2(t)$ grow without limit, whilst if $\omega_1 < 0$ then they remain bounded – indeed they go to zero if the immigration rates $\alpha_1 = \alpha_2 = 0$. Specifically, in the absence of immigration it follows from (9.46) that if $\xi_1\xi_2 > v_1 v_2$ and $\xi_1, \xi_2 < 0$ then the total mean population size decreases, if $\xi_1\xi_2 = v_1 v_2$ and $\xi_1, \xi_2 < 0$ then it remains constant, otherwise it increases.

For example, consider $\lambda_1 = \lambda_2 = 2$, $\mu_1 = 1$ and $\mu_2 = 3$, so that if $v_1 = v_2 = 0$ then $m_1(t) \to \infty$ and $m_2(t) \to 0$. As the migration rate v_1 increases, colony 1 will continue to have a net positive growth rate ($\xi_1 = \lambda_1 - \mu_1 - v_1 > 0$) until $v_1 = 1$, after which point ξ_1 goes negative. Hence as $\xi_2 = -1 - v_2 < 0$ for all v_2, on writing $v_1 = 1 + \varepsilon$ for $v_1 > 1$ we see that the above condition for the total mean population size to decay to zero reduces to $\xi_1\xi_2 > v_1 v_2$, i.e.

$$
(2 - 1 - (1 + \varepsilon))(2 - 3 - v_2) > (1 + \varepsilon)v_2
$$

i.e.

$$
v_2 < \varepsilon.
$$

So if $v_1 > 1$ (i.e. colony 1 has net growth rate $\lambda_1 - \mu_1 - v_1 < 0$) and $v_2 > \varepsilon$, then migrants from colony 1 to colony 2 return sufficiently fast to ensure that $m_1(t)$, and hence $m_2(t)$, will still grow without limit.

This example shows that if the intrinsic growth rates $\lambda_1 - \mu_1 > 0$ and $\lambda_2 - \mu_2 < 0$, then an overall increase or decrease in mean population size will depend purely on the values of the migration parameters v_1 and v_2. If the v_i are such that the $m_i(t)$ ultimately remain constant we say that we have a state of 'critical migration'. The associated condition $\xi_1 \xi_2 = v_1 v_2$ and $\xi_1, \xi_2 < 0$ reduces to

$$v_1 = (\lambda_1 - \mu_1)[1 - \{v_2/(\lambda_2 - \mu_2)\}] \tag{9.47}$$

(i.e. $v_1 = 1 + v_2$ in our example).

*9.5.2   *Variances and covariances*

Expressions may also be derived for the variances $V_{11}(t) = \text{Var}[X_1(t)]$, $V_{22}(t) = \text{Var}[X_2(t)]$ and covariance $V_{12}(t) = \text{Cov}[X_1(t), X_2(t)]$ (Renshaw, 1973a). Note, however, that these second-order moments reflect availability across *all* possible realizations of the process at the *specific time t*. They do not provide information on the variability *over time* along a *single* realization (which is what is usually observed in practice); simulation provides the key to understanding here.

Since general expressions for the $V_{ij}(t)$ are cumbersome, we shall just quote results for the spatially homogeneous model in which

$$\lambda_1 = \lambda_2 = \lambda, \quad \mu_1 = \mu_2 = \mu, \quad v_1 = v_2 = v \text{ and } \quad \alpha_1 = \alpha_2 = 0, \tag{9.48}$$

i.e. the transition rates are independent of geographic location. In vector notation

$$
\begin{aligned}
(V_{11}(t), &\, V_{12}(t), V_{22}(t)) \\
&= \tfrac{1}{4}(a_1 + a_2)(\lambda + \mu)(\lambda - \mu)^{-1}[\exp\{2(\lambda - \mu)t\} \\
&\quad - \exp\{(\lambda - \mu)t\}](1, 1, 1) \\
&\quad + \tfrac{1}{2}(a_1 - a_2)(\lambda + \mu)(\lambda - \mu)^{-1}[\exp\{2(\lambda - \mu - v)t\} \\
&\quad - \exp\{(\lambda - \mu - 2v)t\}](1, 0, -1) \\
&\quad + \tfrac{1}{4}(a_1 + a_2)(\lambda + \mu + 4v)(\lambda - \mu - 4v)^{-1}[\exp\{2(\lambda - \mu - 2v)t\} \\
&\quad - \exp\{(\lambda - \mu)t\}](1, -1, 1).
\end{aligned}
\tag{9.49}
$$

So for large t the correlation coefficient $\rho(t) = V_{12}(t)/\{V_{11}(t)V_{22}(t)\}^{\frac{1}{2}}$ is given by

$$\rho(t) \sim \begin{cases} 1 & \text{if } \lambda > \mu \\ 4v\lambda/(\mu^2 + 4v\mu - \lambda^2) & \text{if } \lambda < \mu. \end{cases} \tag{9.50}$$

In the first case any positive v causes the two expanding population sizes to develop more or less in step; whilst in the second ρ ranges from 0 at $v = 0$

(two independent colonies) to near λ/μ when v is very large (two closely connected colonies).

9.5.3   *Three special cases*

Although the presence of the birth rate λ_i renders direct mathematical solution of the probabilities $p_{ij}(t) = \Pr(X_1(t) = i, X_2(t) = j)$ intractable under scheme (9.42), information on approximate probabilities can still be extracted. Before doing this, however, it is instructive to see where the mathematical difficulties arise (Renshaw, 1973b, 1986).

Model 1 $(\lambda_1 = \lambda_2 = 0)$
Since birth is excluded the $\{p_{ij}(t)\}$ can be calculated directly. Of particular interest is that in equilibrium colonies 1 and 2 behave as though they are isolated from each other, having immigration rates of $(\alpha_2 v_2 + \alpha_1 \theta_2)$ and $(\alpha_1 v_1 + \alpha_2 \theta_1)$, respectively, and a common death rate of $(\theta_1 \theta_2 - v_1 v_2)$, where $\theta_i = \mu_i + v_i$.

Model 2 $(\lambda_1 = v_2 = 0)$
Colony 1 behaves as a simple immigration–death process with parameters α_1 and $\mu_1 + v_1$, respectively, and migration only takes place from colony 1 to colony 2. In equilibrium the colonies again act as though they are independent; this time colony 1 behaves as a simple immigration–death process with parameters α_1 and $\mu_1 + v_1$, whilst colony 2 behaves as a simple immigration–birth–death process with parameters $\alpha_2 + \alpha_1 v_1/(\mu_1 + v_1)$, λ_2 and μ_2, respectively.

Model 3 $(\lambda_2 = v_2 = 0)$
The assumption of one-way migration ($v_1 > 0$ but $v_2 = 0$) is retained, but now $\lambda_2 = 0$ instead of $\lambda_1 = 0$. Thus colony 1 develops as a simple immigration–birth–death process with death rate $\mu_1 + v_1$, whilst colony 2 develops as an immigration–death process having a time-dependent immigration rate. This process is considerably more difficult to analyse than the previous two, and even the equilibrium solution does not possess a simple structure.

Solution of these three models involves a successively increasing order of difficulty, and even though model 3 is in principle a very simple process, the solution for the probabilities $\{p_{ij}(t)\}$ is sufficiently complex to hide the underlying properties of the process from view. Not only are attempts to develop exact probability solutions for more complex birth–death type processes (e.g. $\lambda_1, \lambda_2 > 0$) therefore futile, but corresponding analyses of potentially more interesting spatial processes, such as epidemic, predator–prey and competition models, will be even more hopeless.

9.5.4   *Approximate probabilities*

Since such mathematical difficulties are associated with birth, and not immigration or death, one way forward is to develop approximate solutions by changing the probability of a birth in colony i in the small time interval $(t, t + dt)$ from $\lambda_i X_i(t)\, dt$ to $\lambda_i m_i(t)\, dt$. As this new probability depends on the mean number, and not on the actual number, of individuals present in colony i at time t, it corresponds to a time-dependent immigration rate. Approximate probabilities $\{\tilde{p}_{ij}(t)\}$ can now be evaluated (Renshaw, 1973a), and provided that $X_i(t)$ and $m_i(t)$ are not too dissimilar the results should be reasonably good.

Unfortunately, the general question of exactly when and where this technique performs well is a difficult one to tackle. Numerical or simulation comparisons provide the only practical means of assessing individual situations. Nevertheless, let us consider one particular case in some detail, and then mention an alternative procedure which does at least yield correct means and variances.

For the spatially homogeneous model (9.48) with initial values $X_1(0) = 1$, $X_2(0) = 0$ the approach yields

$$\tilde{p}_{ij}(t) = [\tfrac{1}{2} e^{-\mu t}(1 + e^{-2\nu t})]^i [\tfrac{1}{2} e^{-\mu t}(1 - e^{-2\nu t})]^j \psi^{i+j-1}(i!\,j!)^{-1}$$
$$\times [\psi(1 - e^{-\mu t}) + i + j]\exp\{-\psi e^{-\mu t}\} \tag{9.51}$$

where $\psi = \exp(\lambda t) - 1$. Unless t is quite small, this simplifies to give

$$\tilde{p}_{ij}(t) \simeq (i!\,j!)^{-1}[\tfrac{1}{2} e^{(\lambda - \mu)t}]^i [1 + (i + j)\, e^{-\lambda t}]\exp\{-e^{(\lambda - \mu)t}\}. \tag{9.52}$$

Thus for fixed total population size $i + j = k$, and large time t, $\tilde{p}_{ij}(t)$ is proportional to $(i!\,j!)^{-1}$, i.e.

$$\tilde{p}_{i,k-i}(t) = \binom{k}{i}(\tfrac{1}{2})^k.$$

So given k, individuals are distributed binomially (with parameter $\tfrac{1}{2}$) between the two colonies, as would be expected since the model is homogeneous.

This technique of replacing a random variable by its mean value is of general applicability and is not restricted to the special case considered here. For example, it also applies to models of virus–antibody interaction, bacterium–phage interaction, and two-sex birth–death processes (Morgan and Hinde, 1976). One way of assessing its accuracy is to examine the probability of extinction $p_{00}(t)$ for the homogeneous model (9.48). If $\lambda_1 = \lambda_2$

and $\mu_1 = \mu_2$ then migration does not affect the probability of extinction, and so it follows directly from (2.53) that with $X_1(0) + X_2(0) = 1$

$$p_{00}(t) = [\mu - \mu \exp\{-(\lambda - \mu)t\}]/[\lambda - \mu \exp\{-(\lambda - \mu)t\}]. \quad (9.53)$$

This exact value may now be compared with the approximation (9.52), namely that for appropriately large t

$$\tilde{p}_{00}(t) \simeq \exp\{-e^{-(\lambda-\mu)t}\}. \quad (9.54)$$

These two forms are clearly quite different in structure. In particular,

$$\begin{array}{lll}
\tilde{p}_{00}(\infty) = 0, & p_{00}(\infty) = \mu/\lambda & (\lambda > \mu) \\
\tilde{p}_{00}(\infty) = e^{-1}, & p_{00}(\infty) = 1 & (\lambda = \mu) \\
\tilde{p}_{00}(\infty) = 1, & p_{00}(\infty) = 1 & (\lambda < \mu).
\end{array} \quad (9.55)$$

This comparison suggests that even in this simple case the approximation approach should be treated rather gingerly, especially if $\lambda \geqslant \mu$. The reason is that although extinction is possible in the original process if $\lambda > \mu$, it is impossible in the approximating one because the birth rates are proportional to the two increasing means. So the technique really does change the nature of the process quite fundamentally.

Indeed, the approximation causes a considerable disturbance to the variance-covariances $V_{ij}(t)$ even though the means $m_i(t)$ are left unaltered. For example, in the homogeneous model with $\lambda > \mu$ all three exact moments $V_{ij}(t)$ are of order $\exp\{2(\lambda - \mu)t\}$, whilst the approximations $\tilde{V}_{11}(t)$ and $\tilde{V}_{22}(t)$ grow much more slowly at rate $\exp\{(\lambda - \mu)t\}$ and $\tilde{V}_{12}(t)$ actually decays to zero at rate $\exp\{-2\mu t\}$. So can a different form of approximation be found which leaves both first- and second-order moments intact?

Now we have already used the negative binomial approximation for the single-colony birth–death process (Section 3.9) since we can choose the parameters to ensure that it yields the correct mean and variance. Thus an obvious way of extending result (3.74) to two colonies is to write

$$p_{ij}(t) \simeq \binom{h_1 + i - 1}{i} p_1^{h_1} q_1^i \times \binom{h_2 + j - 1}{j} p_2^{h_2} q_2^j \quad (9.56)$$

where $p_k = m_k/V_{kk}$, $q_k = 1 - p_k$ and h_k is the nearest integer to $m_k^2/(V_{kk} - m_k)$ for $k = 1, 2$. This assumed independence between the colonies means that the associated covariance term must be zero, so (9.56) should be a good approximation when the migration rates are small relative to the birth and death rates, i.e. when the colonies are 'slightly connected'. A generalization of (9.56) which yields the correct covariance does not seem to be readily available.

*9.5.5   *Conditions for ultimate extinction*

On a theoretical note, exact necessary and sufficient conditions for ultimate extinction to occur *can* be determined, namely that

$$A + B \geqslant C \tag{9.57}$$

where $A = v_1 + \mu_1 - \lambda_1$, $B = v_2 + \mu_2 - \lambda_2$ and $C = \sqrt{\{(A-B)^2 + 4v_1v_2\}}$ (Puri, 1968). Moreover, if $v_2 = 0$ then on starting with $X_1(0) = 1$, $X_2(0) = 0$ the probability of ultimate extinction is given by

$$p_{00}(\infty) = \begin{cases} (\tfrac{1}{2}\lambda_1)[\theta_1 - \sqrt{\{\theta_1^2 - 4\lambda_1(\mu_1 + v_1 q)\}}] & (\lambda_2 > \mu_2) \\ (\mu_1 + v_1)/\lambda_1 & (\lambda_2 \leqslant \mu_2; \lambda_1 > \mu_1 + v_1) \\ 1 & (\lambda_2 \leqslant \mu_2; \lambda_1 \leqslant \mu_1 + v_1) \end{cases} \tag{9.58}$$

where q is the smaller of 1 and μ_2/λ_2, and $\theta_1 = \lambda_1 + \mu_1 + v_1$. Aksland (1975) and Helland (1975) extend these results to the general N-colony situation.

For example, consider the two-colony process with $\lambda_1 = \lambda_2 = 1.0$, $v_1 = v_2 = 0.05$ and different death rates $\mu_1 = 1.1$ and $\mu_2 = 0.9$. Then $A = 0.15$, $B = -0.05$ and $C = \sqrt{(0.05)} = 0.2236$. Thus as $A + B = 0.1 < C = 0.22$, it follows from (9.57) that ultimate extinction is not certain. If $v_1 = 0.5$ and $v_2 = 0$, then since $\lambda_2 = 1.0 > \mu_2 = 0.9$ the first expression in (9.58) applies. So as $q = \min[1, (0.9/1.0)] = 0.9$ and $\theta_1 = 2.15$, we have $p_{00}(\infty) = 0.97$.

9.5.6   *Simulation*

As we have already seen how to simulate the simple birth–death process (Section 2.3.4) and a spatial Lotka–Volterra process (Section 7.2), simulation of the two-colony birth–death–migration process presents no new problems. Let the population sizes be $\{X_1(t), X_2(t)\}$ at time t, and choose two uniform pseudo-random numbers $0 \leqslant Z_1, Z_2 \leqslant 1$. Then on denoting

$$R = (\lambda_1 + \mu_1 + v_1)X_1 + (\lambda_2 + \mu_2 + v_2)X_2,$$

the next event is

$X_1 \to X_1 + 1$	if $0 \leqslant Z_1 \leqslant \lambda_1 X_1/R$	$(= R_1)$	(birth)
$X_2 \to X_2 + 1$	if $R_1 < Z_1 \leqslant R_1 + \lambda_2 X_2/R$	$(= R_2)$	(birth)
$X_1 \to X_1 - 1$	if $R_2 < Z_1 \leqslant R_2 + \mu_1 X_1/R$	$(= R_3)$	(death)
$X_2 \to X_2 - 1$	if $R_3 < Z_1 \leqslant R_3 + \mu_2 X_2/R$	$(= R_4)$	(death)
$X_1 \to X_1 - 1, X_2 \to X_2 + 1$	if $R_4 < Z_1 \leqslant R_4 + v_1 X_1/R$	$(= R_5)$	(migration)
$X_1 \to X_1 + 1, X_2 \to X_2 - 1$	otherwise		(migration),

whilst the time (s) to this event is

$$s = -[\log_e(Z_2)]/R.$$

We now replace t by $t + s$ and repeat the cycle as often as required.

Figures 9.8a and b show two realizations with $\lambda_1 = \lambda_2 = 1$, $\nu_1 = \nu_2 = 0.05$, $\mu_1 = 1.1$, $\mu_2 = 0.9$ and $X_1(0) = 10$, $X_2(0) = 0$ which exhibit markedly different behaviour. In the first, $X_1(t) = 0$ by time $t = 5$, though at time $t = 2.8$ colony 1 starts sending migrants to colony 2 which then subsequently develops in its own right. In turn this colony sends migrants to colony 1 to restart the process there. Surprisingly, colony 1 then experiences a surge in population growth, in spite of the net growth rate $\lambda_1 - \mu_1 - \nu_1 = -0.15$ being negative. A subsequent population crash is closely followed by the collapse of colony 2 and hence extinction, even though $\lambda_2 - \mu_2 - \nu_2 = 0.05$ is positive. In contrast, the second simulation never sees colony 1 empty, and both colonies soon show continued upward growth in line with deterministic prediction.

Figure 9.9 shows the times to extinction for 100 such realizations, under the assumption that if a realization does not become extinct by time $t = 100$ then

Figure 9.8. Two realizations of $\{X_1(t)\}$ (———) and $\{X_2(t)\}$ (---) for a simple two-colony birth–death–migration process with $\lambda_1 = \lambda_2 = 1$, $\nu_1 = \nu_2 = 0.05$ and $\mu_1 = 1.1$, $\mu_2 = 0.9$ showing: (a) extinction, (b) non-extinction.

(a)

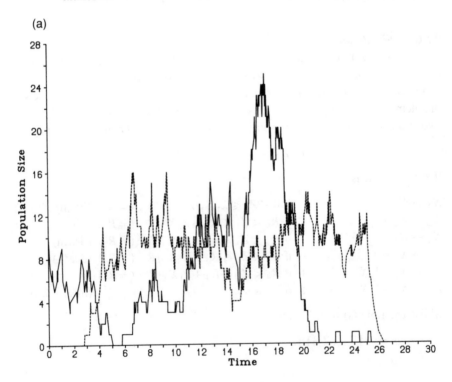

it never will. The distribution is clearly highly skew with the bulk of the extinction times occurring early on before colony 2 has had time to become properly established. As 15 of the 100 realizations reach $t = 100$ this yields the simulated estimate $p_{00}(\infty) \simeq 0.85$.

Since colony 2 dominates colony 1 we can also obtain a theoretical estimate of this probability. Colony 1 ceases to have much importance once colony 2 has become established, so as far as the extinction result (9.58) is concerned we may suppose that colony 2 has death rate $\mu_2 + \nu_2 = 0.95$ and zero migration rate. Hence as $\lambda_2 = 1 > 0.95$ and $\theta_1 = \lambda_1 + \mu_1 + \nu_1 = 2.15$, we take the first line of (9.58) for $X_1(0) = 1$, $X_2(0) = 0$ and obtain

$$p_{00}(\infty) \simeq (\tfrac{1}{2})[2.15 - \sqrt{\{2.15^2 - 4[1.1 + (0.05)(0.95)]\}}] = 0.9849.$$

Thus with $X_1(0) = 10$

$$p_{00}(\infty) \simeq 0.9849^{10} = 0.86,$$

which is in almost exact agreement with our simulated estimate.

The extinction results (9.57) and (9.58) are clearly of considerable practical use, as indeed are the means (9.45). However, the variance-covariances $\{V_{ij}(t)\}$

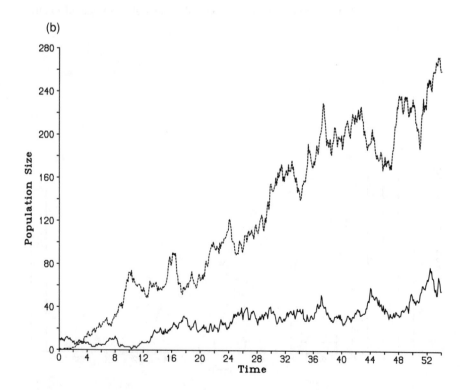

(b)

and population size probabilities $\{p_{ij}(t)\}$ are of less practical interest because they incorporate different types of behaviour pattern. When ultimate extinction is not certain, realizations may or may not become extinct; whilst both extinction and the onset of sustained exponential growth can occur over a wide range of times. The $\{V_{ij}(t)\}$ and $\{p_{ij}(t)\}$ therefore just provide a crude summary of overall behaviour. Only simulation can show what may actually happen within a single realization.

9.6   Stepping-stone models (2)

Interest in the effect of migration between geographic regions is widespread; two examples being the spread of disease between cities in the USSR (Bailey, 1975), and the spread of epidemics around the coastline of Iceland (Cliff, Haggett, Ord and Versey, 1981). The mathematical difficulty we have just experienced with the probability structure of the two-colony birth–death–migration (BDM) process highlights the extent to which solutions to these potentially more complex situations will not only be more difficult to obtain but will also yield less readily to interpretation. Thus

Figure 9.9. Times to extinction for 100 realizations of the model of Figure 9.8.

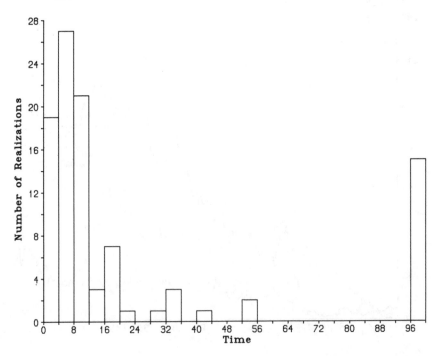

although general genetic/epidemiological/competition/predator–prey, etc., processes are fundamentally of much greater interest than our rather naive BDM-process, for reasons of mathematical tractability we shall have to remain with it for the moment.

Whilst exact results for the means, variances and covariances of population size can be obtained for the N-colony BDM-process in which migration occurs between any pair of colonies (Renshaw, 1972), the solutions are in an opaque vector-matrix form. Some spatial structure must be provided if more transparent solutions are to be obtained, and the simplest approach is to preclude large jumps so that individuals may migrate between nearest-neighbours only.

Suppose there are N colonies arranged in a line, labelled by the integers $i = 1, \ldots, N$ (Figure 9.10). For simplicity we have taken common birth rates (λ) and death rates (μ), with migration rates of v_1 to the right and v_2 to the left. On noting the net inflow and outflow from each colony, we see that the equations for the mean population sizes $\{m_i(t)\}$ are given for $i = 2, \ldots, N - 1$ by

$$\mathrm{d}m_i(t)/\mathrm{d}t = \underset{\text{(net gain in } i)}{(\lambda - \mu - v_1 - v_2)m_i(t)} + \underset{\text{(migr. } i-1 \text{ to } i)}{v_1 m_{i-1}(t)} + \underset{\text{(migr. } i+1 \text{ to } i)}{v_2 m_{i+1}(t)} \qquad (9.59)$$

together with

$$\mathrm{d}m_1(t)/\mathrm{d}t = (\lambda - \mu - v_1)m_1(t) + v_2 m_2(t) \qquad (9.60)$$

and

$$\mathrm{d}m_N(t)/\mathrm{d}t = (\lambda - \mu - v_2)m_N(t) + v_1 m_{N-1}(t). \qquad (9.61)$$

9.6.1   *Special case of $v_2 = 0$*

For ease of solution suppose that $v_1 = v$ and $v_2 = 0$, so that migration can occur only to the right, and assume that the initial population at time $t = 0$ consists of just one individual in colony 1. Then (9.60) solves directly to give

$$m_1(t) = \exp\{(\lambda - \mu - v)t\},$$

whence equation (9.59) with $i = 2$ becomes

$$\mathrm{d}m_2(t)/\mathrm{d}t = (\lambda - \mu - v)m_2(t) + v \exp\{(\lambda - \mu - v)t\}.$$

Figure 9.10. The N-colony stepping-stone model of population growth.

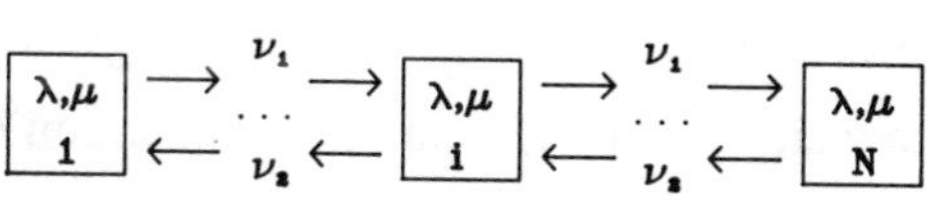

On writing this equation in the form

$$\frac{d}{dt}[m(t)\exp\{-(\lambda - \mu - v)t\}] = v$$

and integrating, we have

$$m_2(t) = \exp\{(\lambda - \mu - v)t\}(vt)$$

since $m_2(0) = 0$. Successively solving (9.59) with $i = 3, 4, \ldots$ then leads to

$$m_i(t) = \exp\{(\lambda - \mu - v)t\}(vt)^{i-1}/(i-1)! \quad (i = 1, \ldots, N-1), \qquad (9.62)$$

i.e. for large N

$$m_i(t) = (\text{total mean pop. size}) \times (\text{Poisson } (v)).$$

If $\lambda > \mu$ then this exponentially growing population will spread itself over an ever-increasing number of colonies. To see how far the population travels in time t let us parallel the diffusion approach of Section 9.3.

Since migration now takes place between discrete sites, rather than through a continuous medium, let us slightly alter our definition of wavefront $R(t)$ by defining $R + 1$ to be where we expect to find exactly one individual at time t, i.e. by

$$m_{R+1}(t) = 1. \qquad (9.63)$$

(Note that virtually any wavefront definition we choose gives rise to the same final result (9.66).) Then when N is large (9.62) gives

$$\exp\{(\lambda - \mu - v)t\}(vt)^R/R! = 1.$$

So on using Stirling's formula, namely

$$R! \sim \sqrt{(2\pi)}\,e^{-R}R^{R+\frac{1}{2}}, \qquad (9.64)$$

and taking logarithms, we have

$$(\lambda - \mu - v)t + R\log_e(vt) \simeq \tfrac{1}{2}\log_e(2\pi) - R + (R + \tfrac{1}{2})\log_e(R)$$

i.e.

$$(\lambda - \mu - v)t - R\log_e(R/vt) + R \simeq \text{constant} + \tfrac{1}{2}\log_e(R). \qquad (9.65)$$

Now suppose that R increases with constant velocity c, so that $R = ct$ (as in (9.29)). Then on dividing both sides of (9.65) by t and letting t become large, we have

$$\mu + v - \lambda = c - c\log_e(c/v). \qquad (9.66)$$

The required (asymptotic) velocity c is therefore a root of equation (9.66).

*9.6.2   *Velocities for $v_2 > 0$*

Analogous results can be obtained for the general stepping-stone model, though the mathematics is easier if the number of colonies is unlimited in both directions, i.e. if there are no barriers to movement. Then

$$m_i(t) = \exp\{(\lambda - \mu - v_1 - v_2)t\}(v_1/v_2)^{i/2}I_i[2t\sqrt{(v_1 v_2)}]$$
$$(-\infty < i < \infty) \quad (9.67)$$

(see Bailey, 1968, and Renshaw, 1974, 1977). The function $I_i(x)$ is called the 'modified Bessel function of the first kind'. Like its running mate $K_v(x)$ (see expression (9.32)), $I_i(x)$ has nice mathematical properties, a lot is known about it, and it is easy to compute numerically (e.g. by using NAG-routines). It frequently turns up in mathematical ecology, and really deserves the popular acclaim accorded to the Poisson and Binomial functions. For the record, $I_i(x)$ is the coefficient of z^i in the expansion of $\exp\{\frac{1}{2}x(z + z^{-1})\}$, namely

$$I_i(x) = (\tfrac{1}{2}x)^i \sum_{r=0}^{\infty} (\tfrac{1}{4}x^2)^r/[(i + r)! r!]. \quad (9.68)$$

The analogous equation to (9.66) for the velocity c is

$$v_1 + v_2 + \mu - \lambda = (c^2 + 4v_1 v_2)^{\frac{1}{2}} - c \log_e\{[c + (c^2 + 4v_1 v_2)^{\frac{1}{2}}]/(2v_1)\} \quad (9.69)$$

(Renshaw, 1977). Suppose $\lambda > \mu$ so that the total mean population size is increasing, and $v_1 > v_2$ (say) so that there is a general drift to the right. Then since the spreading population has a left and a right wavefront, equation (9.69) will have two solutions c_L and c_R, and it may be shown that the left wavefront moves right or left depending on whether

$$\lambda - \mu < (v_1^{\frac{1}{2}} - v_2^{\frac{1}{2}})^2 \quad \text{or} \quad \lambda - \mu > (v_1^{\frac{1}{2}} - v_2^{\frac{1}{2}})^2, \quad (9.70)$$

respectively.

*9.6.3   *Comparison of diffusion and stepping-stone velocities*

To compare the stepping-stone velocities with those derived from the diffusion approach, we first note that the parameters (9.20) take the values

$$a = v_1 - v_2 \quad \text{and} \quad D^2 = v_1 + v_2. \quad (9.71)$$

Thus the (diffusion) population density (9.22) is

$$f(x, t) = \{2\pi(v_1 + v_2)t\}^{-\frac{1}{2}} \exp\{-[x - (v_1 - v_2)t]^2/2(v_1 + v_2)t\}. \quad (9.72)$$

Since the total mean population size at time t is $\exp\{(\lambda - \mu)t\}$, we may define the wavefront to be at $R(t)$ where

$$1 = \exp\{(\lambda - \mu)t\} f(R; t), \tag{9.73}$$

i.e. on taking logarithms, where

$$0 \simeq (\lambda - \mu)t + \text{const.} - \tfrac{1}{2}\log_e(t) - \{[R - (v_1 - v_2)t]^2/2(v_1 + v_2)t\}.$$

On dividing this equation by t, putting $R = ct$, and then letting t become large, we have

$$0 = (\lambda - \mu) - \{[c - (v_1 - v_2)]^2/2(v_1 + v_2)\}$$

which yields the (diffusion) solutions

$$c_{\text{diff}} = (v_1 - v_2) \pm \sqrt{\{2(\lambda - \mu)(v_1 + v_2)\}}. \tag{9.74}$$

If $\lambda > \mu$ and $v_1 > v_2$ then the left wavefront moves right only if

$$(v_1 - v_2) - \sqrt{\{2(\lambda - \mu)(v_1 + v_2)\}} > 0$$

i.e.

$$\lambda - \mu < (v_1 - v_2)^2/2(v_1 + v_2), \tag{9.75}$$

which is different from the stepping-stone result (9.70).

Indeed, since under the stepping-stone model a migrating individual makes jumps of size $+1$ and -1, whilst the diffusion model is based upon infinitesimally small steps of size $\pm \delta x$, we may ask whether the two sets of velocities (9.69) and (9.74) are at all compatible. A little algebra shows that they are, but only if the net growth rate $(\lambda - \mu)$ is small in comparison to the sum of the migration rates $(v_1 + v_2)$. Otherwise, similar migration patterns give rise to substantially different velocities in the two cases.

9.6.4   *Further results*

Results akin to those already constructed for the two-colony process can be developed for variance-covariances, exact probabilities $(\lambda = 0)$ and approximate probabilities $(\lambda > 0)$ (see Renshaw, 1986, for details). Let us just briefly quote two particularly interesting features before moving on.

First, if there are a finite number (N) of colonies, then for large t in the symmetric migration case $(v_1 = v_2 = v)$

$$m_i(t) \sim N^{-1} \exp\{(\lambda - \mu)t\} \tag{9.76}$$

irrespective of the position (s) of the initial individual. Whilst if emigration occurs at rate v from the two end colonies $(i = 1, i = N)$, then

$$\begin{aligned} m_i(t) \sim\ & [2/(N + 1)] \sin[i\pi/(N + 1)] \sin[s\pi/(N + 1)] \\ & \times \exp\{(\lambda - \mu - 2v)t\} \exp\{2vt \cos[\pi/(N + 1)]\} \end{aligned} \tag{9.77}$$

(Renshaw, 1972). So the boundary conditions at the two end colonies play a major role in determining the structure of the means $\{m_i(t)\}$. In (9.76) the means are all asymptotically equal, whilst (9.77) shows that inclusion of end emigration causes the $m_i(t)$ to vary almost sinusoidally, both with i and s.

Second, considerable progress has recently been made in the probabilistic analysis of spatial genetic structures. Though such models are outside the scope of this text, it is worth noting that they are directly analogous to spatial population processes. Shiga (1985), for example, allows each colony to comprise a population of $2N$ genes at a single locus. Evolution through time is assumed to arise both by migration between colonies, and by mutation, selection, and random sampling drift within each colony. For one particular process the stepping-stone model of population dynamics appears as a dual process of the stepping-stone model of population genetics, with the simple transition rates (9.36) being given by

$$\alpha_i(\mathbf{n}) = \lambda n_i, \quad \beta_i(\mathbf{n}) = \mu n_i + \beta n_i(n_i - 1) \quad \text{and} \quad \gamma_{ij}(\mathbf{n}) = g_{i-j} n_i.$$

This defines a linear birth–migration model with a severe, non-linear, mortality component. If $\mu = 0$ then the population can never become extinct, and the equilibrium distribution is conjectured to be a product of individual Poisson distributions, namely

$$\pi(\mathbf{n}) = \prod_i \exp\{-\lambda/\beta\}(\lambda/\beta)^{n_i}/(n_i!). \tag{9.78}$$

If $\mu > 0$ then there exists a critical parameter λ_c such that if $\lambda < \lambda_c$ then the whole population is bound to become extinct, whilst if $\lambda > \lambda_c$ then the population size may explode as $t \to \infty$.

9.7   An application to the spread of *Tribolium confusum*

An excellent example of how the results of this chapter can be applied is provided by an analysis of data obtained by Neyman, Park and Scott (1956) in an experiment to determine the spatial distribution of adult flour beetles. A $10 \times 10 \times 10$-inch cubic container was filled with fresh flour, and on the surface of the flour 2257 adult *Tribolium confusum* beetles were placed in a one-to-one sex ratio. The container was kept in a dark incubator at an approximately constant temperature of 29 °C and humidity of 70%. In order to equalize possible gradients of temperature within the incubator, the container was periodically rotated. After four months, at which time the population totalled 73 009 individuals excluding eggs, the contents of each of the 1000 individual $1 \times 1 \times 1$-inch cubes were lifted and examined.

Table 9.1 represents the second layer from the top of the experimental cube and shows the distribution of adult males and females within the 100 constituent small cubes. The general character of this distribution in other

Table 9.1. Densities of adult flour beetles in the second layer of the cubic container.
The upper and lower figures in each cell refer to the density of females and males,
respectively (taken from Neyman, Park and Scott, 1956).

18.8	22.7	16.1	12.4	12.8	13.8	11.0	14.4	11.8	16.7
28.7	24.2	14.2	12.0	11.1	15.3	14.1	18.1	20.2	23.0
10.1	11.2	9.6	7.9	7.3	7.3	5.1	4.3	6.9	11.9
14.8	6.9	3.8	3.1	5.3	3.5	3.2	4.9	8.0	16.0
7.9	6.5	5.7	4.4	3.5	3.6	4.4	5.9	8.4	10.4
13.3	4.6	2.0	1.7	1.0	1.9	1.9	1.9	3.5	9.8
9.9	6.9	5.4	5.8	3.2	2.9	3.1	5.2	7.0	8.7
9.5	4.5	1.5	0.7	1.3	1.3	0.2	1.0	1.7	7.7
10.9	6.6	3.0	2.6	4.0	4.2	3.9	3.5	5.8	17.8
12.5	4.6	4.6	2.2	2.0	2.5	3.9	1.6	2.2	12.2
6.0	7.8	7.2	2.4	1.0	3.5	4.0	4.0	6.2	16.5
8.0	4.5	2.4	1.8	1.3	3.3	1.8	2.4	3.5	19.4
10.5	6.5	1.2	5.7	4.7	4.2	5.4	3.7	9.7	14.2
10.6	3.5	5.1	2.0	0.4	1.9	2.3	0.9	3.0	12.2
14.9	10.8	4.9	3.9	4.0	3.5	5.8	4.3	7.3	13.3
15.8	5.9	2.8	2.7	0.8	1.4	2.8	1.7	3.8	11.8
18.3	10.4	8.5	5.3	5.6	4.8	6.4	8.3	8.2	12.1
26.6	8.9	3.4	4.6	2.7	3.5	2.5	2.4	4.9	18.6
19.1	12.6	15.8	14.7	12.8	11.4	12.9	12.3	11.5	17.1
35.2	14.1	17.0	17.2	21.8	13.9	14.9	12.2	14.1	34.5

layers is very similar, though the average density of beetles shows a marked
decrease from the top of the container downwards. The distributions for
males and females are different in that the density for the females falls steadily
towards the centre of the square from a relatively low maximum at the
corners, whilst the density for the males falls more sharply from a high
maximum at the corners and is substantially constant in the central portion.
However, both sexes show a gradual increase in density towards the edges and
along the edges towards the corners.

9.7.1   *Possible models*

A number of random-walk models have been tried in an effort to
obtain a distribution which possesses these characteristics. Sherman showed
that a bounded random walk over a square lattice could produce a
concentration of beetles at the edges and in the corners, but could not
account for the gradual increase in density towards the edges. Sherman
(1956) investigated the corresponding one-dimensional random walk and

showed that distributions more like the one observed could be obtained by using a rather strange boundary condition, in which after striking the boundary a beetle remains there a randomly distributed time and is then instantaneously placed a finite distance within the region of motion. Cox and Smith (1957) take a completely different approach to obtain a distribution of the type observed for motion within a circular region by assuming: (i) within the region a beetle follows a straight path; (ii) at the boundary with probability p it returns along its original path; or (iii) with probability $1 - p$ it moves along the boundary before choosing a new direction of motion. They tried to extend their results to the case of the square boundary, and though unsuccessful theoretically they were able to obtain an approximate empirical result.

9.7.2   *A stepping-stone approach*

Now whilst it is extremely unlikely that a beetle will follow straight paths within the flour, the Brownian motion models may be criticized on the grounds that they imply that the path of a beetle is completely irregular. Renshaw (1980) develops a simple stepping-stone model which not only produces a spatial pattern closely resembling the observed data, but also balances these two extremes of complete irregularity and straight-line paths. He assumes, with biological justification, that: beetles migrate over a set of lattice points $i, j = 0, \ldots, n$; they may change their direction of motion at each point; and that the net growth rates are higher on the sides and in the corners than in the interior. In this two-dimensional case an interior position may be reached from one of four nearest-neighbours, whilst side and corner positions may be reached only from one of three and two nearest-neighbours, respectively. Thus the sides, and especially the corners, may well provide a protected environment. Moreover, whilst it is known that for several organisms jostling between population members reduces fecundity (notably *Drosophila melanogaster*; Pearl, 1932), there is also indirect evidence (based on reports of Boyce, 1946, and Rich, 1956) that the loss in the number of eggs hatched (fertility) increases as the number of contacts increases. Thus if we assume that the reduction in fertility caused by increased local population size is counterbalanced by the increased chance of two beetles meeting and mating, then we may also reasonably assert that the net growth rates in the model will remain approximately constant over time.

Let the birth and death rates be λ and μ in the interior, λ' and μ' on the sides, and λ'' and μ'' in the corners. Denote the migration rate between nearest-neighbours by v, and let $m_{ij}(t)$ ($i, j = 0, \ldots, n$; $t = 0, 1, \ldots$) represent the number of beetles in colony (i, j) at the discrete time point t. Then on studying the net inflow and outflow of beetles between adjacent squares (illustrated in Figure 9.11), we have the deterministic equations

$$m_{ij}(t+1) = \begin{cases} (1 + \lambda - \mu - 4v)m_{ij}(t) + vH_{ij}(t) & \text{(interior)} \\ (1 + \lambda' - \mu' - 3v)m_{ij}(t) + vH_{ij}(t) & \text{(sides)} \\ (1 + \lambda'' - \mu'' - 2v)m_{ij}(t) + vH_{ij}(t) & \text{(corners)} \end{cases} \quad (9.79)$$

where

$$H_{ij}(t) = m_{i+1,j}(t) + m_{i-1,j}(t) + m_{i,j+1}(t) + m_{i,j-1}(t).$$

We put $m_{ij}(t) = 0$ for any points (i,j) lying outside the square region $i, j = 0, \ldots, n$.

Suppose we assume that the $\{m_{ij}(t)\}$ eventually all grow at the same rate ω, so that for large t

$$m_{ij}(t+1) = \omega m_{ij}(t). \quad (9.80)$$

Then the interior solution (Renshaw, 1980) has the form

$$m_{ij} = K \cosh[\tfrac{1}{2}(n - 2i)\theta] \cosh[\tfrac{1}{2}(n - 2j)\theta] \quad (9.81)$$

for some scaling constant K, where

$$\cosh(\theta) = (\omega - 1 - \lambda + \mu + 4v)/4v \quad (9.82)$$

denotes $\tfrac{1}{2}(e^\theta + e^{-\theta})$. The side and corner equations (9.79) are equivalent to the conditions

$$(\lambda'' - \mu'') - (\lambda - \mu) = 2[(\lambda' - \mu') - (\lambda - \mu)]$$
$$= 2v\{2\cosh(\theta) - 1 - \cosh[\tfrac{1}{2}(n-2)\theta]/\cosh(\tfrac{1}{2}n\theta)\}, \quad (9.83)$$

i.e. the excess growth activity at the corners must be twice that at the sides.

Figure 9.11. Parameter structure of the stepping-stone model for Neyman, Park and Scott's (1956) flour beetle experiment.

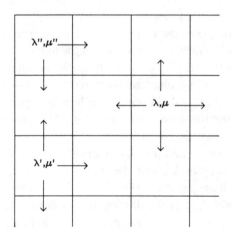

To fit this model to Neyman, Park and Scott's data we first note that the initial population of 2257 adult beetles grew into 73 009 individuals excluding eggs, and we shall assume that all of these were alive at the time of counting. Now the sums of the densities in the two 10×10 matrices (Table 9.1) are 838.4 (females) and 782.5 (males). So if we assume that the corresponding values for the other nine horizontal layers are in the same ratio, then we have the following matrix of population numbers:

	Female	Male
Day 1	1 128.5	1 128.5
Day 120	37 763	35 246

Taking the four-month period of the experiment to be 120 days, the daily rates of increase may therefore be estimated by

$$\omega = (37\ 763/1128.5)^{1/120} = 1.029\ 68 \quad \text{(females)}$$
$$\omega = (35\ 246/1128.5)^{1/120} = 1.029\ 09 \quad \text{(males)}.$$

Since the variance of both the male and female densities increases substantially as we move away from the centres of the two 10×10 data matrices, a transformation of the data is clearly required if we are to avoid over-emphasizing the boundary values. A square-root transformation provides a satisfactory balance between interior, edge and corner densities (Renshaw, 1980), whence solution (9.81) leads to the least-squares estimates

$$\hat{\theta}(\text{female}) = 0.34 \quad \text{and} \quad \hat{\theta}(\text{male}) = 0.51$$

for θ. These values highlight the striking difference in spatial behaviour between the two sexes.

Substituting the above estimates for θ and ω into equations (9.82) and (9.83) with $n = 9$ yields

$$\lambda - \mu = 0.0297 - 0.2334\nu \qquad \lambda - \mu = 0.0291 - 0.5316\nu$$
$$\lambda' - \mu' = 0.0297 + 0.1405\nu \qquad \lambda' - \mu' = 0.0291 + 0.1230\nu$$
$$\lambda'' - \mu'' = 0.0297 + 0.5144\nu \qquad \lambda'' - \mu'' = 0.0291 + 0.7776\nu$$

for males and females, respectively. So unless ν is very small we require the net growth rate at the corners to be considerably larger than that at either the edges or the interior; the restriction $\lambda - \mu > 0$ clearly limits the maximum permissible sizes of both $\lambda' - \mu'$ and $\lambda'' - \mu''$. If further (independent) information were available on ν, then the above relations would yield explicit values for the growth rates themselves.

9.7.3 *Discussion*

One possible criticism of this approach is that we are fitting a two-dimensional model to a three-dimensional distribution, for beetles in interior cells may migrate to one of six nearest-neighbours and not just four. However, a parallel argument to the above yields the three-dimensional solution

$$m_{ijk} = L \cosh[\tfrac{1}{2}(n - 2i)\theta] \cosh[\tfrac{1}{2}(n - 2j)\theta] \cosh[\tfrac{1}{2}(n - 2k)\theta] \quad (9.84)$$

where $i, j, k = 0, \ldots, n$ and L is an appropriate scaling constant. In particular, when $k = 1$ (corresponding to the second layer) solution (9.84) is identical to (9.81) since $L \cosh[\tfrac{1}{2}(n - 2)\theta]$ is independent of both i and j. Thus the data for the second layer may indeed be analyzed as though it were obtained from a two-dimensional experiment.

Another point of possible contention is that since the data relate to 10×10 one-inch squares we have automatically chosen $i, j = 0, \ldots, 9$ even though the true distance between neighbouring lattice points, assuming of course that the model holds, will most likely be considerably less than one inch. Fortunately, it turns out (Renshaw, 1980) that dividing each one-inch square into, say, $h \times h$ sub-squares, and allowing migration to occur only between adjacent sub-squares, does not affect the total mean number in each one-inch square. Thus the distribution (9.81) is independent of the chosen scale of migration. Note that this equivalence is a purely deterministic one and it does not follow that the corresponding stochastic motions are compatible.

One important consequence of this equivalence is that it holds true even when $h \to 0$ and the process turns into Brownian motion. So in the steady state the stepping-stone and Brownian motion models yield the same mean population sizes.

The associated one-dimensional equations to (9.79) were, in effect, developed by Usher and Williamson (1970), though they were based on different modelling assumptions. At each time point t a proportion p of the beetles migrate to a nearest-neighbour and the remainder stay where they are, the birth and death rates of the 'movers' and 'stayers' being λ_m, μ_m and λ_s, μ_s, respectively. So the mean population sizes $\{m_i(t)\}$ for $i = 1, \ldots, n - 1$ are given by the equations

$$m_i(t + 1) = \underbrace{(1 - p)(1 + \lambda_s - \mu_s)m_i(t)}_{\text{(stayers)}}$$

$$+ \underbrace{(\tfrac{1}{2}p)(1 + \lambda_m - \mu_m)(m_{i-1}(t) + m_{i+1}(t))}_{\text{(movers)}}, \quad (9.85a)$$

whilst at the boundaries

$$m_0(t + 1) = (1 - \tfrac{1}{2}p)(1 + \lambda_s - \mu_s)m_0(t)$$
$$+ (\tfrac{1}{2}p)(1 + \lambda_m - \mu_m)m_1(t)$$
$$m_n(t + 1) = (1 - \tfrac{1}{2}p)(1 + \lambda_s - \mu_s)m_n(t)$$
$$+ (\tfrac{1}{2}p)(1 + \lambda_m - \mu_m)m_{n-1}(t). \quad (9.85b)$$

To compare these equations with the equivalent ones for Renshaw's model, namely

$$m_0(t+1) = (1 + \lambda' - \mu' - \nu)m_0(t) + \nu m_1(t) \qquad \text{(end)}$$
$$m_i(t+1) = (1 + \lambda - \mu - 2\nu)m_i(t) + \nu m_{i-1}(t) + \nu m_{i+1}(t) \quad \text{(interior)} \quad (9.86)$$
$$m_n(t+1) = (1 + \lambda' - \mu' - \nu)m_n(t) + \nu m_{n-1}(t) \qquad \text{(end)},$$

we equate coefficients of $m_i(t)$ and obtain

$$(1-p)(1 + \lambda_s - \mu_s) = (1 + \lambda - \mu - 2\nu) \quad \text{(interior)}$$
$$(1 - \tfrac{1}{2}p)(1 + \lambda_s - \mu_s) = (1 + \lambda' - \mu' - \nu) \quad \text{(ends)} \qquad (9.87)$$
$$(\tfrac{1}{2}p)(1 + \lambda_m - \mu_m) = \nu \qquad \text{(all)}.$$

Equations (9.85) and (9.86) are therefore identical provided that

$$(\lambda - \mu) = (1 - p)(\lambda_s - \mu_s) + p(\lambda_m - \mu_m)$$

and

$$(\lambda' - \mu') = (1 - \tfrac{1}{2}p)(\lambda_s - \mu_s) + (\tfrac{1}{2}p)(\lambda_m - \mu_m). \qquad (9.88)$$

Thus at least two plausible models (and doubtless many more) give rise to equations of the same form. So it is not possible to infer model structure from these *Tribolium* population data without further biological information.

9.8   Simulation of the diffusion and stepping-stone processes

The equilibrium 'cosh' solution (9.81) for the mean proportion of individuals $m_{ij}(t)$ in colony (i, j) is quite appealing in its simplicity, especially since this result may be derived from both the stepping-stone and diffusion models. However, this deterministic equivalence does not carry over to stochastic behaviour. We shall therefore now simulate both types of model, using comparable parameter values so that direct comparison can be made between them. To avoid unnecessary complexity we shall just consider one-dimensional processes.

9.8.1   *The diffusion process*

Suppose we take the step and time increments to be $\delta x = 0.1$ and $\delta t = 0.01$, respectively, with $p = q = \frac{1}{2}$ (zero drift). Then we see from (9.20) that the variance parameter

$$D^2 = \sigma^2(1) = 4pq(\delta x^2)/\delta t = 1.$$

Let the individual birth rate $\lambda = 0.2$, and for convenience let the death rate $\mu = 0$.

Since the number of available sites can be very large, it makes computational sense to discard the usual procedure of recording colony sizes, and storing, instead, the position $\{X(k)\}$ of each individual (labelled by $k = 1, \ldots, M$). Let $\{Z_i\}$ be a sequence of uniform pseudo-random numbers on the interval $(0, 1)$, and suppose that there is just one individual at time $t = 0$, sited at $x = 0$. That is we write $X(1) = 0$. Then at time $t = \delta t$ we

(i) have a birth if $Z_1 \leqslant \lambda\delta t = 0.002$, in which case we put $X(1) = X(2)$ $= 0$, otherwise

(ii) let $X(1) = -\delta x$ or $+\delta x$ according as $Z_2 \leqslant \frac{1}{2}$ or $Z_2 > \frac{1}{2}$, respectively.

As time develops we therefore have the computational procedure:

set population sizes MNEW $= M = 1$

10 cycle over times $t = \delta t, 2\delta t, 3\delta t, \ldots$

20 cycle over individuals $I = 1, \ldots, M$

if $Z_i \leqslant \lambda\delta t$ write $X(\mathrm{MNEW} + 1) = X(I)$ and put MNEW $= \mathrm{MNEW} + 1$

if not, then $X(I) \rightarrow X(I) - \delta x$ or $X(I) \rightarrow X(I) + \delta x$ according as $Z_{i+1} \leqslant \frac{1}{2}$ or $Z_{i+1} > \frac{1}{2}$

Figure 9.12. Simulated diffusion process with $\lambda = 0.2$, $\mu = 0$, $\delta x = 0.1$ and $\delta t = 0.01$, shown at times $t = 0, 20, 40$ (———) and $t = 10, 30$ (– – –).

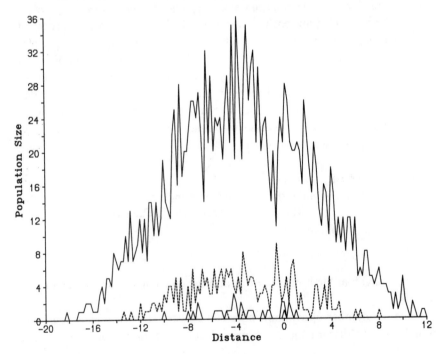

20 continue
    set M = MNEW
10 continue.

Figure 9.12 shows a simulated realization of this diffusion process at times $t = 0, 10, \ldots, 40$. Since an individual's mean time between births is $1/\lambda = 5$, the process can wander quite considerably before becoming spatially established. For example, we see from (9.19) that the position variance $\sigma^2(t) = t$, so after say 10 time units the 95% confidence interval for the position of the initial individual is $\pm 1.96\sqrt{(10)}$, i.e. ± 6.2. Our simulation shows two individuals present at $t = 10$, sited at $x = -3.4$ and -4.8, so the centre of the population distribution has already moved away from zero. By $t = 20$ the population is still spread fairly thinly, lying over the range -10.0 to 1.0. However, exponential growth now takes hold and by $t = 40$ the population is well established, though the density varies considerably between neighbouring positions.

Here the distance increment δx takes the fairly large value 0.1. If δx is much smaller, say 0.001, then with $D^2 = \delta x^2/\delta t = 1$ we require $\delta t = 10^{-6}$ which is hardly computationally efficient! A better approach is to simulate the distance x_i individual i travels between giving birth, together with the associated inter-birth time t_i. By simulating successive values of (x_i, t_i) for $i = 1, 2, \ldots$ we can then construct the spatial-temporal development of the family which issues from the original individual (Figure 9.13).

9.8.2 *The stepping-stone process*

The stepping-stone process $\{X_i(t)\}$ can be simulated by using the Lotka–Volterra approach of Section 7.2.1. We simply replace the two-species

Figure 9.13. Family development of a single individual showing inter-birth times (t_i) and distances (x_i), with $i(j)$ denoting parent (offspring).

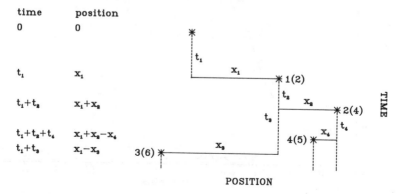

transition probabilities (7.4) by the one-species probabilities

(i) $\Pr[X_i(t + h) = X_i(t) + 1] = \lambda X_i(t)h$ (birth in colony i)

(ii) $\Pr[X_i(t + h) = X_i(t) - 1$ and $X_{i-1}(t + h) = X_{i-1}(t) + 1]$
$= v_2 X_i(t)h$ (migration from i to $i - 1$)

(iii) $\Pr[X_i(t + h) = X_i(t) - 1$ and $X_{i+1}(t + h) = X_{i+1}(t) + 1]$
$= v_1 X_i(t)h$ (migration from i to $i + 1$). (9.89)

To ensure compatibility with our diffusion example we need $\sigma^2(1) = v_1 + v_2$
$= 1$, i.e. $v_1 = v_2 = \frac{1}{2}$.

Figure 9.14 shows such a realization. Though the overall shape is similar to the diffusion realization (Figure 9.12), it is far smoother. This is to be expected, since after an even number of steps the diffusion process can occupy the positions..., $-2\delta x, 0, 2\delta x, 4\delta x, ...,$ and so when $\delta x = 0.1$ five neighbouring diffusion sites relate to just one stepping-stone colony.

9.8.3   *Comparison of velocities*

Both the stepping-stone and diffusion processes may send out

Figure 9.14. Simulated stepping-stone process with $\lambda = 0.2$, $\mu = 0$, $v_1 = v_2 = 0.5$ shown at times as for Figure 9.12.

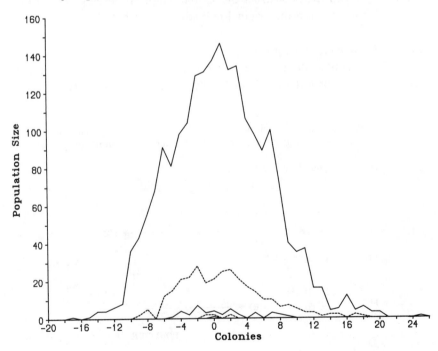

forerunners a short distance ahead of the main population wave, the effect of which is to complicate our concept of wavefront $R(t)$ and thereby velocity c. Indeed, $R(t)$ and c are now random variables since they will vary between realizations. If the stochastic wavefront is defined to be the position of the furthest right individual at time t, then one way of excluding forerunners is to use the position of say the fifth-furthest individual instead. However, for velocity purposes the difference between wavefront definitions disappears as t becomes large, since any forerunners will generally remain close to the pursuing pack.

We see from (9.69) that when $\lambda = 0.2$, $\mu = 0$ and $v_1 = v_2 = 0.5$, the (deterministic) stepping-stone velocity c is the solution of the equation

$$0.8 = (c^2 + 1)^{\frac{1}{2}} - c \log_e\{c + (c^2 + 1)^{\frac{1}{2}}\}.$$

Using standard numerical techniques (the quickest being trial-and-error on a programmable calculator!) gives

$$c = 0.6424,$$

whilst from (9.74) the (deterministic) diffusion velocity

$$c_{\text{diff}} = \sqrt{\{2(\lambda - \mu)(v_1 + v_2)\}} = \sqrt{(0.4)} = 0.6325.$$

This value is only marginally lower than c, so the net growth rate $\lambda - \mu = 0.2$ is small enough in relation to $v_1 + v_2 = 1$ for these two velocities to be compatible (see the comment after (9.75)).

As $t = 40$ is too early for either process to have become properly established, any comparison between simulated and theoretical velocities must be crude. Nevertheless, on noting that

$$\text{population spread} \simeq 2 \times \text{velocity} \times \text{time},$$

we see that between times $t = 20$ and 40 our two stochastic realizations give:

diffusion   – spread rises from 11 to 30, so
        velocity $\simeq (30 - 11)/(2 \times 20) = 0.48$;
stepping-stone – spread rises from 14 to 42, so
        velocity $\simeq (42 - 14)/(2 \times 20) = 0.70$.

The second estimate, in particular, is quite close to the deterministic value $c = 0.64$.

9.9   Spatial predator–prey processes revisited

The diffusion and stepping-stone simulations shown in Figures 9.12 and 9.14 are spatial variants of the simple birth–death process in which individuals develop independently of each other, regardless of the total number present. In reality, birth, death, and even migration rates are density-dependent. We might therefore wish to replace the birth–death component by a logistic process, or prevent the migration of an individual to a site already occupied by a member of the same species. These, and a whole host of other possibilities, are easy to simulate since they involve only minor modifications to the basic program which simulates the diffusion and stepping-stone structures.

The study of spatial interaction between two or more species has recently received considerable attention in the ecological literature. Dubois (1975), for example, proposed an extension of the Lotka–Volterra predator–prey model in the context of marine populations. He considered spatial diffusion of both phytoplankton (prey) and herbivorous zooplankton (predator), and a stochastic simulation of his model by Dubois and Monfort (1978) showed that it led to the spontaneous emergence of very strong spatial heterogeneities similar to those observed in the sea (see Levin, 1978, for further references to this work). To date diffusion principles underlie the great majority of scenarios for population spread, and a wide variety of these are described in an absorbing review by Okubo (1980).

Before we show how two-species diffusion processes operate let us first return to the stepping-stone model. Whilst a diffusion approach may be adequate for describing predator–prey interactions over a continuous region, such as the sea or a large homogeneous area of land, in other situations potential colonization sites may have well-defined local boundaries. Examples include a group of oceanic islands, and ponds interconnected by small waterways, and MacArthur and Wilson (1967) provide a fascinating biogeographical account of the importance of stepping-stone islands to the success and rate of spread of population dispersal. A major question to ask is: to what extent does the behaviour of a spatial process change when the diffusion and stepping-stone mechanisms of migration are interchanged? To a large extent the answer is unknown.

9.9.1   *The spatial Volterra model*

Whilst the general deterministic equations (7.3) for the Lotka–Volterra stepping-stone model are displayed in the context of Huffaker's spatial orange–mite experiment, this present chapter examines spatial spread in the context of wavefronts and velocities. To combine these two approaches let us retain the stepping-stone structure and replace (7.3) by the nearest-neighbour equations

$$dM_i/dt = M_i(\lambda - \alpha N_i) - M_i(u_1 + u_2) + (M_{i-1}u_1 + M_{i+1}u_2)$$
$$dN_i/dt = N_i(-\mu + \beta M_i) - N_i(v_1 + v_2) + (N_{i-1}v_1 + N_{i+1}v_2),$$

$$(9.90)$$

where the migration rates from i to $i + 1$, $i - 1$ are u_1, u_2 for prey and v_1, v_2 for predators.

Suppose we wish to investigate the effect of introducing predators into a previously predator-free environment. Then since the number of prey at any one site will be limited by local resources, we need to introduce a prey carrying capacity K. So replace λM_i in equations (9.90) by

$$\lambda M_i(1 - M_i/K). \tag{9.91}$$

Let predators be introduced into a single site at time $t = 0$, say colony 0. Then not only will a predator–prey cycle commence within this colony, but as the system of colonies is spatially connected the predators will spread out in a stochastic wave.

As equations (9.90) and (9.91) apply over all colonies $i = \ldots, -1, 0, 1, \ldots$ and are therefore difficult to handle, let us confine our attention to a small region around the wavefront. Then since the development at the wavefront is only affected by individuals close to it, we might anticipate that the (deterministic) waveprofile there will assume a constant form once the process has had sufficient time to become established.

This is easily demonstrated for the one-species birth–death–migration process with $v_1 = v$ and $v_2 = 0$, for which result (9.62) gives

$$m_i(t) = \exp\{(\lambda - \mu - v)t\}(vt)^i/(i!)$$

over $i = 0, 1, 2, \ldots$. With large t and velocity of propagation c, the deterministic wavefront is situated at ct. Hence

$$m_{ct}(t) = \exp\{(\lambda - \mu - v)t\}(vt)^{ct}/(ct)! = 1,$$

whilst at a distance s from the wavefront

$$\begin{aligned} m_{s+ct}(t) &= \exp\{(\lambda - \mu - v)t\}(vt)^{s+ct}/(s + ct)! \\ &= [(ct)!/(vt)^{ct}] \times [(vt)^{s+ct}/(s + ct)!]. \end{aligned}$$

On using Stirling's formula (9.64), this expression reduces to

$$m_{s+ct}(t) \simeq (ev/c)^s. \tag{9.92}$$

Now it follows from (9.66) that $c > ve$ if $\lambda - \mu - v > 0$, i.e. if the left-most colony (0) has a positive net growth rate. So result (9.92) shows that in this situation the waveprofile decreases geometrically at rate ev/c as s increases.

Corresponding expressions to (9.62) for $M_i(t)$ and $N_i(t)$ are not available, so Renshaw (1982) develops a 'travelling-wave' approach by first writing

$$x(s) = M_{s+ct}(t); \quad y(s) = N_{s+ct}(t) \tag{9.93}$$

and then looking for solutions of the form

$$x(s) = a_0 + a_1 e^{-s} + a_2 e^{-2s} + \cdots; \quad y(s) = b_0 + b_1 e^{-s} + b_2 e^{-2s} + \cdots$$

$$(9.94)$$

for suitable constants $\{a_r\}$ and $\{b_r\}$. Substituting for $x(s)$ and $y(s)$ from (9.93) and (9.94) into equations (9.90) and (9.91) leads to a method in which the velocity c can be found together with successive values of (a_0, b_0), (a_1, b_1), ... Far ahead of the wavefront there are no predators and K prey (K is the prey carrying capacity), so

$$a_0 = K \quad \text{and} \quad b_0 = 0;$$

whilst it may be shown that

$$a_1 = -\alpha K/[c\theta + \lambda + u_1(1 - e^{\theta}) + u_2(1 - e^{-\theta})] \quad \text{and} \quad b_1 = 1.$$

Since numerical studies suggest that the sequence (a_1, b_1), (a_2, b_2), ... converges *very* fast to zero, the first-order approximation

$$x(s) \simeq K + a_1 e^{-\theta s} \quad \text{and} \quad y(s) \simeq e^{-\theta s} \tag{9.95}$$

is sufficient for most practical purposes. Here the parameter θ is given by

$$e^{\theta} = (1/2v_1)[c + (c^2 + 4v_1 v_2)^{\frac{1}{2}}], \tag{9.96}$$

where the velocity c is the solution of the equation

$$v_1 + v_2 + \mu - \beta K = (c^2 + 4v_1 v_2)^{\frac{1}{2}} - c \log_e\{[c + (c^2 + 4v_1 v_2)^{\frac{1}{2}}]/(2v_1)\}. \tag{9.97}$$

Note that this equation just involves the predator parameters; prey have influence only through the carrying capacity K. This is to be expected, since near the wavefront there will be very few predators and even these will have had little time in which to deplete the prey population. So in the immediate vicinity of the wavefront the overall birth rate of predators will be very close to βK. It is therefore not surprising to discover that result (9.97) is identical to expression (9.69) for the velocity of propagation of a simple birth–death–migration process with birth rate βK, death rate μ and migration rates v_1, v_2.

Behind the wavefront predators and prey interact to form a system of linked stochastic cycles around the equilibrium prey and predator values M_i^* and N_i^*. For algebraic simplicity suppose that the carrying capacity K is large in comparison with the equilibrium prey values, so that $1/K$ can be replaced by zero. Then placing $dM_i/dt = dN_i/dt = 0$ in equations (9.90) gives

$$M_i^* = \mu/\beta \quad \text{and} \quad N_i^* = \lambda/\alpha. \tag{9.98}$$

To investigate small departures from equilibrium we may parallel the non-spatial approach of Chapter 6 by writing

$$M_i(t) = M_i^* + h_i(t) \quad \text{and} \quad N_i(t) = N_i^* + p_i(t). \tag{9.99}$$

On assuming that cycles are of sufficiently small amplitude for us to disregard the product $h_i p_i$, equations (9.90) then reduce to

$$\begin{aligned}
\mathrm{d}h_i/\mathrm{d}t &= -(\alpha\mu/\beta)p_i - u_1(h_i - h_{i-1}) - u_2(h_i - h_{i+1}) \\
\mathrm{d}p_i/\mathrm{d}t &= (\beta\lambda/\alpha)h_i - v_1(p_i - p_{i-1}) - v_2(p_i - p_{i+1}).
\end{aligned} \tag{9.100}$$

The general solution to these equations is given in Renshaw (1982) and involves complicated power-series expansions. However, for some parameter values this solution may be considerably simplified, whilst for others it yields to approximation techniques.

For example, suppose that at time $t = 0$ all colonies are in equilibrium except for colony 0 which is perturbed by a sudden influx of predators, so that $h_i(0) = p_i(0) = 0$ $(i \neq 0)$ and $h_0(0) = 0$, $p_0(0) = p$. Then for the rather unusual case in which predators and prey have the same migration rates $v_1 = u_1$ and $v_2 = u_2$, the general solution simplifies for large t and relatively small i to

$$\begin{aligned}
h_i(t) &\simeq -(p\alpha/2\beta)\{\lambda\pi t(v_1 v_2)^{\frac{1}{2}}/\mu\}^{-\frac{1}{2}}(v_1/v_2)^{i/2} \sin[t\sqrt{(\lambda\mu)}] \\
&\quad \times \exp\{-t(v_1^{\frac{1}{2}} - v_2^{\frac{1}{2}})^2\} \\
p_i(t) &\simeq (p/2)\{\pi t(v_1 v_2)^{\frac{1}{2}}\}^{-\frac{1}{2}}(v_1/v_2)^{i/2} \cos[t\sqrt{(\lambda\mu)}] \\
&\quad \times \exp\{-t(v_1^{\frac{1}{2}} - v_2^{\frac{1}{2}})^2\}.
\end{aligned} \tag{9.101}$$

Thus under these conditions both $h_i(t)$ and $p_i(t)$ change geometrically with i at rate $\sqrt{(v_1/v_2)}$. Moreover, all colonies are in phase and attenuate at the same rate $t^{-\frac{1}{2}}\exp\{-t(v_1^{\frac{1}{2}} - v_2^{\frac{1}{2}})^2\}$. Note that if $v_1 = v_2 = u_1 = u_2$ then decay is no longer exponential but occurs at the relatively slow rate $t^{-\frac{1}{2}}$.

As a second example, if $\lambda\mu$ is large in comparison with $(v_1 - u_1)$ and $(v_2 - u_2)$, then an approximate solution may be developed which shows that predators and prey no longer cycle exactly $\pi/2$ out of phase.

Although the deterministic predator–prey cycles decay to zero in both of these cases as t increases, stable cycles may be generated by slightly modifying the spatial arrangement of the colonies. For example, if migration is allowed only to the right, then not only will any cycle present in colony 0 be maintained indefinitely, but this cycle will also govern the behaviour of all other colonies. In the simplest case with $u_1 = u_2 = v_2 = 0$, for $i > 0$

$$\begin{aligned}
h_i(t) &= -(\alpha\mu/\xi\beta)pr^i \sin(\xi t - i\delta) \\
p_i(t) &= pr^i \cos(\xi t - i\delta)
\end{aligned} \tag{9.102}$$

where

$$\xi = \sqrt{\{\lambda(\mu + v_1)\}}, \quad r = \sqrt{\{(\mu + v_1)/(\mu + v_1 + \lambda)\}}$$

and

$$\tan(\delta) = \sqrt{\{\lambda/(\mu + v_1)\}}.$$

Thus for large t the deterministic behaviour of the system is governed solely by the initial influx of p predators; the result being a train of linked elliptical cycles around the equilibrium values, in which the amplitude decreases geometrically with i at rate r and the phase-lag increases linearly at rate δ (Renshaw, 1982). When v_1 is small, $r \sim \sqrt{\{\mu/(\mu + \lambda)\}}$ and $\tan(\delta) \sim \sqrt{(\lambda/\mu)}$; whilst as v_1 becomes large, $r \sim 1$ and $\delta \sim 0$, the system behaving in the limit as $v_1 \to \infty$ as though the spatial separation of colonies can be ignored.

Though we have applied this approach to just one particular form of spatial predator–prey model, the technique should be applicable to many other spatial systems with two or more interacting species.

9.9.2   *A spatial diffusion model*

The spatial predator–prey system developed in Chapter 7 is centred around Huffaker's experiments in which he tried to maintain predator–prey cycles by making the spatial arrangement of colonies (oranges) increasingly complex. At a far more simplistic level simulation of a Lotka–Volterra stepping-stone system (Section 7.2) shows that early extinction becomes increasingly unlikely as the number of colonies rises. Now comparison of Figures 9.12 and 9.14 shows that whilst population numbers in neighbouring stepping-stone colonies are closely related, those in neighbouring diffusion sites are far more variable. This intuitively suggests that a diffusion predator–prey process might have an even better chance of sustaining stochastic cycles than a stepping-stone process; when population numbers are low a few sites may still contain 'spikes' of individuals ready to start the process off again.

Support for this conjecture is provided by Ziegler (1977), though he employs a migration scheme in which a species can migrate between nearest-neighbours only when one cell is at its maximum value and the other is empty. By combining analytic and simulation methods he shows that two forms of indefinite persistence of populations are possible over a wide range of parameter settings, even though all the cells tend to rapid extinction in isolation. In one form the population distribution is homogeneous, but in the other it consists of dynamically stable patches and waves.

For demonstration purposes consider a circular configuration of 250 sites spaced $\delta x = 0.1$ units apart at positions $x = 0, 0.1, 0.2, \ldots, 24.9$. Suppose that predators are constantly on the move searching for prey, i.e. they follow the Brownian scheme of Section 9.2 with time increment $\delta t = 0.01$, and that at each discrete time point $t = 0, \delta t, 2\delta t, \ldots$ a prey may either stay where it is or else move a distance δx to the right or left. Otherwise the usual Lotka–

Volterra rules apply. Denote $\{X_i(t), Y_i(t)\}$ to be the stochastic number of prey and predators at time t and site $i = x/\delta x = 0, \ldots, 249$, and consider the following transition rates:

$$\begin{aligned}
\Pr(X_i \to X_i + 1) &= 0.1X_i\,\delta t &&\text{(prey birth)}\\
\Pr(X_i \to X_i - 1) &= 0.05X_iY_i\,\delta t &&\text{(prey death)}\\
\Pr(Y_i \to Y_i + 1) &= 0.01X_iY_i\,\delta t &&\text{(predator birth)} &&(9.103)\\
\Pr(Y_i \to Y_i - 1) &= 0.05Y_i\,\delta t &&\text{(predator death)}\\
\Pr(X_i \to X_i - 1, X_{i+1} &\to X_{i+1} + 1)\\
&= \Pr(X_i \to X_i - 1, X_{i-1} \to X_{i-1} + 1)\\
&= 0.04X_i\,\delta t &&\text{(prey migration)};
\end{aligned}$$

together with

$$\text{Pr(each predator moves one step left/right)} = 0.5$$
$$\text{(predator diffusion).} \quad (9.104)$$

Note that because of the circularity sites 0 and 250 are equivalent.

A simple routine for simulating this process is

```
10   cycle over t = δt, 2δt, ...
20   cycle over i = 0, ..., 249
20   determine events (9.103)
30   cycle over i = 0, ..., 249
30   determine diffusion events (9.104)
10   continue.
```

Figure 9.15 shows simulated total prey and predator population sizes at times $t = 0, 1, \ldots, 500$, starting with an initial configuration of 5 prey and 1 predator at each site. Taken purely by itself, this plot of total population numbers exhibits behaviour even more conservative than that of the stochastic non-spatial Volterra process (Section 6.2), since prey and predators undergo *damped* cycles. Now the non-spatial part of (9.103) corresponds to the basic Lotka–Volterra model, and we have seen earlier that this process exhibits *unstable* stochastic cycles which quickly lead to extinction of one of the species. So the addition of spatial mobility renders the unstable (non-spatial) stochastic process stable. This behavioural difference highlights the danger of ignoring the spatial component of a model purely for mathematical convenience, as often happens in studies of total population numbers which relate to a region substantially greater than a single individual's territory.

In the notation of the non-spatial equations (6.3), our example parameter values are $r_1 = 0.1$, $r_2 = 0.05$, $b_1 = 0.05$ and $b_2 = 0.01$. The equilibrium values at individual sites are therefore $M_i^* = r_2/b_2 = 5$ and $N_i^* = r_1/b_1 = 2$, whence over all 250 sites the total equilibrium values are $M^* = 1250$ and N^* = 500. Now we see from Figure 9.15 that the prey have a peak value of 2683 at $t = 114$, reducing to 1942 three peaks later at $t = 384$. This yields the estimated period $T_{est} = 270/3 = 90$, which happens to be in almost perfect agreement with the theoretical period of oscillation $T = 2\pi/\sqrt{(r_1 r_2)} = 88.9$ calculated under the assumption of a (non-spatial) Lotka–Volterra model.

The corresponding Volterra process exhibits damped deterministic cycles (as required) with period of oscillation $T_V = 4\pi/\theta$, where from (6.35)

$$\theta = \sqrt{(-\Delta)} = \sqrt{\{4r_2 b_1 N^* - (cM^*)^2\}} = 0.141\ 32.$$

Thus $T_V = 88.9$, which again is in excellent agreement with the estimated value $T_{est} = 90$. Hence as far as this particular simulation is concerned, the deterministic non-spatial Volterra model provides a good summary of the stochastic spatial Lotka–Volterra diffusion process. However, continuing the simulation beyond $t = 500$ to $t = 2000$ showed that the prey maxima did *not* attenuate towards M^* but varied quite widely in the range 1250 to 2000, with no evidence whatsoever that the cycles would not be sustained indefinitely.

Figure 9.15. Simulated total prey (———) and predator (– – –) population sizes for the Lotka–Volterra spatial diffusion process with transition rates (9.103) and (9.104), shown at times $t = 0, 1, \ldots, 500$.

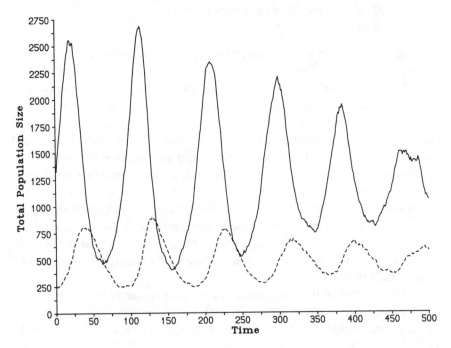

Thus the initial damping is clearly misleading, and shows just how hazardous drawing inferences from medium-length simulation experiments can be!

Figure 9.16 shows the spatial dispersion of prey and predators at the following times.

(i) $t = 65$:   the total prey population is near a minimum; prey are dispersed unevenly throughout the region with 148 of the 250 sites containing either 0 or 1 prey and 5 sites containing more than 10 prey.

(ii) $t = 90$:   the localized patches of prey, together with newly emergent ones, develop further, being relatively unimpeded by the predator population which is now at a minimum; just one site contains as many as 4 predators and 97 sites contain none.

(iii) $t = 115$:   these localized predator–prey processes have now developed to their full prey potential; note the strong patchy pattern with 52 sites still containing either 0 or 1 prey.

Figure 9.16. Spatial dispersion of prey (———) and predators (– – –) over the 250 sites for the simulation of Figure 9.15, shown at times (i) $t = 65$, (ii) $t = 90$, (iii) $t = 115$ and (iv) $t = 130$.

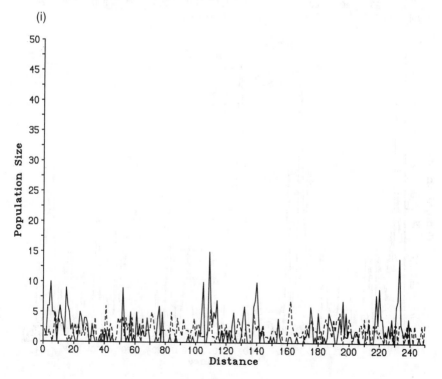

(i)

(ii)

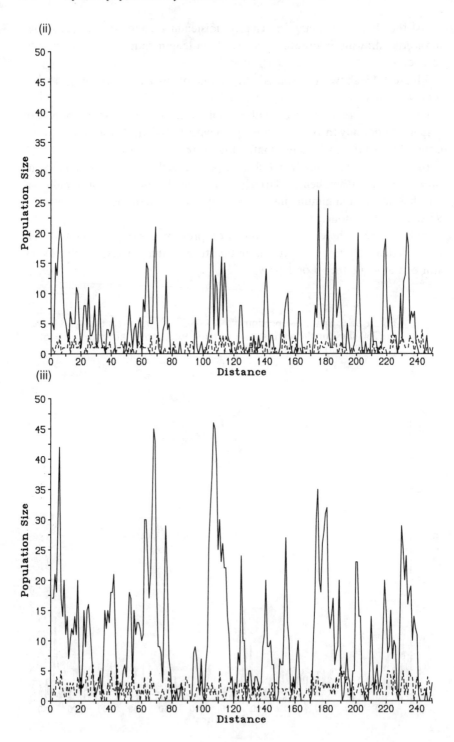

(iii)

(iv) $t = 130$:   the predators now reach their maximum population size, but unlike prey the constant movement of predators ensures that they fill the region fairly uniformly; the decline in prey produces a prey-pattern similar to that in (ii).

Though the behaviour shown in Figure 9.16 is very spiky, far smoother patches can be produced by modifying the transition probability structure (9.103) and (9.104). Dubois and Monfort (1978), for example, produced a simulation run (Figure 9.17) showing the spontaneous emergence of a large single patch of prey which they claim to be like the real situation observed in the sea. This patch then splits and moves off as two separate patches, due to an increasing predator population at its centre.

Mimura and Murray (1978) show that when the diffusion of prey is small compared with that of the predator then stable heterogeneity can persist indefinitely, which is exactly in line with our simulation results above. However, if the diffusion of both species is sufficiently large then there can be no spatial structure within a bounded region; the faster a species moves around the less chance it has of staying in a locality long enough to make its

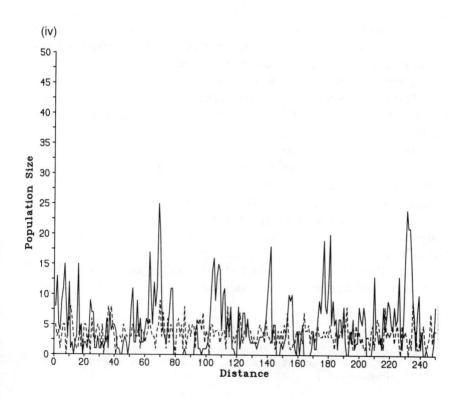

presence felt there. The theoretical analysis of spatial heterogeneity (patchiness) in a homogeneous environment is clearly an important and potentially rewarding area of study.

9.10   Turing's model for morphogenesis

In reality we are usually denied the mathematical nicety of unlimited space, and so investigations into the effect of interaction between nearest-neighbours in a spatial process are often plagued by the presence of edge-effects. Even in one dimension the assumptions governing the development at the two end-points can dramatically affect development right through the whole region, as demonstrated by the difference in behaviour between results (9.76) and (9.77) for the birth–death–migration process. Indeed, in the two-dimensional *Tribolium* situation (Section 9.7) it is the (mathematical) manipulation of the edge conditions themselves (see Figure 9.11) that enables the model to mimic the data successfully.

At a practical level, Ford (1975) describes several glasshouse experiments which examine the effects of between-plant competition on *Tagetes patula* L. (marigolds). Seeds were planted in a regular triangular pattern in wooden boxes, and to minimize edge-effects within a box sides were attached when the plants reached a height of 150 mm, these sides being progressively heightened to keep pace with plant growth. However, this solution is not ideal. For as Renshaw (1984) remarks, if the side of a box is regarded as a row of pseudo-plants then these pseudo-plants would be completely dependent; yet the alternative of ignoring plants sited near the edges would involve the loss of too much information. The extent of this loss can be seen by supposing that 400 plants are arranged on a 20 × 20 square grid: 76 plants (19%) lie on the edges, whilst 144 plants (36%) lie within one row of the edges.

An obvious way of avoiding edge-effects in a one-dimensional experiment is to arrange for sites to lie on a ring. Turing (1952), in a classic pioneering paper, developed elegant mathematical solutions arising from such a circular configuration, and this current section is devoted to a study of his main results. Although his analyses specifically relate to a system of chemical substances, called morphogens, which react together and diffuse through tissue, his approach is of general applicability.

Suppose the ring contains N cells, and that there are just two morphogens of interest. Note that the ensuing results can be extended to deal with any

Figure 9.17. Simulation run by Dubois and Monfort (1978) showing the spontaneous emergence of a patch of prey (———) at times $t = 5, 6.4$ and 6.7; predators are shown as (– – –). (Reproduced by permission of Physica-Verlag.)

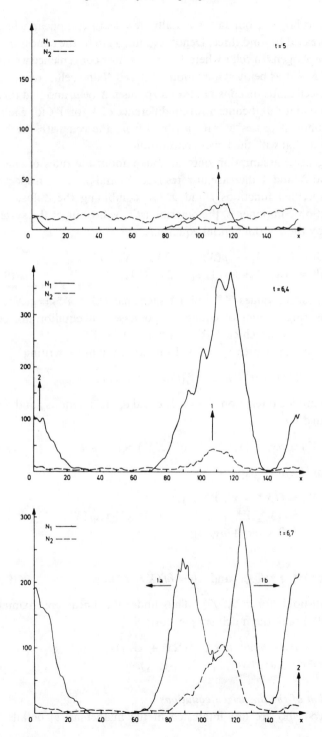

number of morphogens, but no essentially new features appear when the number is increased beyond three. Denote X_r and Y_r to be the concentrations of the two morphogens in cell r where $1 \leqslant r \leqslant N$. For convenience we regard cell 0 and cell $N + 1$ as being synonymous with cell N and cell 1, respectively. Let the cell-to-cell diffusion constant for morphogen X be μ, and that for Y be v. This means that for unit concentration difference of X (or Y), for each cell-pair the morphogen passes at rate μ (or v) from the cell with the higher concentration to that with the lower concentration.

The most general assumption one can make about the rates of chemical reaction is that X and Y increase at rates $f(X, Y)$ and $g(X, Y)$, respectively, for some unspecified functions f and g. On combining the diffusion and chemical reaction components, the deterministic behaviour of the system is then defined by the $2N$ differential equations

$$dX_r/dt = f(X_r, Y_r) + \mu(X_{r+1} - 2X_r + X_{r-1})$$
$$dY_r/dt = g(X_r, Y_r) + v(Y_{r+1} - 2Y_r + Y_{r-1}). \tag{9.105}$$

If there exist positive values X^* and Y^* such that $f(X^*, Y^*) = g(X^*, Y^*) = 0$, then this ring system of equations possesses an equilibrium, either unstable or stable, in which each $X_r = X^*$ and $Y_r = Y^*$.

Consider small departures x_r and y_r from equilibrium by writing

$$X_r(t) = X^* + x_r(t) \quad \text{and} \quad Y_r(t) = Y^* + y_r(t).$$

Then for constant local reaction rates a, b, c and d, the functions f and g may be approximated by

$$f(X_r, Y_r) \simeq ax_r + by_r \quad \text{and} \quad g(X_r, Y_r) \simeq cx_r + dy_r. \tag{9.106}$$

[Since on using Taylor's expansion,

$$f(X_r, Y_r) = f(X^* + x_r, Y^* + y_r)$$
$$\simeq f(X^*, Y^*) + x_r \, \partial f(X^*, Y^*)/\partial x_r + y_r \, \partial f(X^*, Y^*)/\partial y_r$$
$$= 0 + ax_r + by_r$$

where

$$a = \partial f(X^*, Y^*)/\partial x_r \quad \text{and} \quad b = \partial f(X^*, Y^*)/\partial y_r. \tag{9.107}$$

A similar result holds for $g(X_r, Y_r)$.] Thus under this linear approximation equations (9.105) take the much simpler form

$$dx_r/dt = ax_r + by_r + \mu(x_{r+1} - 2x_r + x_{r-1})$$
$$dy_r/dt = cx_r + dy_r + v(y_{r+1} - 2y_r + y_{r-1}). \tag{9.108}$$

*9.10.1 *Solution of the linearized equations*
Readers who are not interested in the mathematical details sur-

rounding the solution of equations (9.108) should move on directly to Section 9.10.2; the following brief sketch is provided for those who are. More than one approach exists, but the one used by Turing is particularly neat. Let i denote $\sqrt{(-1)}$, and make the following (Fourier) transformation of the x_r and y_r

$$u_r = (1/N) \sum_{s=1}^{N} x_s \exp\{-2\pi irs/N\}$$

$$v_r = (1/N) \sum_{s=1}^{N} y_s \exp\{-2\pi irs/N\}. \tag{9.109}$$

Now a routine mathematical exercise shows that

$$x_r = \sum_{s=0}^{N-1} u_s \exp(2\pi irs/N)$$

$$y_r = \sum_{s=0}^{N-1} v_s \exp(2\pi irs/N). \tag{9.110}$$

Whence substituting for x_r and y_r from (9.110) into equations (9.108) yields, after a little algebra,

$$du_s/dt = [a - 4\mu \sin^2(\pi s/N)]u_s + bv_s$$
$$dv_s/dt = [d - 4v \sin^2(\pi s/N)]v_s + cu_s. \tag{9.111}$$

The sine terms occur because $\exp(iz) = \cos(z) + i \sin(z)$.

These equations are quite manageable since, unlike equations (9.108), each pair contains just two variables (u_s and v_s). Their solution has the standard form

$$u_s = A_s \exp(p_s t) + B_s \exp(p_s' t)$$
$$v_s = C_s \exp(p_s t) + D_s \exp(p_s' t), \tag{9.112}$$

where the constants A_s, B_s, C_s and D_s satisfy the relations

$$A_s[p_s - a + 4\mu \sin^2(\pi s/N)] = bC_s$$
$$B_s[p_s' - a + 4\mu \sin^2(\pi s/N)] = bD_s \tag{9.113}$$

and p_s, p_s' are the roots of the equation

$$[p - a + 4\mu \sin^2(\pi s/N)][p - d + 4v \sin^2(\pi s/N)] = bc. \tag{9.114}$$

Substituting these results back into (9.110), and replacing the local variables x_r, y_r by X_r, Y_r, then gives

$$X_r = X^* + \sum_{s=1}^{N} [A_s \exp(p_s t) + B_s \exp(p_s' t)] \exp\{2\pi irs/N\}$$

$$Y_r = Y^* + \sum_{s=1}^{N} [C_s \exp(p_s t) + D_s \exp(p_s' t)] \exp\{2\pi irs/N\}. \tag{9.115}$$

9.10.2 *A spatial predator–prey example*

Expressions (9.115) provide the general solution to equations (9.105) under the assumption that departures from equilibrium are sufficiently small for the functions $f(X, Y)$ and $g(X, Y)$ to be replaced by their linear approximations (9.106). When this assumption holds true the result is *completely general* in its applicability, and therein lies its power. Morphogenesis is clearly just one of a whole host of situations for which it may be used. Moreover, as an alternative to a ring of separate cells (i.e. a stepping-stone model) one might prefer to consider a continuous ring of tissue (i.e. a diffusion model), yet this involves making only slight modifications to the above results (for details see Turing, 1952).

To illustrate the simplicity of determining appropriate values for the constants a, b, c and d in approximation (9.106), consider the spatial Lotka–Volterra probabilities (7.4) for a ring of N cells (Figure 7.3a shows $N = 3$). With X_r and Y_r now denoting the number of prey and predators, respectively, in cell r, the general equations (7.3) take the form

$$dX_r/dt = X_r(r_1 - b_1 Y_r) + \mu(X_{r+1} - 2X_r + X_{r-1})$$
$$dY_r/dt = Y_r(-r_2 + b_2 X_r) + \nu(Y_{r+1} - 2Y_r + Y_{r-1}) \tag{9.116}$$

where μ and ν denote the prey and predator migration rates to neighbouring cells. So here

$$f(X, Y) = X(r_1 - b_1 Y) \quad \text{and} \quad g(X, Y) = Y(-r_2 + b_2 X)$$

with $X^* = r_2/b_2$ and $Y^* = r_1/b_1$. We therefore see from (9.107) that

$$a = r_1 - b_1 Y^* = 0 \quad \text{and} \quad b = -b_1 X^* = -r_2 b_1/b_2$$
$$c = b_2 Y^* = r_1 b_2/b_1 \quad \text{and} \quad d = -r_2 + b_2 X^* = 0. \tag{9.117}$$

9.10.3 *Types of behaviour*

Although expression (9.115) is indeed the required mathematical solution to equations (9.105), from a biological viewpoint it is too opaque to be of much use as it stands and so we clearly need to investigate its properties. The guts of Turing's analysis can be understood once we appreciate that any general exponential term, say $\exp(kt)$, gives rise to three basic types of behaviour as t increases:

(i) if k is real and positive then $\exp(kt)$ grows indefinitely large;
(ii) if k is real and negative then $\exp(kt)$ decays to zero;
(iii) if k is imaginary (i.e. $k = il$ for real l) then

$$\exp(kt) = \exp(ilt) = \cos(lt) + i\sin(lt),$$
and so $\exp(kt)$ oscillates.

Note that the same interpretation of (9.115) holds for both the stepping-stone and diffusion models. In the latter situation the migration rates μ and v are replaced by the diffusion rates μ' and v' where

$$\mu = \mu'(N/2\pi\rho)^2 \quad \text{and} \quad v = v'(N/2\pi\rho)^2, \tag{9.118}$$

ρ being the radius of the ring.

After some time has elapsed solution (9.115) becomes dominated by those terms for which the corresponding p_s (say $p_{\tilde{s}}$) has the largest real part. Two distinct situations arise depending on whether this dominant value is real or complex. In the *stationary* case there are $\tilde{s}$ stationary waves arranged around the ring, and since the coefficients $A_{\tilde{s}}$ and $C_{\tilde{s}}$ are in a definite ratio (given by (9.113)), the pattern for one morphogen defines that for the other. Moreover, provided there is genuine instability the waves become more pronounced as time progresses. In the *oscillatory* case the interpretation is similar except that waves are not stationary; two wave trains move around the ring in opposite directions. Turing split these two situations into the following six sub-cases.

(a) *Stationary case with extreme long wavelength*
For example, if $\mu = v = \frac{1}{4}$, $b = c = 1$ and $a = d$, then $p_s = a - \sin^2(\pi s/N) + 1$. This expression is real, positive, and takes its maximum value (for $0 \leqslant s \leqslant N/2$) when $s = 0$. The contents of all the cells are the same; there is no resultant flow from cell to cell and so each behaves as though it were isolated. The general condition for this sub-case to occur is that

   (i) $bc > 0$, or
   (ii) $bc < 0$ and $(d - a)/\sqrt{(-bc)} > (\mu + v)/\sqrt{(\mu v)}$, or
   (iii) $bc < 0$ and $(d - a)/\sqrt{(-bc)} < -2$.

In addition, the condition for instability is that either $bc > ad$ or $a + d > 0$.

(b) *Oscillatory case with extreme long wavelength*
If $\mu = v = \frac{1}{4}$, $b = -c = 1$ and $a = d$, then $p_s = a - \sin^2(\pi s/N) \pm i$ which is complex and has greatest real part when $s = 0$. Each cell behaves as though it were isolated, with an oscillatory departure from equilibrium. The general condition is that

$$bc < 0 \quad \text{and} \quad -2 < (d - a)/\sqrt{(-bc)} < 4\sqrt{(\mu v)}/(\mu + v).$$

Instability occurs if, in addition, $a + d > 0$.

(c) *Stationary waves of extreme short wavelength*
This sub-case occurs, for example, if $v = 0$, $\mu = 1$, $d = I$ (say), $a = I - 1$ and $b = -c = 1$. Here

$$p_s = I - \tfrac{1}{2} - 2\sin^2(\pi s/N) + \sqrt{\{(2\sin^2(\pi s/N) + \tfrac{1}{2})^2 - 1\}},$$

and is greatest when $\sin^2(\pi s/N)$ is greatest, i.e. when $s = N/2$ which corresponds to a very high frequency oscillation of wavelength 2. If N is even then population sizes in cells two apart are similar, but those in neighbouring cells are distinctly different. However, if N is odd then this arrangement is impossible, and the difference in population size between neighbouring cells varies around the ring, from zero at one point to a maximum at the point diametrically opposite. The general condition is that

$$bc < 0 \quad \text{and} \quad \mu' > v' = 0,$$

with instability if, in addition, $a + d > 0$.

(*d*) *Stationary waves of finite length*
This is the sub-case of greatest biological interest. It occurs, for example, if $a = I - 2$, $b = 2.5$, $c = -1.25$, $d = I + 1.5$, $\mu' = 1$, $v' = \frac{1}{2}$ and $\mu/\mu' = v/v' = (N/2\pi\rho)^2$. Write

$$U = (N/\pi\rho)^2 \sin^2(\pi s/N).$$

Then with these special values equation (9.114) can be written as

$$(p-I)^2 + (\tfrac{1}{2} + (3U/2))(p-I) + \tfrac{1}{2}(U - \tfrac{1}{2})^2 = 0,$$

which has a solution $p = I$ if $U = \frac{1}{2}$. Hence if the radius of the ring (ρ) is chosen to ensure that for some integer $\tilde{s}$

$$\tfrac{1}{2} = U = (N/\pi\rho)^2 \sin^2(\pi\tilde{s}/N), \tag{9.119}$$

then there will be $\tilde{s}$ stationary waves around the ring since every other p_s has a real part smaller than I. If ρ is chosen so that $\tilde{s}$ is not an integer, then the actual number of waves is one of the two nearest integers to $\tilde{s}$. The general condition for this sub-case to occur is that

$$bc < 0 \quad \text{and} \quad \frac{4\sqrt{(\mu'v')}}{\mu' + v'} < \frac{d - a}{\sqrt{(-bc)}} < \frac{\mu' + v'}{\sqrt{(\mu'v')}}, \tag{9.120}$$

with instability occurring if in addition (for $v' \leqslant \mu'$)

$$d\sqrt{(\mu'/v')} - a\sqrt{(v'/\mu')} > 2\sqrt{(-bc)}. \tag{9.121}$$

When $v' > \mu'$ we just interchange the morphogen parameters.

The remaining two possibilities require the presence of at least three morphogens; with just one morphogen the only possibility is (a).

(*e*) *Oscillatory case with medium wavelength*
Two sets of travelling waves may move around the ring, one travelling clockwise and the other anticlockwise.

(f) *Oscillatory case with extreme short wavelength*
Neighbouring cells oscillate nearly 180° out of phase.

The linear approximation used in the derivation of solution (9.115) clearly has a fundamental impact on the validity of the above results in specific situations. For example, if quadratic terms are included in the Taylor series expansion of $f(X^* + x_r, Y^* + y_r)$ then it is possible for an instability to become catastrophic, in that wave growth makes the whole system more unstable than ever. With this warning in mind, Turing's conclusions for the two-morphogen sub-cases may be summarized as follows.

(a) If a change in the reaction rates causes the equilibrium to become unstable then each cell will drift away from its equilibrium value, with the drift in neighbouring cells, but not far away ones, likely to be in the same direction. This might account for dappled colour patterns in certain animals, though if morphogenesis is to be regarded as an explanation then dappling must occur when the foetus is only a few inches long since the morphogens could not diffuse over larger distances.

(b) This case is similar to (a), except that departure from equilibrium is oscillatory.

(c) Any drift from equilibrium occurs in opposite directions in contiguous cells; no biological examples of this are claimed to be known.

(d) There is a stationary wave pattern around the ring, with no change in time apart from a slow increase in amplitude. The peaks of the waves are uniformly spaced. Though it is difficult to find direct applications because isolated rings of tissue are rare, systems that have the same kind of symmetry as a ring are extremely common and it is to be expected that under appropriate conditions stationary waves may well develop on these bodies. Consider, for example, *Hydra*, which resembles a sea-anemone but lives in fresh water and has from about five to ten tentacles. When part of a *Hydra* is cut off, it rearranges itself to form a completely new organism. At one stage in its development this organism still has circular symmetry, and forms a tube open at the head end and closed at the other end. Later on this symmetry disappears; an appropriate stain will highlight a number of patches on the widened head end where the tentacles will subsequently appear (Child, 1941). Applying the morphogen theory, we may suppose that these patches occur at the lobes of a developing stationary wave system.

9.10.4   *Application to a spatial Volterra system*
Although Turing talks purely in terms of morphogenesis, the surrounding theory and resulting conclusions are completely general in their applicability. So to conclude this chapter let us consider a spatial Volterra

predator–prey system represented by the deterministic equations

$$dX_r/dt = X_r(r_1 - c_1X_r - b_1Y_r) + \mu(X_{r+1} - 2X_r + X_{r-1})$$
$$dY_r/dt = Y_r(-r_2 + b_2X_r) + \nu(Y_{r+1} - 2Y_r + Y_{r-1}). \qquad (9.122)$$

These equations are the same as the Lotka–Volterra system (9.116), except that the logistic parameter c_1 has been included to avoid the risk of a local prey population explosion in the absence of nearby predators. The functions f and g in the general equations (9.105) now have the form

$$f(X, Y) = X(r_1 - c_1X - b_1Y) \quad \text{and} \quad g(X, Y) = Y(-r_2 + b_2X).$$

Thus the equilibrium values are

$$X^* = r_2/b_2 \quad \text{and} \quad Y^* = (r_1b_2 - r_2c_1)/b_1b_2,$$

whilst the linear coefficients (9.107) become

$$a = -c_1X^*, \quad b = -b_1X^*, \quad c = b_2Y^* \quad \text{and} \quad d = 0.$$

Conditions (9.120) and (9.121) of sub-case (d) therefore translate into

$$-b_1b_2X^*Y^* < 0 \quad \text{and} \quad \frac{4\sqrt{(\mu\nu)}}{\pi + \nu} < \frac{c_1X^*}{\sqrt{(b_1b_2X^*Y^*)}} < \frac{\mu + \nu}{\sqrt{(\mu\nu)}} \qquad (9.123)$$

with instability if, for $\nu \leqslant \mu$,

$$2\sqrt{(\mu/\nu)} < c_1X^*/\sqrt{(b_1b_2X^*Y^*)}. \qquad (9.124)$$

For given equilibrium values X^*, Y^* and migration parameters μ, ν, it is apparent from (9.123) that choice of c_1 is important in determining whether a stationary wave system exists. Moreover, as the difference between μ and ν increases, so does the admissible choice for the other parameter values. Since if $\mu = \nu$ then both ends of the second inequality (9.123) equal 2 and so it cannot be satisfied, whilst if ν approaches zero then these two ends approach the all-embracing values of zero and infinity. Also, we see from (9.124) that instability occurs if $2\sqrt{(\mu/\nu)} < [(\mu/\nu) + 1]/\sqrt{(\mu/\nu)}$, i.e. if $(\mu/\nu) < 1$. As this condition contradicts the earlier requirement that $\mu \geqslant \nu$, instability cannot therefore occur. This result is hardly surprising, for not only does each cell have a stable equilibrium value, but any local absence of prey or predators will eventually be filled by migrants.

Stochastic interpretation in such situations is bound to be difficult since in the absence of instability any patterns that are created by natural stochastic effects will soon decay. For example, consider a stochastic realization of system (9.122) on a ring of 50 cells with $r_1 = 1.5$, $b_1 = 0.1$, $c_1 = 0.01$, $r_2 = 0.25$, $b_2 = 0.01$, $\mu = 1.0$ and $\nu = 0.01$. The equilibrium values are $X^* = 25$

and $Y^* = 12.5$, and for an isolated, predator-free colony the prey have a carrying capacity of $r_1/c_1 = 150$. Condition (9.123) is satisfied, being $-0.31 < 0$ and $0.396 < 0.447 < 10.1$, and so this particular model corresponds to the stable version of sub-case (d). With the uniform, near-equilibrium, starting configuration of $X_r(0) = 25$ and $Y_r(0) = 12$ ($r = 1, \ldots, 50$), we see for the simulation shown in Figure 9.18 that by time $t = 2.5$ the process has produced a relatively dense patch of prey centred around cell 11, with two low-density regions around cells 1 and 19. By time $t = 5$ these two regions still remain, but the patch at 11 is already waning and a new one is emerging diametrically opposite to it.

9.10.5   *Simulated stochastic waves*

Such an example is hardly enthralling, since all it really shows is the presence of short-term patchiness around the ring. Spectacular effects, such as

Figure 9.18. Stochastic realization of the Volterra ring system

$$\mathrm{d}X_r/\mathrm{d}t = X_r(1.5 - 0.01X_r - 0.1Y_r) + (X_{r+1} - 2X_r + X_{r-1})$$
$$\mathrm{d}Y_r/\mathrm{d}t = Y_r(-0.25 + 0.01X_r) + 0.01(Y_{r+1} - 2Y_r + Y_{r-1}),$$

showing the spatial distribution of prey at times $t = 2.5$ (— — —) and $t = 5$ (———), together with the equilibrium value $X^* = 25$ (– – –).

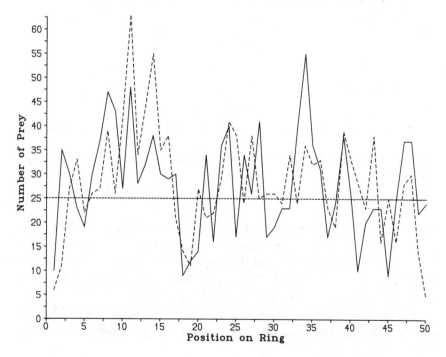

Hydra's ability to regenerate, are associated with instability, and since we have already shown that our Volterra predator–prey process is inherently stable let us return to the general two-species interaction process (6.1). Here the functions f and g in equations (9.105) take the form

$$f(X, Y) = X(r_1 + a_{11}X + a_{12}Y)$$

and

$$g(X, Y) = Y(r_2 + a_{21}X + a_{22}Y), \tag{9.125}$$

so the equilibrium values X^*, Y^* are the solutions of the equations

$$r_1 + a_{11}X^* + a_{12}Y^* = 0 = r_2 + a_{21}X^* + a_{22}Y^*. \tag{9.126}$$

The coefficients (9.107) are therefore given by

$$a = (r_1 + a_{11}X^* + a_{12}Y^*) + a_{11}X^* = a_{11}X^*, \quad b = a_{12}X^*$$
$$d = (r_2 + a_{21}X^* + a_{22}Y^*) + a_{22}Y^* = a_{22}Y^*, \quad c = a_{21}Y^*. \tag{9.127}$$

We shall now use this non-linear process (9.105) and (9.125) to demonstrate a stochastic version of Turing's example of sub-case (d), viz.

$$a = I - 2, \quad b = 2.5, \quad c = -1.25, \quad d = I + 1.5,$$
$$\mu' = 1 \quad \text{and} \quad v' = 0.5. \tag{9.128}$$

Inverting expressions (9.127) gives

$$a_{11} = (I - 2)/X^*, \quad a_{12} = 2.5/X^*$$
$$a_{21} = -1.25/Y^*, \quad a_{22} = (I + 1.5)/Y^*, \tag{9.129}$$

whilst from (9.126)

$$r_1 = -a_{11}X^* - a_{12}Y^* = 2 - I - 2.5(Y^*/X^*)$$
$$r_2 = -a_{21}X^* - a_{22}Y^* = -1.5 - I + 1.25(X^*/Y^*). \tag{9.130}$$

So when I is near zero and X^*/Y^* is near one, we have the configuration

$$\begin{pmatrix} - & | & - & + \\ - & | & - & + \end{pmatrix}$$

(using the notation of Chapter 6). In biological terms this corresponds to the rather strange situation in which both species undergo a basic death process, the X-species eats both itself and the Y-species, and the Y-species cultivates both itself and the X-species. The system clearly makes most sense when applied to Turing's (1952) chemical reaction process.

Suppose we wish five stationary waves to appear around a ring of $N = 50$ cells. Then from the deterministic result (9.119) we see that the radius ρ of the ring must satisfy

$$\tfrac{1}{2} = (50/\pi\rho)^2 \sin^2(5\pi/50)$$

i.e.

$$(\pi\rho)^2 = 477.46.$$

Whence from (9.118) the required migration parameters are

$$\mu = \mu'(50/2\pi\rho)^2 = 1.3090$$

and

$$v = v'(50/2\pi\rho)^2 = 0.6545.$$

Figure 9.19. Stochastic realization of the ring system

$$dX_r/dt = X_r(-0.3 - 0.22X_r + 0.25Y_r) + 1.3090(X_{r+1} - 2X_r + X_{r-1})$$
$$dY_r/dt = Y_r(-0.05 - 0.125X_r + 0.13Y_r) + 0.6545(Y_{r+1} - 2Y_r + Y_{r-1})$$

showing the spatial distribution of the X-species at times (a) $t = 1.7$, (b) $t = 2.4$ and (c) $t = 5.0$.

(a)

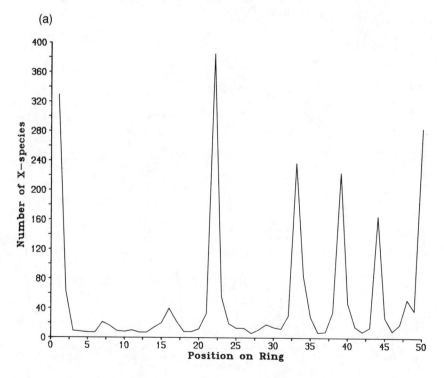

(b)

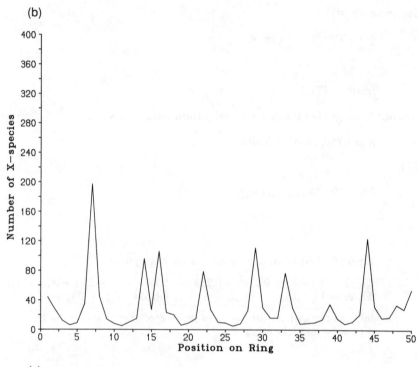

(c)

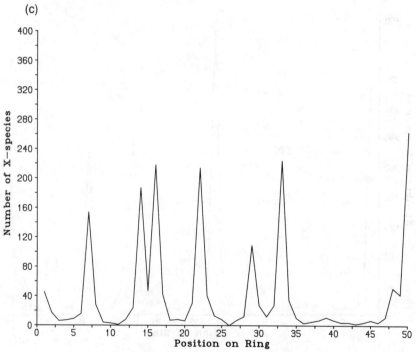

Now in the absence of any X-species, the Y-species in an isolated cell will grow explosively at rate

$$\mathrm{d}Y/\mathrm{d}t = g(0, Y) = Y(r_2 + a_{22}Y) \simeq a_{22}Y^2.$$

Thus even in the stable spatial situation we may anticipate the occurrence of large $Y(t)$-values somewhere around the ring before nearby $X(t)$-values can themselves grow large enough to control them. For example, on choosing $I = -0.2$ and selecting $X^* = Y^* = 10$, the parameter values (9.129) become

$$a_{11} = -0.22, \quad a_{12} = 0.25, \quad r_1 = -0.3$$
$$a_{21} = -0.125, \quad a_{22} = 0.13, \quad r_2 = -0.05.$$

Condition (9.120) is satisfied, being $1.89 < 1.98 < 2.12$, though the instability condition (9.121) is not, being $1.92 > 2$. So we are just inside the stability range of sub-case (d). The result of this near instability is that starting from the uniform equilibrium configuration $X_r(0) = Y_r(0) = 10$ $(r = 1, \ldots, 50)$ random effects can soon give rise to large tentacles.

Figure 9.19 shows the number of X-species for a simulation run in which five large stochastic waves are created as early as time $t = 1.7$ (Figure 9.19a), these being centred at cells $50/1$, 22, 33, 39 and 44. In addition there are also minor blips sited around cells 7, 16 and 29. Since the process is stable these amplitudes do not grow indefinitely, but wax and wane as time progresses. By time $t = 2.4$, all five large amplitudes have substantially decreased and the three blips have increased (Figure 9.19b). The resulting new wave structure is still in the process of development at time $t = 5$ (Figure 9.19c), except that the peak at cell 44 has been replaced by the regrowth of cell 50.

The power of the deterministic analysis to predict general qualitative stochastic features, such as five stationary tentacles, is especially good when one reflects that the local approximation on which it is based is outrageously flouted in this particular instance. The theory is valid only when X and Y are near 10, yet at time $t = 3.7$ in the above run $X_{14} = 649$ and $Y_{14} = 563$! Deterministic analysis is clearly a powerful procedure in the study of complex spatial situations, provided that predicted features are validated by a parallel stochastic simulation exercise.

10

Epidemic processes

We have concentrated on two-species models of competition and predation because they are biologically relevant. The danger in expanding to other areas of application is that the total possible number of theoretical models that can be constructed is inexhaustible, and virtually all will give birth to monsters of purely mathematical imagination. It is therefore refreshing to note that only a slight variation on the basic Lotka–Volterra predator–prey theme gives rise to a whole new major field of study, namely the modelling of infectious diseases.

10.1 Introduction

Consider the Lotka–Volterra equations (6.3), namely

$$dN_1/dt = N_1(r_1 - b_1 N_2)$$
$$dN_2/dt = N_2(-r_2 + b_2 N_1), \tag{10.1}$$

and make the following correspondence:

$N_1(t)$ (no. of prey)$\rightarrow$
$\qquad\qquad x(t)$ (no. of individuals susceptible to a disease),
$N_2(t)$ (no. of predators)$\rightarrow$
$\qquad\qquad y(t)$ (no. of individuals infected by the disease).

Then on taking $r_1 = 0$, $b_1 = b_2 = \beta$, and $r_2 = \gamma$, equations (10.1) may be written in the 'epidemic' form

$$dx/dt = -\beta xy$$
$$dy/dt = \beta xy - \gamma y. \tag{10.2}$$

In brief, β is the infection rate, xy is the number of possible contact-pairs between susceptibles and infectives, and γ is the death (or removal) rate of infectives.

The fundamental difference between these two sets of equations lies in their interpretation. In (10.2) the 'death' of a susceptible automatically gives rise to the 'birth' of an infective, since it involves a transfer of state for the same individual, whilst in (10.1) prey do not become predators but merely act as food for them.

Though equations (10.2) are naive descriptors of epidemic development, they do generate useful qualitative predictions about possible modes of behaviour. Indeed, Kendall (1965) remarks that, 'I consider the ultimate justification for this sort of work to be the fact that precisely similar grossly over-simplified models are talked about (in words, if not in symbols)... by all those who are professionally concerned with epidemics, their natural history, and their prevention'. Bailey (1975) presents a full account of the way in which mathematical modelling can provide scientific insight into the mechanics of epidemic and endemic processes; a comprehensive list of references can be found through Isham (1988). Here we shall just draw on a few of these results to generate further discussion into the deterministic versus stochastic and non-spatial versus spatial comparisons that we have developed earlier.

The importance of epidemiology cannot be overstated. To quote Bailey's figures, in fourteenth century Europe there were 25 million deaths out of a population of 100 million from the Black Death alone, with whole towns and villages being virtually annihilated; the Aztecs lost half their population of $3\frac{1}{2}$ million from smallpox; whilst around 20 million people died in the world pandemic of influenza in 1919. Today, the over-riding concern is the spread of AIDS, whilst vast numbers of people remain affected by less media-conscious diseases such as malaria (350 million living in endemic areas), schistosomiasis (200 million), filariasis (250 million) and hookworm disease (450 million).

The total burden of human misery and suffering that results from communicable disease is immense, and any understanding that modelling techniques can bring to alleviate this terrible state of affairs is truly important. Light can be shed on the life-cycle of the parasite, the transmission and spread of an infectious disease, the nature of threshold population densities above which an epidemic can flare up, and methods for optimal immunization and control. In all of these matters mathematical and statistical studies have an essential role to play. Indeed, bearing in mind the great variety of infectious diseases that currently affect our planet, and how relatively few theoretical inroads have been made into understanding most of them, there is clearly a vast field of worthwhile and directly applicable research just waiting to take off for those with the interest and imagination to embark upon it.

10.2 Simple epidemics

In the simplest type of epidemic model infection spreads by contact between members of the community, and infected individuals are not removed from circulation by recovery, isolation or death (so $\gamma = 0$ in equation (10.2)). Ultimately, all individuals susceptible to the disease must therefore become infected. Although these assumptions are indeed 'simple', they are nevertheless applicable to some milder types of infection.

Suppose we have a homogeneously mixing group of $n + 1$ individuals, and that at time $t = 0$ the epidemic starts off with just one infected individual. The remaining n individuals are all assumed to be susceptible to the infection. Then since there are $x(t)y(t)$ possible contact-pairs between susceptibles and infectives at time t, each of which can turn a susceptible into an infective at rate β, it follows that the deterministic rate of decline of susceptibles is given by

$$dx(t)/dt = -\beta x(t)y(t). \tag{10.3}$$

Clearly

$$x(t) + y(t) = n + 1.$$

On eliminating $x(t)$ from equation (10.3), we see that the rate of increase of infectives is given by

$$dy(t)/dt = \beta y(t)[n + 1 - y(t)], \tag{10.4}$$

which is identical to the deterministic logistic equation (3.20) with $r = \beta(n + 1)$ and $s = \beta$. Note, however, that these two processes are not stochastically equivalent since the number of infectives can never decrease. Use of the logistic solution (3.22) immediately yields

$$y(t) = (n + 1)/[1 + n \exp\{-(n + 1)\beta t\}], \tag{10.5}$$

which implies that the number of remaining susceptibles is

$$x(t) = n + 1 - y(t) = n(n + 1)/[n + \exp\{(n + 1)\beta t\}]. \tag{10.6}$$

10.2.1   *The epidemic curve*

In practice a major feature of interest is the rate (u) at which new infectives occur, namely

$$u = \frac{dy}{dt} = -\frac{dx}{dt} = \beta xy = \frac{\beta n(n + 1)^2 \exp\{-(n + 1)\beta t\}}{[1 + n \exp\{-(n + 1)\beta t\}]^2}. \tag{10.7}$$

This function is called the epidemic curve and reaches its maximum value when $du/dt = 0$, i.e. when $\beta t = [\log_e(n)]/(n + 1)$. At this point $u_{max} = \frac{1}{4}\beta(n + 1)^2$, and as $x = y = \frac{1}{2}(n + 1)$ the epidemic is half over.

From a stochastic viewpoint the apparently simple nature of this process proves deceptive. In the general notation of Section 3.1 the birth and death rates of the number of infectives $Y(t)$ are

$$B(Y) = \beta Y(n + 1 - Y) \quad \text{and} \quad D(Y) = 0, \tag{10.8}$$

and the quadratic nature of $B(Y)$ produces a lot of algebraic hassle in spite of the zero death rate (see Bailey, 1975, for details). Expressions for the probabilities $\{\Pr(Y(t) = y)\}$ can be obtained by means of a rather complicated argument, but they are too tortuous to present here. We shall therefore just state the result for the stochastic mean $\mu(t)$, namely that for even n (Haskey, 1954)

$$\mu(t) = \beta \sum_{j=1}^{n/2} \frac{n!}{(n-j)!(j-1)!} \left\{ (n-2j+1)^2 \beta t + 2 - (n-2j+1) \sum_{k=j}^{n-j} k^{-1} \right\}$$
$$\times \exp\{-\beta j(n-j+1)t\} \tag{10.9}$$

(the result for odd n is slightly different).

Though computation of expression (10.9) does require care because of potential round-off error in the main summation, comparison between the deterministic and stochastic situations is particularly easy when $n = 10$ since exact results for $\mu(t)$ and $w(t) = d\mu(t)/dt$ are already known. In particular

$$w(t) = \beta\{e^{-10\beta t}(8100\beta t - 3156\tfrac{1}{14}) + e^{-18\beta t}(79\,380\beta t - 20\,650\tfrac{1}{2})$$
$$+ e^{-24\beta t}(216\,000\beta t - 38\,931\tfrac{3}{7}) + e^{-28\beta t}(211\,680\beta t - 4032)$$
$$+ e^{-30\beta t}(37\,800\beta t + 66\,780)\} \tag{10.10}$$

(Mansfield and Hensley, 1960, provide tables of $w(t)$ for $n = 5, 6, \ldots, 40$ and $\beta = 1$). On changing the time-scale to $s = \beta t$ all the above results become dimensionless as far as β is concerned, and Figure 10.1 shows the deterministic and stochastic epidemic curves (10.7) and (10.10), respectively, for the $n = 10$ case. Whilst both peak at about the same place the deterministic curve rises higher than the stochastic one and then falls away more quickly.

For much larger, and hence more realistic, values of n it is clearly easier to work with $u(t)$ in (10.7) rather than $w(t)$ which is obtained by differentiating (10.9). Yet if the difference between them is substantial then use of $u(t)$ might lead to wrong interpretation about the progress of the epidemic. An alternative to using the algebraic solution (10.9) is to simulate say 100 realizations of the epidemic and then use their average as an approximation to the curve $\mu(t)$. Once this simulated mean curve is smoothed, the gradient provides an approximation to the epidemic curve $w(t)$.

This technique has an added bonus, since the envelope of the simulations tells us how variable the progress of the epidemic can be. Moreover, since the number of infectives can never decrease, the simulation program just involves a minor adaptation of that used for the simple birth process (Section 2.1.3). The length of time (Z_i) that the population is exactly of size i is exponentially distributed with parameter λi for the pure birth process and $\beta i(n + 1 - i)$ for the epidemic process, so we simply replace λ by $\beta(n + 1 - i)$ in the program BIRTH (listed in the Appendix to Chapter 2).

*10.2.2 *Duration time*

We can exploit this similarity with the birth process still further to determine the duration time (T) of the epidemic, i.e. the time taken for all n susceptibles to become infected. On paralleling results (2.20) and (2.21) we see that

$$T = Z_1 + Z_2 + \cdots + Z_n \tag{10.11}$$

has mean value

$$E(T) = \sum_{i=1}^{n} E(Z_i) = (1/\beta) \sum_{i=1}^{n} \frac{1}{i(n+1-i)}$$

$$= \frac{1}{(n+1)\beta} \sum_{i=1}^{n} \left\{ \frac{1}{i} + \frac{1}{n+1-i} \right\} = \frac{2}{(n+1)\beta} \sum_{i=1}^{n} i^{-1}. \tag{10.12}$$

Similarly,

$$\mathrm{Var}(T) = \frac{1}{\beta^2} \sum_{i=1}^{n} \left\{ \frac{1}{i(n+1-i)} \right\}^2$$

Figure 10.1. Comparison of normalized (i.e. $\beta = 1$) deterministic (———) and stochastic (– – –) epidemic curves for $n = 10$.

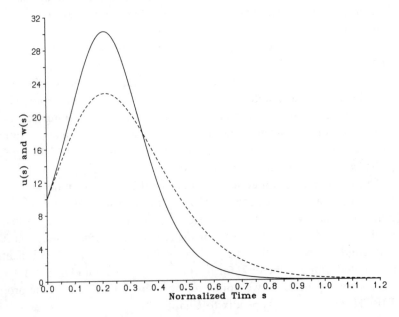

which, on splitting into partial fractions, reduces to

$$\text{Var}(T) = \frac{2}{(n+1)^2\beta^2}\left\{\frac{2}{(n+1)}\sum_{i=1}^{n}i^{-1} + \sum_{i=1}^{n}i^{-2}\right\}. \qquad (10.13)$$

So if n is large, on using the results

$$1 + \frac{1}{2} + \frac{1}{3} + \cdots + \frac{1}{n} \sim \gamma + \log_e(n) \qquad (2.22)$$

and

$$1 + \frac{1}{2^2} + \frac{1}{3^2} + \cdots + \frac{1}{n^2} \simeq \frac{\pi^2}{6}, \qquad (2.27)$$

where $\gamma = 0.577$ is Euler's constant, we see that

$$E(T) \sim [2/(n+1)\beta][\gamma + \log_e(n)] \sim [2\log_e(n)]/n\beta \qquad (10.14)$$

and

$$\text{Var}(T) \sim [2/(n+1)^2\beta^2]\{[2/(n+1)][\gamma + \log_e(n)] + (\pi^2/6)\}$$
$$\sim \pi^2/3n^2\beta^2. \qquad (10.15)$$

Hence the coefficient of variation

$$CV(T) = \sqrt{\{\text{Var}(T)\}}/E(T) \sim \pi/[2\sqrt{3}\log_e(n)]. \qquad (10.16)$$

Thus whilst the mean duration time approaches zero quite quickly as n increases, the coefficient of variation decreases at a much slower pace. For example, with $\beta = 1$ we have

n	10^2	10^3	10^4
$E(T)$	0.10	0.015	0.0020
$CV(T)$	0.20	0.13	0.10

The larger the initial number of susceptibles, the quicker the epidemic is completed, with the growth in maximum infection rate $\frac{1}{4}\beta n^2$ (when $i = \frac{1}{2}n$) far outstripping the total number of susceptibles n to be infected. Moreover, since the coefficient of variation remains moderately large it follows that quite substantial differences in epidemiological behaviour can occur between separate, but otherwise identical, groups of individuals. The practical implication of this is that considerable care is required when ascribing unexpected results to say abnormal virulence or infectiousness, for they could be due to purely chance fluctuations.

10.3   General epidemics

Suppose we now allow infectives to be removed from circulation by isolation or death. Although this new feature does not lead to a completely general epidemic which can take account of migration, the geographical arrangement of possible infection sites, loss of immunity, latent period of infection, etc., it does give rise to a model which is sufficiently realistic to be useful.

Let $z(t)$ denote the number of infected individuals who have been removed from a community of *total* size n. Then

$$x(t) + y(t) + z(t) = n \tag{10.17}$$

for all $t \geqslant 0$, whilst equations (10.2) expand to

$$\begin{aligned} \mathrm{d}x/\mathrm{d}t &= -\beta xy \\ \mathrm{d}y/\mathrm{d}t &= \beta xy - \gamma y \\ \mathrm{d}z/\mathrm{d}t &= \gamma y. \end{aligned} \tag{10.18}$$

For convenience define $\rho = \gamma/\beta$ as the *relative removal rate*.

At the start of the epidemic let $x(0) = x_0$, $y(0) = y_0$ and $z(0) = 0$. Usually y_0 will be small and so x_0 will be near n. Now we see from (10.18) that at time $t = 0$

$$\mathrm{d}y/\mathrm{d}t = \beta y_0(x_0 - \rho), \tag{10.19}$$

so an epidemic can only build up (i.e. $\mathrm{d}y/\mathrm{d}t > 0$) if $x_0 > \rho$. Thus $x_0 = \rho$ defines a (deterministic) threshold density of susceptibles below which an epidemic cannot develop, since infectives are removed at a faster rate than new infectives can be produced.

*10.3.1   *The epidemic curve*

To analyze this model further, we first eliminate y by dividing $\mathrm{d}x/\mathrm{d}t$ by $\mathrm{d}z/\mathrm{d}t$ to form

$$\mathrm{d}x/\mathrm{d}z = -x/\rho.$$

This equation integrates to give

$$x = x_0 \exp\{-z/\rho\}, \tag{10.20}$$

whence the last of equations (10.18) can now be written purely as a function of z, namely

$$\mathrm{d}z/\mathrm{d}t = \gamma[n - x_0 \exp\{-z/\rho\} - z]. \tag{10.21}$$

Unfortunately it does not appear that equation (10.21) can be solved directly, though an approximate solution can be obtained by assuming that z/ρ

remains small. In this case

$$e^{-z/\rho} = 1 + \frac{(-z/\rho)}{1!} + \frac{(-z/\rho)^2}{2!} + \cdots \simeq 1 - (z/\rho) + \tfrac{1}{2}(z/\rho)^2,$$

whence

$$dz/dt \simeq \gamma\{n - x_0 + [(x_0/\rho) - 1]z - (x_0/2\rho^2)z^2\}. \tag{10.22}$$

Equation (10.22) is solvable by standard methods, and yields

$$z(t) = (\rho^2/x_0)\{(x_0/\rho) - 1 + \alpha \tanh(\tfrac{1}{2}\alpha\gamma t - \phi)\} \tag{10.23}$$

where $\tanh(\theta)$ denotes $(e^\theta - e^{-\theta})/(e^\theta + e^{-\theta})$. Here

$$\alpha = \sqrt{\{[(x_0/\rho) - 1]^2 + 2x_0 y_0/\rho^2\}}, \tag{10.24}$$

and since $z(0) = 0$, ϕ is determined from

$$\tanh(\phi) = [(x_0/\rho) - 1]/\alpha.$$

The epidemic curve, now defined in terms of the removals $z(t)$ and not the infectives $y(t)$, is therefore obtained by differentiating (10.23) to form

$$dz/dt = (\gamma\alpha^2\rho^2/2x_0) \operatorname{sech}^2(\tfrac{1}{2}\alpha\gamma t - \phi) \tag{10.25}$$

where $\operatorname{sech}(\theta)$ denotes $2/(e^\theta + e^{-\theta})$. Expression (10.25) describes a symmetric bell-shaped curve centred around the maximum value of $\gamma\alpha^2\rho^2/2x_0$ at time $t = 2\phi/\alpha\gamma$, and successfully mimics the way in which in many epidemics the daily number of new reported cases first climbs to a peak and then decays.

10.3.2   *The deterministic threshold theorem*

The total size of the epidemic, i.e. the total number of removals when there are no longer any infectives left, is obtained by letting $t \to \infty$ in (10.23), whence

$$z(\infty) = (\rho^2/x_0)[(x_0/\rho) - 1 + \alpha]. \tag{10.26}$$

Now if $2x_0 y_0/\rho^2$ is much smaller than $[(x_0/\rho) - 1]^2$, then (10.24) reduces to $\alpha \simeq (x_0/\rho) - 1$ (for $x_0 > \rho$). Thus

$$z(\infty) \simeq 2\rho[1 - (\rho/x_0)]. \tag{10.27}$$

Suppose that x_0 is bigger than ρ, and write

$$x_0 = \rho + v \qquad (v > 0). \tag{10.28}$$

Then (10.27) gives

$$z(\infty) \simeq 2\rho v/(\rho + v) \simeq 2v, \tag{10.29}$$

and so the initial number of susceptibles $\rho + v$ is eventually reduced to $\rho - v$, i.e. to a value as far below the threshold as it was initially above. This is the celebrated *threshold theorem* of Kermack and McKendrick (1927), and it clearly provides considerable insight into the mechanics governing the outbreak of epidemic disease. Unfortunately, this remarkable paper attracted little initial attention, and twenty years elapsed before its worth was fully appreciated.

The reliability of this theorem obviously depends on the accuracy of the approximations used to obtain it. Now an elegant argument by Kendall (1956) shows that expression (10.23) for $z(t)$ represents the exact number of removals for an epidemic process in which the infection rate β is assumed to depend on z, namely

$$\beta(z) = 2\beta/[(1 - z/\rho) + (1 - z/\rho)^{-1}].\qquad(10.30)$$

It follows that although $\beta(0) = \beta$, $\beta(z) < \beta$ when $0 < z < \rho$. So Kermack and McKendrick's result consistently underestimates the infection rate β, and hence the total size of the epidemic. If, for example, $z(\infty) = 0.2\rho$, then $\beta[z(\infty)] = 0.98\beta$ and so the approximations have negligible effect. However, once $z(\infty) > 0.4\rho$ then $\beta[z(\infty)] < 0.88\beta$ and the underestimation becomes far more substantial.

10.3.3   *The stochastic threshold theorem*

A far more important consideration is whether a deterministic threshold is appropriate at all. For the idea that an outbreak can or cannot occur as the population of susceptibles switches from being just above to just below a threshold value is clearly fanciful. What seems far more likely to happen is that the *probability* that an outbreak occurs will change.

Let $X(t)$ and $Y(t)$ denote the stochastic number of susceptibles and infectives, respectively, at time t; the number of removals $Z(t)$ is automatically given by $n - X(t) - Y(t)$. Then in the small time interval $(t, t + h)$ we have the transition probabilities

$$\Pr\{(X, Y) \to (X - 1, Y + 1)\} = \beta X Y h \quad \text{(infection)}$$
$$\Pr\{(X, Y) \to (X, Y - 1)\} = \gamma Y h \quad\text{(removal)}.\qquad(10.31)$$

The usual argument for constructing the equations for the probabilities

$$p_{ij}(t) = \Pr\{X(t) = i, Y(t) = j\}$$

leads to

$$p_{ij}(t + h) = \beta h(i + 1)(j - 1)p_{i+1, j-1}(t) + \gamma h(j + 1)p_{i, j+1}(t)$$
$$+ [1 - (\beta ij + \gamma j)h]p_{ij}(t).\qquad(10.32)$$

Whence dividing both sides of (10.32) by h and then letting $h \to 0$ yields the set of differential equations

$$dp_{ij}(t)/dt = \beta(i + 1)(j - 1)p_{i+1, j-1}(t) - (\beta i + \gamma)jp_{ij}(t)$$
$$+ \gamma(j + 1)p_{i, j+1}(t). \tag{10.33}$$

These equations are extremely difficult to work with, and are still taxing the imagination of applied probabilists who are trying to develop new ways of handling them. However, numerical solutions can be obtained fairly easily by the crude means of evaluating (10.32) over successive time points $t = 0, h, 2h, \ldots$ for appropriately small h.

A different theoretical approach is clearly required, and comes from noting from (10.31) that with one initial infective the transition rates βXY, which successively equal $\beta(n - 1)$, $2\beta(n - 2)$, $3\beta(n - 3), \ldots$, lie close to $n\beta Y$ for small Y. Thus in the opening stages of the epidemic the number of infectives $Y(t)$ behaves like a simple birth–death process with parameters $\lambda = n\beta$ and $\mu = \gamma$. Now we know from (2.54) that for this associated process the probability of ultimate extinction is given by

$$q = \begin{cases} \mu/\lambda = \gamma/n\beta = \rho/n & \text{if } \lambda > \mu, \text{ i.e. if } n > \rho \\ 1 & \text{if } \lambda \leqslant \mu, \text{ i.e. if } n \leqslant \rho. \end{cases} \tag{10.34}$$

Moreover, if $\lambda > \mu$ and the population has managed to grow to size a, then the probability of extinction from then on, namely $(\mu/\lambda)^a$, is substantially reduced. So the future of the epidemic is determined in its opening stages, when its similarity with the simple birth–death process is strongest. We therefore have the basic *stochastic threshold theorem*:

(i) if $n \leqslant \rho$ then a major outbreak cannot occur,
(ii) if $n > \rho$ then a minor or major outbreak occurs with probability ρ/n and $1 - \rho/n$, respectively.

Note that if a infectives are initially present, instead of just one, then these probabilities change to $(\rho/n)^a$ and $1 - (\rho/n)^a$.

Whittle (1955) strengthens this result by determining the probability that an epidemic of not more than a given *intensity* takes place, where the intensity of an epidemic is defined to be the proportion of the total number of susceptibles who eventually contract the disease.

10.3.4 *Stochastic realizations*

Mathematicians have found considerable enjoyment in developing the theory of epidemics, and readers interested in seeing the many and varied lines of mathematical arguments employed are well advised to consult

Bailey's (1975) excellent compendium of results. For example, the average size of the epidemic, excluding the initial a infectives, is approximately given by

(i) $an/(\rho - n)$ (if $n < \rho$)

(ii) $(\rho/n)^a[a\rho/(n - \rho)] + [1 - (\rho/n)^a](r - a)$ (if $n > \rho$)   (10.35)

where r is the unique positive root of the equation

$$a - r + n[1 - \exp(-r/\rho)] = 0. \tag{10.36}$$

These results are easy to compute numerically (just solve (10.36) by trial and error) and provide a useful indication of how many susceptibles may eventually become infected. Moreover, if n is large then the average duration time of the epidemic is given very simply by

$$T_{ave} \simeq \gamma^{-1} \log_e(a + n). \tag{10.37}$$

To see how individual realizations behave we have to simulate the process, and this is particularly easy since all that is required is a minor modification to the *single*-species program GENONE (Section 3.4 and Appendix to Chapter 3). Although there are three types of individual, we see from (10.31) that only two types of event are possible. So on letting $\{W\}$ denote a sequence of uniform pseudo-random variables in the range 0 to 1, we have:

(i) if $W \leqslant \beta X Y/(\gamma Y + \beta X Y) = \beta X/(\gamma + \beta X)$ then $X \to X - 1$, $Y \to Y + 1$ (infection), otherwise $Y \to Y - 1$, $Z \to Z + 1$ (removal);

(ii) the inter-event times $= -[\log_e(W)]/\{Y(\gamma + \beta X)\}$.   (10.38)

Consider a general epidemic with parameter values $\beta = 0.01$ and $\gamma = 0.5$, starting from $X(0) = 99$ and $Y(0) = a = 1$. Since the threshold value $\rho = \gamma/\beta = 50$ is substantially less than $n = 100$, Kermack and McKendrick's result (10.29) suggests that major outbreaks will be fairly severe. Moreover, the stochastic result (10.34) shows that these have an even chance of occurring since the extinction probability $q = \rho/n = 0.5$. To determine the average size ($\bar{w}$) of the epidemic we first solve equation (10.36) to obtain $r = 81.4$, whence result (10.35ii) tells us that $\bar{w} = r/2 = 40.7$. From (10.37) the average duration time over all outbreaks is $T_{ave} \simeq 2\log_e(101) = 9.2$, so T_{ave} for a major outbreak is roughly double this at 18.4.

Two such simulated epidemics are shown in Figure 10.2 and illustrate the striking difference between major and minor outbreaks. The minor outbreak soon fizzles out and results in a total of only seven removals with a maximum of five infectives being present at any one time. In contrast, the major outbreak results in three-quarters of the susceptible population becoming infected, with the number of available infectives peaking at 21. So this general epidemic process really does hammer home just how different the stochastic and deterministic approaches can be, for the latter does not predict this dual behaviour at all.

To assess the accuracy of our theoretical results this process was simulated 100 times and both the duration times (Figure 10.3a) and the final epidemic sizes (Figure 10.3b) were recorded. Of the 100 runs 49 and 51 result in minor and major outbreaks, respectively, in almost exact agreement with the 50–50 split predicted theoretically. Minor duration times range from 0.02 to 6.14, and major duration times from 12.1 to 39.6. The latter have a median value of 19 which is very close to our crude theoretical average of 18.4, though the mean value 21 is slightly higher (due to the skewness of the distribution). The histogram of final epidemic size highlights the great difference between minor and major outbreaks. Of the 49 minor ones, 37 result in no new infection at all, and the remaining 12 range from two to five removals. So for this process a simulation with seven removals (as in Figure 10.2) is unusually large as minor outbreaks go. Apart from the isolated value 24 (there always has to be

Figure 10.2. Number of infectives (———) and removals (— — —) for a major outbreak of a simulated general epidemic with $\beta = 0.01$ and $\gamma = 0.5$, together with the number of infectives (– – – –) for a minor outbreak.

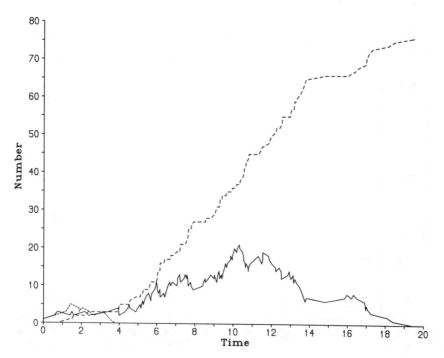

one that's awkward!), all of the remaining 50 major outbreaks have $57 \leqslant z(\infty) \leqslant 96$, i.e. a loss of just over half the population to almost total annihilation. The agreement between the average epidemic size (major and minor) of 41.9 and the theoretical value $\bar{w} = 40.7$ is surprisingly good, all of which illustrates just how useful simple, albeit approximate, mathematical results can be in helping to describe the outcome of a process.

10.4   Recurrent epidemics

For the general epidemic we have shown that either a minor outbreak occurs which then swiftly dies away, or else there is a major build-up of infectives which then slowly subsides. This description is clearly useful for rare diseases, like (present-day) plague or smallpox, since any outbreak that does occur can be regarded as a single phenomenon. However, with more common diseases like measles, chicken-pox, influenza, diptheria, etc., there are periodic flare-ups with infection being sustained at a low level

Figure 10.3. Results from 100 simulations of the epidemic process shown in Figure 10.2: (a) duration times, (b) final sizes.

(a)

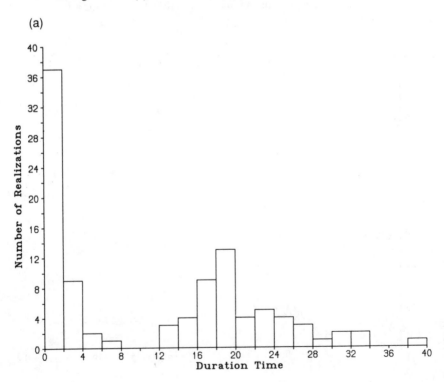

inbetween times by a gradual spread to new susceptibles. These incomers can either be new recruits to the population or previously infected individuals who have lost their immunity. To be strictly accurate such diseases are really *endemic* rather than epidemic, since they are always with us, but as the flare-ups often tend to be oscillatory in nature we shall regard them as *recurrent*.

Bartlett (1960) gives the monthly notifications (Table 10.1) of measles and chicken-pox in Philadelphia for 1941–43, the statistics for New York City, Baltimore and other large towns being similar in character. Comparison of the two diseases shows a much more stable seasonal rhythm for chicken-pox, in contrast with the more violent measles outbreaks. Such phenomena appear to be universal; a plot of the observed measles notifications for Bristol, UK, for 1945–55 shows exactly the same recurrent unstable behaviour (Bartlett, 1957b).

He observed that in a study of nineteen towns in England and Wales ranked according to population size N, the mean period between successive measles flare-ups varied with $N^{-\frac{1}{2}}$. Moreover, the persistence of the oscillations also depended on N. He noted sustained oscillations in two

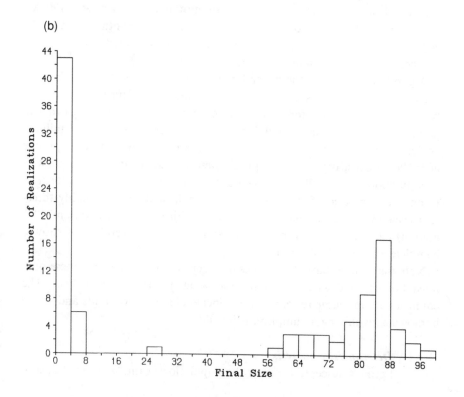

(b)

Table 10.1. Measles and chicken-pox monthly notifications, Philadelphia (Bartlett, 1960)

Month	Measles			Chicken-pox		
	1941	1942	1943	1941	1942	1943
Jan.	2906	62	4923	705	1021	556
Feb.	4770	93	4759	555	991	439
Mar.	6991	132	3583	720	1167	461
Apr.	5457	222	1428	827	1144	432
May	2203	194	1198	582	974	550
June	527	135	821	739	531	548
July	77	70	235	79	92	214
Aug.	8	38	10	25	16	80
Sept.	7	44	14	22	18	32
Oct.	10	275	22	87	124	160
Nov.	19	1122	23	360	227	345
Dec.	16	2770	22	730	387	585

towns (Birmingham and Manchester) of more than 650 000 inhabitants; occasional fade-out (Bristol, Hull and Plymouth) in three towns with 180 000 to 420 000 inhabitants; and invariable fade-out in all of the other 14 towns he considered (fewer than 115 000 inhabitants). A suggested threshold for measles is *c.* 250 000 inhabitants.

A good illustration of the dependence of periodicity on N is provided by Figure 10.4 which shows the number of reported cases of measles per month between 1945 and 1970 for four countries, arranged in decreasing order of population size. In the United States, with a population of 210 million, epidemic peaks occur every year (note the dramatic reduction in amplitude after 1964 because of vaccination programmes). In the United Kingdom, with a population of 56 million, peaks occur every two years. The pattern in Denmark, population 5 million, is more complicated with a roughly three-year cycle in the second half of the period. Whilst in contrast, Iceland (0.2 million) has only 8 peaks in the 25-year period, and several years show a complete absence of infection.

Note that unlike Bartlett's studies on a series of urban centres, these data must not be used as input to the following mathematical analysis. The countries are too large to represent distinct epidemiological units and so the homogeneous-mixing assumption is invalid.

10.4.1 *Deterministic analysis*

Let new susceptibles enter the population at rate α. Then the basic

Figure 10.4. Number of reported cases of measles per month between 1945 and 1970 for four countries, arranged in descending order of population size. (Reproduced from Cliff, Haggett, Ord and Versey, 1981, by permission of Cambridge University Press.)

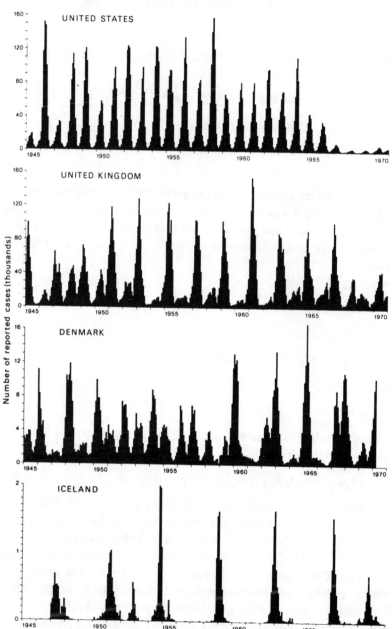

deterministic equations (10.2) become

$$dx/dt = -\beta xy + \alpha$$
$$dy/dt = \beta xy - \gamma y. \tag{10.39}$$

The effect of small departures from the equilibrium values

$$x^* = \gamma/\beta \quad \text{and} \quad y^* = \alpha/\gamma \tag{10.40}$$

can be studied in exactly the same way as for the competition and predator–prey processes. On writing

$$x = x^*(1 + u) \quad \text{and} \quad y = y^*(1 + v)$$

for small u and v, equations (10.39) yield

$$du/dt = -(\alpha\beta/\gamma)(u + v)$$
$$dv/dt = \gamma u. \tag{10.41}$$

Now these equations are identical to the general two-species equations (6.24) and (6.25) with $k_{11} = k_{12} = -\alpha\beta/\gamma$, $k_{21} = \gamma$ and $k_{22} = 0$, so their solution follows along standard lines as laid down in Section 6.2.2. In particular, on denoting $\sigma = \gamma/\alpha\beta$ and eliminating u, we obtain the second-order differential equation

$$\sigma \frac{d^2v}{dt^2} + \frac{dv}{dt} + \gamma v = 0. \tag{10.42}$$

This has the solution

$$v(t) = v(0) \exp\{-t/2\sigma\} \cos(\xi t), \tag{10.43}$$

where

$$\xi = \sqrt{\{(\gamma/\sigma) - (1/4\sigma^2)\}} \tag{10.44}$$

and the time-origin is suitably chosen to ensure that $D_2 = 0$ in (6.38). Equations (10.41) then give

$$u(t) = \gamma^{-1}dv/dt = -v(0)(\gamma\sigma)^{-\frac{1}{2}}\exp\{-t/2\sigma\} \cos(\xi t - \theta) \tag{10.45}$$

where $\cos(\theta) = 1/\sqrt{(4\gamma\sigma)}$. So both the infective and susceptible populations undergo damped cosine oscillations with period $2\pi/\xi$. Note that the phase difference θ is *not* $\pi/2$, unlike the earlier Lotka–Volterra process.

In his early deterministic work on the periodicity of recurrent outbreaks of measles Soper (1929) took γ^{-1} to be two weeks, since this is the approximate incubation period, and estimated σ from London data to be roughly 68 weeks. So from (10.44) we have $\xi = 0.0854$, giving a period (T) of $2\pi/\xi = 74$ weeks and a peak-to-peak damping factor of $\exp\{-T/2\sigma\} = 0.58$. Thus whilst the assumption of a constant influx of new susceptibles does give rise to epidemic waves with roughly the right period, the swift damping of the infective cycles towards a steady endemic state clearly contradicts the epidemiological facts.

10.4.2   *Stochastic considerations*

However, rather than discarding this model out of hand, let us recall the Volterra predator–prey process (Section 6.2). Although the deterministic Volterra solution exhibits damped cycles (Figure 6.4), its stochastic counterpart can sustain irregular cycles (Figure 6.5). Let us therefore see what happens when we simulate our recurrent epidemic; a simple operation since we just replace the exact probabilities (10.38) by

(i) if $W \leqslant \beta X Y/(\alpha + \gamma Y + \beta X Y)$ $(= A$, say$)$ then $X \to X - 1$,
$Y \to Y + 1$,
if $A < W \leqslant (\gamma Y + \beta X Y)/(\alpha + \gamma Y + \beta X Y)$ then $Y \to Y - 1$,
otherwise $X \to X + 1$;

(ii) the inter-event times $= -[\log_e(W)]/(\alpha + \gamma Y + \beta X Y)$.

$$(10.46)$$

Retaining Soper's values of $\gamma = 0.5$ and $\sigma = 68$, suppose that the equilibrium number of infectives $y^* = 100$. Then from (10.40) $\alpha = \gamma y^* = 50$, whence $\beta = \gamma/\alpha\sigma = 0.000\ 147$ and the equilibrium number of susceptibles $x^* = \gamma/\beta = \alpha\sigma = 3400$. Simulations of this model starting a long way from the equilibrium value (x^*, y^*) generally led to a huge upsurge in the number of infectives and a correspondingly large drop in the number of susceptibles. So when the inevitable infection crash occurred the susceptible population could not grow fast enough to pass the point $x = x^*$ (at which the rate of growth of the infectives, $\beta y(x - x^*)$, becomes positive again) before the infectives died out. Conversely, simulations starting nearer to (x^*, y^*) initially followed a deterministic-type path before wandering about (x^*, y^*) thereafter. If this model is to lead to sustained flare-ups then we therefore need: (i) $(X(0), Y(0))$ to be substantially different from (x^*, y^*); and (ii) the occasional infective entering the system from outside to ensure that the epidemic restarts should the infective population die out.

Figure 10.5 shows one such simulation for an infective immigration rate $\delta = 0.1$, i.e. on average one new infective arrives every 10 weeks. Numbers of infectives are given at weekly intervals over a twenty-year period (1000 weeks), and the simulation starts well away from (x^*, y^*) with $X(0) = 5000$ and $Y(0) = 5$. Seven flare-ups occur during the first 14 years, though the mean cycle period of 97 weeks is considerably higher than the deterministic prediction of 68 weeks since the process takes some time to restart whenever the infective population dies out. After 14 years the simulation undergoes a dramatic change, because during the next infective trough there are now just enough infectives remaining to keep the susceptible population in check. This causes a substantial reduction in the next infective peak, and so the trajectory $\{X(t), Y(t)\}$ passes near enough to the equilibrium state (x^*, y^*) for it to become trapped into undergoing damped cycles towards an endemic level. Although the stochastic peak-to-peak damping factor 0.73 is higher than the deterministic prediction 0.58, it is nevertheless qualitatively similar.

For this situation to be reversed the infective population would have to wander substantially below y^*, i.e. near to temporary extinction. A rough idea of the probability of this event can be obtained by regarding the susceptible population size as being fixed equal to x^*, and then treating $\{Y(t)\}$ as a simple birth–death process with parameters $\lambda = \beta x^* = \gamma$ and $\mu = \gamma$, respectively. Since $\lambda = \mu$, and disregarding the occasional infected immigrant, the probability of infective extinction is given by (2.55), namely

$$p_0(t) = [\tfrac{1}{2}t/(1 + \tfrac{1}{2}t)]^{y^*} = [1 + (2/t)]^{-100}. \tag{10.47}$$

Figure 10.5. A simulated stochastic recurrent epidemic corresponding to the deterministic representation $\mathrm{d}x/\mathrm{d}t = -\beta xy + \alpha$ and $\mathrm{d}y/\mathrm{d}t = \beta xy - \gamma y + \delta$, with $\alpha = 50$, $\beta = 0.000147$, $\gamma = 0.5$ and $\delta = 0.1$.

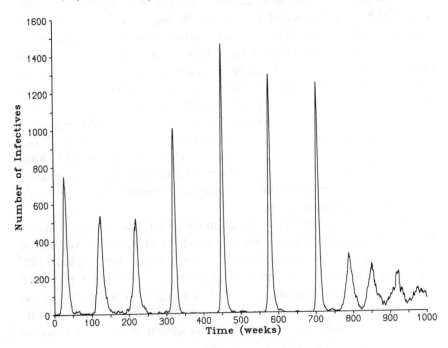

This expression takes the values

t	10	100	1000	10 000
$p_0(t)$	1.21×10^{-8}	0.14	0.82	0.98

with the median time to extinction, given by $p_0(t) = 0.5$, being $t = 288$ (weeks). So on a human time-scale this particular process has the potential for flip-flopping between the epidemic and endemic states.

The behaviour of small fluctuations about the deterministic equilibrium values $x^* = \gamma/\beta$ and $y^* = \alpha/\gamma$ can be determined by examining the variance-covariances of the number of infectives and susceptibles under the assumption that oscillations are maintained indefinitely. Following the approach of Section 6.2.4 (which led to the variance-covariances (6.48)) gives rise to:

$$\sigma_X^2 = x^*(1 + x^*/y^*) = 119\,000, \quad \text{so } \sigma_X = 345;$$
$$\sigma_Y^2 = x^* + y^* = 3500, \quad \text{so } \sigma_Y = 59;$$

and

$$\sigma_{XY} = -x^* = -3400$$

(Bailey, 1975). Unfortunately, the assumption of small fluctuations relegates the applicability of these values to the endemic state. However, the fact that $x^* = 3450 \simeq 10\sigma_X$ and $y^* = 100 < 2\sigma_Y$ does suggest that in the far more interesting flare-up situation early short-term extinction of infectives (susceptibles) is likely (highly unlikely). This ties in with our simulation experience.

10.4.3   *The Lotka–Volterra approach*

Since equations (10.39) represent an extremely simple model it is tempting to try and make them more epidemiologically realistic. For example, we might allow for an incubation period or for seasonal variations in infection rate (see Bailey, 1975). However, such modifications also give rise to damped deterministic cycles, and the wish to obtain undamped cycles suggests considering disease which is not only lethal to all those who contract it but which is also sufficiently virulent to suppress any live births among circulating infectives. Let us therefore suppose that susceptibles, but not infectives, give birth. Then on replacing the immigration term α by the birth term λx, equations (10.39) become

$$dx/dt = -\beta xy + \lambda x$$
$$dy/dt = \beta xy - \gamma y. \tag{10.48}$$

Now these epidemic equations correspond to the Lotka–Volterra equations (6.3), with predators and prey being analogous to infectives and susceptibles, respectively. Since the predator–prey trajectories are a family of closed curves (Figure 6.1) it therefore follows that this particular epidemic process does indeed give rise to sustained deterministic cycles. Moreover, direct use of results (6.12) and (6.14) with $N_1(t) = x(t)$, $N_2(t) = y(t)$, $b_1 = b_2 = \beta$, $r_1 = \lambda$ and $r_2 = \gamma$ yields the undamped solution

$$x(t) = (\gamma/\beta)[1 + A \cos\{t\sqrt{(\gamma\lambda)} + B\}]$$
$$y(t) = (\lambda/\beta)[1 + A\sqrt{(\gamma/\lambda)} \sin\{t\sqrt{(\gamma\lambda)} + B\}]. \tag{10.49}$$

The constants A and B are determined from the initial values

$$x(0) = (\gamma/\beta)[1 + A \cos(B)] \quad \text{and} \quad y(0) = (\lambda/\beta)[1 + A\sqrt{(\gamma/\lambda)} \sin(B)].$$

Note that this model cannot apply to milder diseases such as measles and chicken-pox.

This equivalence does not extend to stochastic solutions, since in the predator–prey context the interaction term βxy gives rise to two separate changes of state (i.e. $X \to X - 1$ and $Y \to Y + 1$ are different events), whilst in the epidemic context the two changes of state occur together (i.e. $X \to X - 1$ and $Y \to Y + 1$ simultaneously). Hence stochastic simulations of these two deterministically equivalent processes will not necessarily give rise to similar behaviour. In particular, it does not automatically follow that stochastic epidemic trajectories will dive onto an axis (see Figure 6.3) resulting in the early extinction of either infectives or susceptibles. Indeed, simulations of the equivalent process to that shown in Figure 10.5 (i.e. $\beta = 0.000\ 147$, $\gamma = 0.5$ and $\lambda = \beta y^* = 0.0147$), but this time with infective immigration rate $\delta = 0$, showed that it was not unusual for six or seven full cycles to occur before the infectives became extinct. In one such run, for example, the infective population peaked six times in the range 200–360, then the seventh peak of 620 proved too much for the process to cope with and the infectives crashed to extinction. Such realizations certainly qualify as genuine recurrent epidemics on a human time-scale; whilst introducing the occasional immigrant infective, as before, enables the process to persist indefinitely.

10.5   An extension to malaria

The simple, general and recurrent epidemic models are mathematically very simple, and possibilities abound for making them more realistic. In particular, much work has already been done based on bacterial models which describe the flow of host individuals between various states of infection (e.g. latent, infectious, immune, carrier, death). These states can be thought of as 'compartments', with an individual's change of state being viewed as a migration from one compartment to another.

Although such compartmental frameworks are well established they need to be handled delicately, since there are two distinct attitudes towards using them. In one the emphasis is theoretical with little or no attempt being made to tie together mathematical results with biological observation. In the other, mathematics is considered purely as a means of deriving biological conclusions which are testable, and any elegant mathematics which is developed *en route* is simply icing on the cake (see Gross, 1986). However well intentioned, it is all too easy to drift from the second attitude to the first, and care must be taken to ensure that the rules for moving between compartments are based far more on biological reality than mathematical expediency.

Fortunately, sufficient is known about certain diseases to enable the construction of fairly accurate models. Cvjetanović (1982), for example, describes compartmental systems for tetanus, typhoid, cholera and diptheria, using 9, 10, 11 and 10 states, respectively. A strong partnership between deterministic theory and stochastic simulation is clearly vital if meaningful conclusions are to be drawn in such complex situations.

10.5.1   *The basic deterministic model*

These four systems are too involved for the purpose of this text. So to illustrate the way in which simple models can be usefully extended let us consider the transmission and maintenance of malaria, since this disease can be modelled as a basic two-state system. The account of Aron and May (1982) is a good one to follow, since it emphasizes the relation between biology and mathematics, focussing both on the role played by the basic reproductive rate of the parasite and on the dynamics of the prevalence of infection among humans and mosquitoes. The basic deterministic model, which incorporates the interaction between the infected human hosts and the mosquito vector population, but little else, may be written in the form

$$dx/dt = \sigma y(1 - x) - rx \qquad (\sigma = abM/N) \tag{10.50}$$

$$dy/dt = ax(1 - y) - \mu y \tag{10.51}$$

where:

> x and y are the proportions of the human and female mosquito populations that are infected;
>
> N and M are the (constant) sizes of the human and female mosquito populations;
>
> a is the bite rate of a single mosquito;
>
> b is the proportion of infected bites that produce an infection;

r is the individual recovery rate for humans;

μ is the individual death rate for mosquitoes.

Equation (10.50) describes how the proportion of humans infected changes: new infectives accrue at a rate which depends on the number of mosquito bites per person per unit time (aM/N), on the probabilities that the biting mosquito is infected (y) and that a bitten human is not infected ($1 - x$), and on the probability that an uninfected bitten person will become infected (b); infected people recover at rate *r*. Similarly, equation (10.51) describes changes in the proportion of mosquitoes infected: gains are proportional to the number of bites per mosquito per unit time (a), and to the probabilities that the biting mosquito is uninfected ($1 - y$) and that the bitten human is infected (x); infected mosquitoes die at rate μ.

Aron and May (1982) are quick to point out that this model is highly simplified; in particular it fails to take account of the different developmental stages of the parasite. Nevertheless, it does contain the essentials of the transmission process and enables distinctions to be made between patterns found in various sets of data from different geographic regions.

Closer inspection of the deterministic equations (10.50) and (10.51) reveals their novelty, for they do not fall within the general structure (6.1) for two interacting species, namely

$$dN_1/dt = N_1(r_1 + a_{11}N_1 + a_{12}N_2)$$
$$dN_2/dt = N_2(r_2 + a_{21}N_1 + a_{22}N_2).$$

Placing $dx/dt = dy/dt = 0$ shows that the equilibrium values (x^*, y^*) are solutions of the equations

$$\sigma y(1 - x) = rx$$
$$ax(1 - y) = \mu y. \tag{10.52}$$

Eliminating *y* gives

$$x^* = (a\sigma - \mu r)/(a\sigma + ar).$$

Hence on writing

$$R = a\sigma/\mu r = a^2 bM/\mu rN \tag{10.53}$$

we see that

$$x^* = (R - 1)/[R + (a/\mu)], \tag{10.54}$$

whence using (10.52) gives

$$y^* = \frac{r(R - 1)}{\sigma[1 + (a/\mu)]} = \left[\frac{R - 1}{R}\right]\left[\frac{a/\mu}{1 + (a/\mu)}\right]. \tag{10.55}$$

The value R is called the 'basic reproductive rate' of the parasite (see MacDonald, 1957), and is essentially the number of secondary cases of infection generated by a single infected individual. If $R < 1$ then the disease will be unable to maintain itself (x^* must be positive for an equilibrium solution to exist). The more R exceeds 1, the greater is the resistance of the infection to eradication (for further references see Aron and May, 1982).

Stability analysis of the equilibrium point (x^*, y^*) proceeds in exactly the same manner as in Chapters 5 and 6 for the competition and predator–prey processes. On writing the two isoclines (10.52) as

(i) $y = rx/\sigma(1 - x)$ and   (ii) $y = ax/(\mu + ax)$

and differentiating, we have

(i) $dy/dx = r/\sigma(1 - x)^2$ and   (ii) $dy/dx = a\mu/(\mu + ax)^2$.

Hence as x rises from 0 to 1, y rises from 0 to (i) ∞ and (ii) $a/(a+\mu)$; whilst dy/dx rises (i) from r/σ to ∞ and (ii) from a/μ to $a\mu/(a + \mu)^2$. For an equilibrium point to exist these two isoclines must cross, and we see (Figure

Figure 10.6. Phase-plane diagram for the malaria model (10.50) and (10.51): (a) R much greater than 1; (b) R just greater than 1; (c) R less than 1.

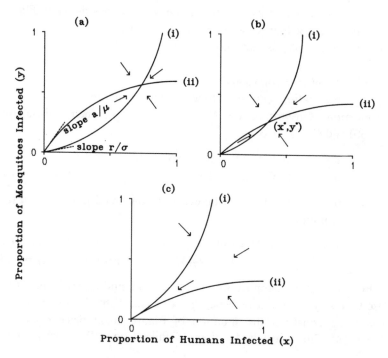

10.6) that for this to happen the initial slope of isocline (ii), a/μ, must exceed that of isocline (i), r/σ. Thus we need $R = (a/\mu) \div (r/\sigma) > 1$. The initial slope a/μ denotes the average number of bites on human hosts made by a mosquito during its lifetime, and if this number is relatively large (Figure 10.6a) then the equilibrium point will lie near to $(1, 1)$ in a fairly deep hole (dynamically speaking). Conversely, if a/μ is fairly small (Figure 10.6b) then the equilibrium point will not only be closer to $(0,0)$ but it will also lie within a shallow, elongated trough.

Comparing these two situations shows that small changes in parameter values will affect the proportion of humans infected in the latter case (a/μ small) more than in the former (a/μ large). Thus a/μ can be treated as an index of stability (MacDonald, 1957). In areas where mosquitoes often bite humans and have relatively long lifespans this index is high and malaria will therefore be endemic (stable malaria); whereas if the opposite holds true then the index is low and the disease will tend to be subject to epidemic outbreaks (unstable malaria).

Finally, if $a/\mu < r/\sigma$ (Figure 10.6c) then all trajectories converge on the origin and so the disease cannot be sustained.

The occurrence of epidemic outbreaks for low a/μ means that the number of human infectives (x) must wander quite appreciably from the equilibrium value (x^*) before the centralizing force towards x^* takes effect. Large outbreaks can therefore only occur if the two isoclines shown in Figure 10.6b remain close together for a large distance above (x^*, y^*), which may clearly be a lot to ask of this particular model.

10.5.2 *Stochastic comments*

Simulation requires us to work with population numbers rather than proportions. On writing $X = Nx$ and $Y = My$ the deterministic equations (10.50) and (10.51) become

$$dX/dt = (\sigma Y/M)(N - X) - rX \quad \text{and}$$
$$dY/dt = (aX/N)(M - Y) - \mu Y, \qquad (10.56)$$

whence the associated stochastic events in the small time interval $(t, t + h)$ have probabilities

$$\Pr(X \to X + 1) = (\sigma Y/M)(N - X)h, \quad \Pr(X \to X - 1) = rXh$$
$$\Pr(Y \to Y + 1) = (aX/N)(M - Y)h, \quad \Pr(Y \to Y - 1) = \mu Yh.$$

For illustration, suppose we choose $a = 0.1$ and $\mu = 0.3$, since this gives rise to a very small but realistic a/μ value (see Table 5.1 of Aron and May, 1982). To construct the situation of Figure 10.6b place $r/\sigma = 2/7$, since $R = a\sigma/\mu r = 7/6$ is just larger than 1 whilst from (10.54) $x^* = 1/9$ is fairly close to zero. Finally, let $N = 1000$, $M = 10\,000$ and $\sigma = 0.05$. On simulating this process over a long time period, $X(t)$ was observed to meander about the equilibrium value $X^* = Nx^* = 111$ over the range 69 to 156. This behaviour contrasts markedly from reality, as portrayed, for example, in Figure 10.7 which shows severe upsurges in the monthly malaria attendance per 1000 population in the upper catchments of the Dedura Oya and Maha Oya basins in the epidemic zone of Sri Lanka.

Bearing in mind the highly simplistic nature of our model there is no reason why close agreement between stochastic realizations and reality should occur, and a greater degree of realism is clearly necessary for this to be achieved. One possible way is to incorporate a fluctuating environment into the model (we have already seen this approach used in a logistic context in Chapter 8). Aron and May (1982), for example, generate large simulated outbreaks by letting the total mosquito population M vary seasonally with a random amplitude. Though they review other extensions to the basic process, detailed specification does not concern us here. What is important is for us to

Figure 10.7. Monthly malaria attendance per 1000 population in the epidemic zone of Sri Lanka (reconstructed from Rajendram and Jayewickreme, 1951).

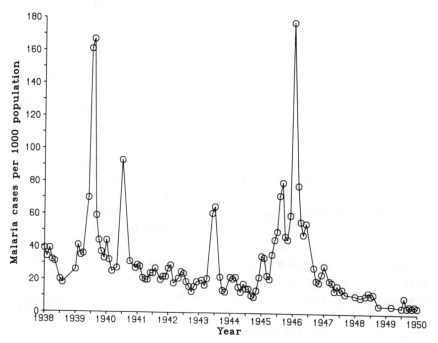

realize that use of a simple model structure enables understanding to be gained into the nature, and thereby possible control, of the disease.

For example, incorporating the incubation period (s) of the parasite within the mosquito puts us back into the time-lag situation of Chapter 4. The basic reproductive rate is now

$$R = (a^2bM/\mu rN) \times \exp\{-\mu s\},$$

and this leads to the important conclusion that imagicides are more effective than larvicides (MacDonald, 1957) – reducing larval recruitment by a factor of n reduces M/N to M/Nn and thus R to $R_1 = R/n$, whilst a corresponding increase in the adult mortality rate from μ to μn changes R to R_2 $= (a^2bM/n\mu rN) \times \exp\{-n\mu s\}$ which decreases exponentially faster than R_1 with increasing n.

10.6   Spatial models

In many communities susceptibles who live a long way from infected individuals have a smaller chance of becoming infected than those who live in close proximity to the disease, and so the basic homogeneous-mixing assumption which underlies the simple and general epidemic models is violated. For example, a few of the passengers arriving at an Icelandic port from Copenhagen in the early summer of 1907 were suffering from measles. One filtered through the quarantine procedure and joined the local population, though the ensuing epidemic outbreak was contained. Unwittingly, an infective was then allowed to travel to Reykjavík, and mixed with crowds gathering for a visit by the King of Denmark. The measles virus was transmitted to some neighbours in the crowd, resulting in an epidemic which took 16 months to spread through the whole island, infecting 7000 people and killing 354.

A rich source of similar situations involving the geographic spread of infection (and other phenomena) is the entertaining text of Cliff, Haggett, Ord and Versey (1981). One such situation is the spread of one of the world's greatest cholera pandemics, the El Tor strain, between 1960 and 1971 (Figure 10.8). This strain was first identified in the bodies of six Muslim pilgrims outside Mecca in 1905, and was later recognized in the 1930s as being endemic in the Celebes which has a largely Muslim population. Little was heard of it for the next 30 years, until in 1961 it suddenly exploded out of the Celebes, reaching India in 1964 (replacing the normal cholera strain which had been endemic in the Ganges delta for centuries), and advancing into central Africa, Russia and Europe by the early 1970s.

10.6.1   *Deterministic spread*

Regardless of whether disease spreads by wind (e.g. stem rust) or direct contact (e.g. measles or AIDS) it will enter different geographic regions at different times, and an idea of the likely velocity of spread can be obtained by exploiting our previous results on spatial birth–death processes. Near the wavefront of an advancing epidemic the number of susceptibles may be assumed to be fairly constant. So the non-spatial deterministic equation for the number of infectives in the simple epidemic case, namely

$$dy(t)/dt = \beta x(t)y(t),$$

may be approximated by the *local* equation

$$dy(t)/dt \simeq (\beta N)y(t). \tag{10.57}$$

Here $x(t)$ and $y(t)$ denote the local density (i.e. the number per unit area) of susceptibles and infectives, respectively.

Now equation (10.57) describes a simple birth process with birth rate $\lambda = \beta N$. So if we allow diffusion to occur at rate b^2 then result (9.29) immediately tells us that the velocity of propagation $v = b\sqrt{(\beta N)}$. In the Discussion to Soper (1929), Halliday suggested that a measles outbreak starting in September took about 24 weeks to spread over the 37 wards of Glasgow. If this area is crudely regarded as a circle of radius 2 miles, then the velocity $v = 1/12$ mile per week. Inserting appropriate values for β and N then gives an estimate of the diffusion parameter $b = 1/[12\sqrt{(\beta N)}]$ (for Glasgow).

Figure 10.8. Spread of the El Tor cholera pandemic, 1960–71.
(Reproduced from Cliff, Haggett, Ord and Versey, 1981 by permission of Cambridge University Press).

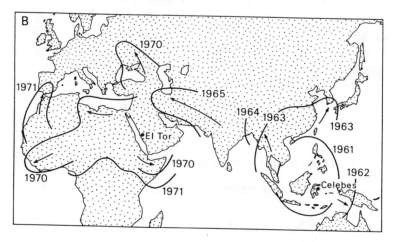

However, even this simple mathematical result has to be treated very carefully, since Halliday comments that as far as Glasgow is concerned all late epidemics collapse suddenly at the end of May, often leaving large tracts of the city unaffected. So an epidemic starting in say February would not be able to cover the whole city, and as a result of this partial immunization measles would recur as an outbreak in the following annual period thus breaking the standard two-year measles cycle.

The same approach can be used when studying the spread of infection through separate geographic regions. Using the simple stepping-stone representation, we see from result (9.66) that with one-way migration the velocity of propagation c is the root of the equation

$$v - \beta N = c - c \log_e(c/v) \tag{10.58}$$

if the infectives have migration rate v and zero death rate.

Note that c is a *deterministic* velocity, whereas in practice we are often interested in the speed of advance of the *stochastic* wave of newly infected colonies. Fortunately the corresponding stochastic velocity can be shown to be no greater than c (Mollison, 1977), and so expression (10.58) yields an upper bound to the velocity of infection. Moreover, for large N it can be shown that

$$c \simeq \beta(N + 1)/\log_e\{\beta(N + 1)/v\} \tag{10.59}$$

which is easy to compute.

That c is an upper bound to the exact deterministic velocity is intuitively reasonable. Since $N \geqslant x(t)$ the approximate force of infection βN generally exceeds, and is never less than, the true force $\beta x(t)$. So the approximation credits the epidemic with more drive than it actually possesses.

A lower bound can be obtained by removing all infectives other than those at the wavefront itself (Faddy and Slorach, 1980). The wavefront of this modified system will advance more slowly than that of the original system because of the absence of infectives behind it. Let r and $N + 1 - r$ denote the number of infectives and susceptibles, respectively, at the wavefront site. Then since in the small time interval $(t, t + h)$ the probability of either a susceptible becoming infected or an infective migrating is $\{\beta(N + 1 - r)r + vr\}h$, it follows that the time T_r to the next event is exponentially distributed with mean

$$E(T_r) = 1/\{\beta(N + 1 - r)r + vr\}.$$

A standard probability argument then leads to the lower bound

$$c_\ell = \beta(N + 1)\bigg/ \sum_{k=0}^{N} (k + v/\alpha)^{-1}. \tag{10.60}$$

As N increases, the ratio c/c_ℓ of the upper and lower bounds approaches 1, with the true velocity being sandwiched between them.

Faddy and Slorach (1980) simulate this epidemic process with 10 susceptibles in each colony and just one initial infective, that being sited in colony 0. In every simulation the wave of infectives settled down quickly to a steady rate of advance, and with $\alpha = 1$ their results

v	Lower bound	Simulated velocity	Upper bound
1	4	5	8
10	14	20	27
20	25	35	43

demonstrate that the simulated velocities do indeed lie within the theoretical bounds.

10.6.2   Stochastic spread

At this point it is worth remembering that all the spatial models we have so far investigated have involved nearest-neighbour migration. In the epidemic context this means infection by direct contact (such as with measles). However, not all disease spreads in this mathematically convenient way. For example, with Black Death people fleeing from infected areas could cross several sites in a single jump; whilst stem rust has a very wide airborne distribution and can take individual jumps of over 500 km. In one-dimensional simulations of such systems Mollison (1972) not only produces interesting graphplots which show the development of wavefronts under different types of jump distribution, but he also demonstrates that in extreme circumstances the wavefront can progress in 'wilder and wilder leaps forward'.

Simulation clearly has considerable potential as a tool for learning about the spread (and thereby the control) of such diseases, an obvious example being the development of foot-and-mouth disease which spreads by both airborne and direct contact mechanisms. Ferret distemper also spreads by both these mechanisms. Kelker (1973) simulates this latter situation by supposing that 100 ferrets are confined in single pens arranged in a 20 × 5 rectangular array in a draughty building; an epidemic is started by placing a single infected ferret in the centre pen of the first row.

Note that simulating epidemic spread in two dimensions can lead to very different results from those in one dimension. Whilst in one dimension infection can only spread between two sites by passing through all the intervening ones, in two dimensions it can bypass directly intervening sites by taking a more circuitous route.

Bartlett (1957b) examined the consequence of including a spatial effect in his recurrent epidemic study by considering a 6×6 square grid of sites, this being the largest layout that was computationally feasible at the time. Infectives could move from their own site to any neighbouring site with a common boundary. Although the threshold value for sustained oscillations was shown to be relatively unaffected by this spatial factor, he demonstrated that persistence of infection in just a few of the sites was sufficient to begin a new epidemic. This persistence characteristic parallels our results for the spatial predator–prey process (Chapter 7).

Whilst Bartlett's prime concern was the study of recurrent epidemics, a detailed simulation study by Bailey (1967) concentrates on the spatial phenomenon itself. He considers a square lattice of size 11×11, with the epidemic being started by a single infective at the centre, and with infection being transmitted to the *eight* neighbouring sites. The epidemic spreads away from the centre until the boundary is reached. A discrete time approach is used, with results being obtained for both the simple epidemic (without removals) and the general epidemic (with removals).

Let the probability of an exposed susceptible becoming infected be p. If it is exposed to r infectives the chance of it not contracting the disease is $(1 - p)^r$, whence the chance that it does become infected is $1 - (1 - p)^r$. Thus when $p = 1$, at the gth generation ($g = 1, 2, \ldots, 5$) the disease completely covers the square with corners at ($\pm g, \pm g$). Since the number of infectives at each site at the gth generation is known, each susceptible in the $(g + 1)$th generation can be examined to see whether it becomes infected or not. Figure 10.9 gives Bailey's results for 25 simulations of a general epidemic, spreading over the 11×11 lattice, with (i) $p = 0.2$ and (ii) $p = 0.8$. They show: (a) the average epidemic curve for each $(2g + 1) \times (2g + 1)$ square; (b) the average epidemic curve for each individual site in these five squares; and (c) the average total number of new cases. We see that when $p = 0.2$ the general epidemic is very localized with its presence being felt less and less as we move away from the centre. Conversely, when $p = 0.8$ the epidemic explodes directly out to the boundary.

With current high-speed computers there is nothing to prevent more highly structured models being developed that contain far more realistic features such as immunization campaigns, quarantining, social and geographical details, and barriers to movement. The general insight that such work would provide could well be quite powerful.

The difference in behaviour between $p = 0.2$ and 0.8 shows the obvious

Figure 10.9. General epidemic on an 11 × 11 square with (i) $p = 0.2$, (ii) $p = 0.8$. (Reproduced from Bailey, 1975, by permission of Edward Arnold.)

(i)

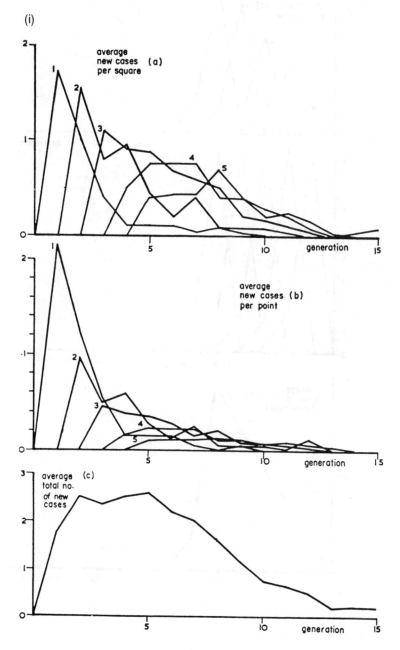

(ii)

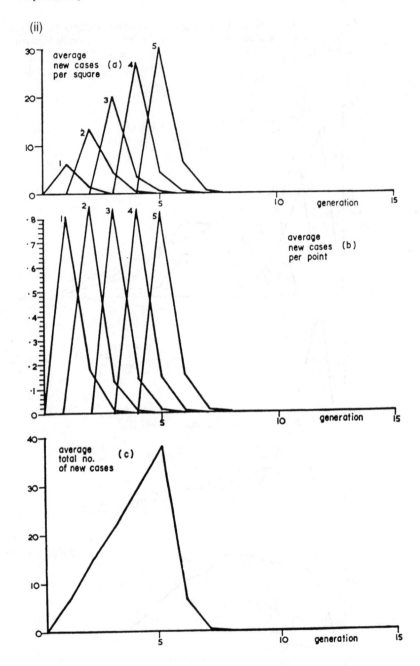

presence of a threshold effect. Indeed, the total number of new cases in 25 simulations, namely

New cases	0–20	21–40	41–60	61–80	81–100	101–200
$p = 0.20$	13	6	5	1	0	0
$p = 0.36$	0	0	0	1	3	21

highlights a very marked change in form between $p = 0.20$ and $p = 0.36$. That a spatial threshold exists has been shown theoretically by D. G. Kendall (see Discussion of Bartlett, 1957b). Let individuals be uniformly distributed across the region with density σ per unit area, and denote the number of individuals in any small area of size A who are susceptible, infectious or removed by $\sigma x A$, $\sigma y A$ and $\sigma z A$, respectively, where the proportions x, y and z sum to one. Then the non-spatial general epidemic equations (10.18) may be written in the spatial form

$$dx/dt = -\beta\sigma x\tilde{y}$$
$$dy/dt = \beta\sigma x\tilde{y} - \gamma y \qquad\qquad (10.61)$$
$$dz/dt = \gamma y,$$

where $\tilde{y}$ is a spatially weighted average of those y-values positioned near enough to the element A to influence any susceptible living there. The ensuing mathematical result is that a *pandemic* affecting every part of the region will occur if and only if the initial population density of susceptibles σ exceeds the (non-spatial) threshold $\rho = \gamma/\beta$. If a pandemic does occur then its severity θ is the unique positive root of the equation

$$\theta = 1 - \exp(-\sigma\theta/\rho). \qquad\qquad (10.62)$$

This means that the proportion of individuals who eventually become infected will be at least θ in any part of the region, no matter how far from the original source of infection. We therefore have a (spatial) pandemic threshold result which corresponds to the (non-spatial) result of Kermack and McKendrick (1927). Moreover, it may be shown that the solution to equation (10.62) is also the severity of the non-spatial epidemic. Thus we have the rather surprising result that inclusion of a spatial effect does not affect the threshold value.

10.6.3   *A carcinogenic growth process*

Even with this relatively simple spatial epidemic model theoretical results are few and far between, and with more complicated processes simulation provides virtually the only means of predicting qualitative behaviour. A good illustration of this is provided by the work of Williams and Bjerknes (1972) on a carcinogenic growth process. The basic biological assumption is that when a basal cell in the epithelium divides, both daughter cells remain in the basal layer and one of the six neighbouring cells (chosen at random) is displaced upwards. Both normal and abnormal cells are allowed

Figure 10.10. Configurations of normal and abnormal cells in the basal layer (Williams and Bjerknes, 1972). The three columns show the configurations when there are, for the first time, 100, 400, 900 and 1600 abnormal cells. In the left column, $k = \infty$; in the middle, $k = 2$; and in the right, $k = 1.1$. (Reprinted by permission of *Nature*, **236**, pp. 19–21. Copyright © 1972 Macmillan Magazines Ltd.)

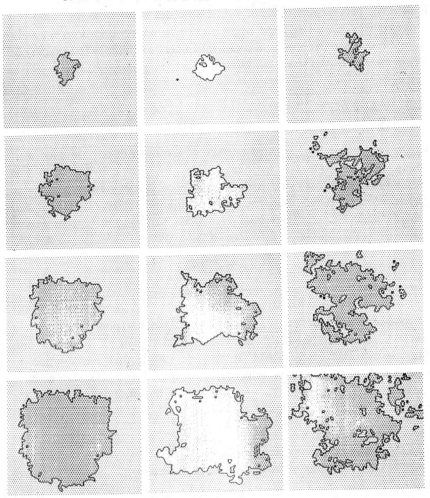

to develop, the latter leading to the development of a tumour. These tumour cells divide at a faster rate than normal ones ($k > 1$ times as fast), and so the clone of developing cancer cells gradually invades the basal layer by pushing non-tumorous neighbours out. Incidentally, if the time variable t is changed to kt and k is allowed to become large, then we obtain a simple spatial epidemic process (Bailey, 1975).

Cells tend to pack into layers in a honeycomb pattern, and so the computer simulations assume a precise honeycomb distribution. Each realization starts with a single abnormal cell in the centre of the honeycomb. The columns of Figure 10.10 show sequences of simulated patterns of the basal layer configuration for $k = \infty$, $k = 2$ and $k = 1.1$ when there are, for the first time, 100, 400, 900 and 1600 abnormal cells in this layer. Normal cells are indicated as dots and abnormal ones as circles (the boundary between normal and abnormal cells has been drawn in).

In the first column k is so large that only the abnormal cells divide and so the normal ones are swiftly pushed out. Note that enclaves of normal cells can occasionally be engulfed by abnormal cells, but that the opposite cannot happen. There can be no abnormal islands by themselves, since as normal cells do not divide they are unable to cut off an abnormal peninsula.

The middle column shows the development of the process when $k = 2$. Half of the cancer patterns (i.e. $1/k$) never take hold; this is one that does. The enclaves of normal cells are now larger and take longer to fill in. Note that there are islands of abnormal cells.

However, it is the third column that is particularly instructive. Here $k = 1.1$ so this sequence of patterns represents the 9% of simulations that ultimately take hold. Now this sequence looks far more invasive than the previous two, which have more the appearance of a carcinoma *in situ*. Thus the model of Williams and Bjerknes implies that 'the infiltrating patterns traditionally associated with cancer growth may be as much due to counter-invasion of the abnormal by the normal cells, as due to invasion of the normal by the abnormal ones'.

11

Linear and branching architectures

A whole new approach is needed if we are to consider the complete route taken by all members of a developing population. For example, we may wish to study the space–time development of an invading species of ant as it reproduces and spreads across a region. The resulting map will resemble a 'tree', with current ant positions corresponding to branch buds, births to branch forks, deaths to branch ends, and the paths between such events to the branches themselves. Though such scenarios are not often encountered in population dynamics, tree-like structures abound in biology, and a way of describing and analyzing them is clearly needed. Obvious examples include lung-airways, neural and arterial networks, and plant rhizome systems; examples of more abstract networks include the concepts of food webs and dominance relations in animal society. MacDonald (1983) provides an excellent overview of this potentially vast subject area. His presentation is eminently readable by mathematician and biologist alike, and provides an ideal starting point for readers wishing to pursue this highly absorbing subject.

In order to find our way around a tree or network (the former implies the absence of closed paths, the latter does not) we need to define an ordering over the connected branches. Fortunately, geographers spent considerable effort in the 1950s and 1960s investigating various possibilities for stream and river networks, and this work has been of considerable benefit to subsequent biological research. The method proposed by Gravelius (1914), for example, involves starting at the river mouth (bole of a root system, trunk of a tree canopy) and mapping out a path upstream. At each junction we select the stream that makes the smaller angle to the direction we approached it from. This process is continued until the main stream has been traced to its source, and then is repeated for each of its tributaries, and *their* tributaries, etc., until the whole system has been covered. A method favoured by many biologists is one proposed earlier by Strahler (1952). This orders in the reverse direction by starting with leaves (stream ends). All twigs contiguous to these leaves are called branches of order 1. When two branches of order m meet, the continuing branch is of order $m + 1$. When branches of orders m and n meet, where m is less than n, the continuing branch is of order n.

These two approaches are illustrated in Figure 11.1, and field studies of such ordering schemes have given rise to various empirical 'laws' of branching structure. For example, Shreve (1966) surveyed data on 246 stream networks and found that the number N_m of branches of Strahler order m obeyed the empirical relation

$$N_m/N_{m+1} = R \quad \text{(a constant)}. \tag{11.1}$$

This implies that a branch of order m will subtend approximately R^{m-1} branches of order 1. Values for R typically lie between 3 and 4. Such mathematical relations can be derived for systems grown under a variety of assumptions, and comparison with results based on known data sets then enables considerable insight to be gained into extremely complex branching systems (with lung-airways N_1 is measured in millions).

As the resulting theoretical considerations go well beyond the scope of this text, we shall now use a pure simulation approach to gain an understanding of three biological growth processes. First we disregard branching, and concentrate solely on trying to detect the direction of spread of creeping willow herb roots from above-ground shoot counts. The second and third studies are far more complex and involve studying the full branching system of Sitka spruce. The former analyzes canopy structure, for which a complete temporal development can be reconstructed from branching measurements taken at a single time. The latter concerns structural root development which, having a far less well-defined structure, does not enable branching maps to be determined through time. Both examples are discussed extensively in Renshaw (1985).

Figure 11.1. The Strahler (Gravelius) ordering scheme.

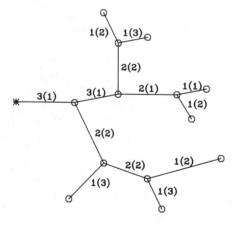

11.1   The spatial distribution of *Epilobium angustifolium* in a recently thinned woodland

Ford and Renshaw (1984) describe a situation in which following thinning of the tree canopy in a woodland south of Edinburgh, Scotland, willow herb (*Epilobium angustifolium* L.) was spreading at the expense of raspberry (*Rubus idaeus* L.) to form a dense pure stand of ground vegetation. As the shoots of willow herb arise on creeping perennial roots (Myerscough, 1980), one might expect a directional pattern at the invasion front (van Andel, 1975), whilst within the established colony growth might fill in between the directional axes.

The problem is to determine: (i) the extent (if any) of this anticipated directional effect of the root pattern; and (ii) the difference between the spatial patterns of willow herb shoots at the invasion front and within the established zone. Since the ground could not be excavated the actual configuration of the root system remains unknown; only surface-based information on shoots could be obtained. Individual shoots of willow herb were therefore counted in two 32 × 32 arrays of contiguous 10 cm sided square quadrats. One array was placed at the advancing front across the interface between the patches of willow herb and raspberry (Figure 11.2a), whilst the other was placed some 2 m behind it within a dense zone of pure willow herb (Figure 11.2b). Growth at the invasion front appears to be more clumped than within the developed stand, though no directional effects are visually apparent in either array.

11.1.1   *Two-dimensional spectral analysis*

An excellent way of analyzing such data is to use two-dimensional

Figure 11.2. Spatial pattern of *Epilobium angustifolium* spreading through a woodland. (a) and (b) are sampled patterns at the invasion front and within the developed stand, respectively. (Reproduced from Ford and Renshaw, 1984, by permission of Kluwer Academic Publishers.)

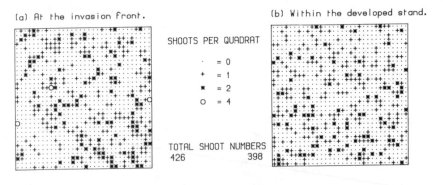

(a) At the invasion front.

(b) Within the developed stand.

SHOOTS PER QUADRAT

  . = 0
  + = 1
  * = 2
  o = 4

TOTAL SHOOT NUMBERS
426   398

spectral analysis, this being a natural extension of the one-dimensional procedure developed in Section 4.5.4. This technique not only yields a comprehensive description of both the structure and scales of pattern in a spatial data set, but it also assumes no structural characteristics in the data prior to analysis. This is particularly advantageous in ecological situations where pattern may exist over a range of scales, and be anisotropic (i.e. have directional components). Programs which perform the analyses may be obtained from the author.

Since we shall be using the approach purely as a means of solving this particular problem, there is no point in discussing it at length here. Readers who are interested in the technique itself should note that Renshaw and Ford (1983) discuss basic theoretical principles together with general problems of estimation, discrimination and interpretation of spectra, and then illustrate them through the analysis of a forest canopy data set. Further illustration through the analysis of simulated competition and invasion patterns is provided by Renshaw and Ford (1984); whilst in addition to the willow herb problem Ford and Renshaw (1984) also investigate the spatial pattern of *Calluna vulgaris* in a regenerating woodland, a data set first studied by Diggle (1981).

Note that although two-dimensional spectral analysis has found extensive application in the *processing* of electronic signals, for example in the reconstruction of data from satellites and the analysis of electron micrographs, few previous attempts have been made to use it as a way of *inferring* process from pattern. Bartlett (1964, 1975) deduced an inhibition process between plants in two stands of Japanese black pine saplings; whilst McBratney and Webster (1981) analyzed yields of wheat grain and straw (data of Mercer and Hall, 1911), and showed the existence of periodic effects attributed to ploughing and an earlier ridge and furrow system.

11.1.2   *A simple cosine wave example*

Since our willow herb problem specifically involves spacing and directional effects, the most convenient way of presenting spectral results is to use a polar representation. Scales of pattern and directional components are then revealed through the 'R-spectrum' and the 'Θ-spectrum', respectively. For example, Figure 11.3a shows the simple cosine wave

$$\cos[2\pi\{(ps/32) + (qt/32)\}] \tag{11.2}$$

sampled at the points $s,t = 1, 2, \ldots, 32$ of a 32×32 square lattice. This function generates a regular wave-pattern directed along the main diagonal, and with $p = q = 4$ gives rise to four complete cycles along each axis. The associated R-spectrum (Figure 11.3b) shows a single large peak in the interval

$5 < r \leqslant 6$; as 8 full waves occur along the diagonal which is of length $32\sqrt{2}$, the exact frequency is $32\sqrt{2}/8 = 5.66$. The corresponding Θ-spectrum (Figure 11.3c) shows the anticipated diagonal alignment through a single large peak in the interval $35° < \theta \leqslant 45°$.

If random noise is added from a Normal distribution with mean 0 and variance 1 to each of the individual 32×32 elements, so that the underlying cosine wave is obscured from view, then the R-spectrum and Θ-spectrum still show large peaks around $r = 5.66$ and $\theta = 45°$. This illustrates the power of the spectral approach; pattern can be detected even when there is no visual suggestion that it might exist.

11.1.3 *Analysis of the willow herb data*

We are now in a position to interrogate the willow herb data. The

Figure 11.3. Spectral analysis of a pure cosine wave with $p = q = 4$: (a) cosine wave; (b) R-spectrum; (c) Θ-spectrum. (Reproduced from Renshaw and Ford, 1984, by permission of Kluwer Academic Publishers.)

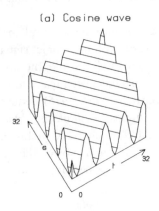

(a) Cosine wave

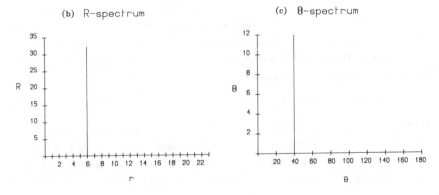

(b) R-spectrum   (c) θ-spectrum

R-spectrum (Figure 11.4a) for the pure zone shows: (i) a peak at $r = 8$–9 (wavelength $\simeq$ 36–40 cm), suggesting regularity at this interval in the organization of shoots along roots; and (ii) weak evidence of a further peak at $r = 18$ (wavelength $\simeq$ 18 cm). The *R*-spectrum for the invasion zone shows large values for $r = 2$–3, indicative of large-scale clumping at the interface, with a shift of feature (i) to $r = 5$–8 (wavelength 40–64 cm); feature (ii) is not represented. The 5% significance levels shown are for individual r (and θ) values; on combining neighbouring values the pure zone peak at $r = 8$–9 turns out to be significant at the 0.1% level (see Ford and Renshaw, 1984, for details). Similarly, the broad peak at $r = 5$–8 for the invasion zone is also highly significant. This shift in the *R*-spectrum between the two stands is particularly interesting. One possible explanation is that as growth takes place there is a more complete exploitation of the ground, competition occurs, and so the apparent large-scale clumping is removed.

Analysis of the Θ-spectrum (Figure 11.4b) for the pure willow herb zone provides strong evidence of a directional component at $\theta = 90°$–$100°$ (significant at the 0.5% level), and although this component is more diffuse than the single peak at $\theta = 70°$ for the advancing front (significant at just over the $2\frac{1}{2}\%$ level), it is stronger. Ford and Renshaw found this result surprising. Although a strong directional component was anticipated to occur close to the invasion front, it was thought (before the analysis!) to be unlikely in the older stand.

Figure 11.4. Spectral analysis of the willow herb patterns of Figure 11.2: (a) *R*-spectrum and (b) Θ-spectrum of invasion (———) and developed (— — —) stands, showing 5% significance levels (- - - - -); $\triangle$ denotes 5% significance levels under randomization of the spectrum. (Reproduced from Ford and Renshaw, 1984, by permission of Kluwer Academic Publishers.)

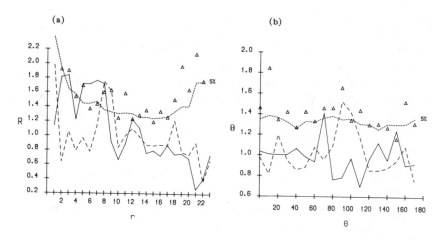

11.1.4   *Simulation of directional spread*

To establish that a directional mechanism of willow herb invasion is consistent with these spectral results, Renshaw and Ford (1984) develop a simple model for plants that spread by underground roots or rhizomes and produce aerial shoots at intervals. First, simulated roots are constructed starting from points placed a constant distance (d) apart along an arbitrary base line. The angle of spread (θ) from this line is taken to be a Normal random variable with mean μ_θ and variance σ_θ^2. So if σ_θ^2 is very small then all the lines are approximately parallel, whilst if σ_θ^2 is very large then all angles (0° to 180°) are equally likely. Aerial shoots are then positioned sequentially along each root a distance m apart, where m is a Normal random variable with mean μ_m and variance σ_m^2. Thus shoot regularity decreases with increasing σ_m^2. Grid counts of aerial shoots on a 32 × 32 array are then made over an area distant from the base line (to eliminate the regularity of start positions) for various combinations of the variances σ_θ^2 and σ_m^2. The values μ_θ = 70° and $\mu_m = d = 3$ are retained throughout.

When $\sigma_\theta^2 = \sigma_m^2 = 0$ the process is purely deterministic and so the pattern consists of parallel lines of equidistant points (Figure 11.5a). There is a peak in the Θ-spectrum (Figure 11.5c) at $\theta = 160°$ which is the direction *across* the roots, and so *their* corresponding direction is at right angles to this, namely 160° − 90° = 70°, as required. Note that the extreme regularity of the pattern means that unintended additional regularity also exists. This is highlighted by the Θ-spectrum which shows:

(i) a strong peak at $\theta = 90°$ – the result of ones in every third (sometimes second) row; and

(ii) conjugate waves angled around $\mu_\theta = 36°$ and 126°.

Similar unwanted effects have been demonstrated in other regular patterns (see Ford, 1976), and failure to appreciate them can lead to wrong interpretation. Great care is therefore needed when working with purely deterministic models, since they may generate more complicated patterns than one might at first imagine.

Only a small positive value of σ_m^2 is needed in order to eliminate these conjugate patterns. When σ_θ^2 is also allowed to be non-zero then the simulated pattern consists of lines in various orientations (in Figure 11.5b, σ_θ^2 = 4.0), and the Θ-spectrum is ideal for determining what the principal orientation is. For example, spectral values for two typical simulations with:

(i) $\sigma_m^2 = 2.5, \sigma_\theta^2 = 4.0$ were 0.63 at 60°, 3.72 at 70°, 1.14 at 80° and 0.44 at 90°;

(ii) $\sigma_m^2 = 1.0, \sigma_\theta^2 = 400$ were 0.89 at 60°, 1.72 at 70°, 1.67 at 80° and 0.69 at 90°.

Remember that these values should be around 1 if the spatial pattern is purely random. Note that in case (ii) the orientation of the simulated roots varies considerably, as does the distance between shoots on the same root, yet the Θ-spectrum is still able to determine the general direction of invasion. Thus even if an ecological data set contains a very weak directional component, it is quite possible that the Θ-spectrum will be able to detect it.

Here we have illustrated the spatial dispersal of non-branching line processes through the study of one specific species, namely *Epilobium angustifolium*, and it is worth reflecting that this type of model can describe

Figure 11.5. Two models of an invasion process: (a) regularly spaced points along parallel angled lines; (b) random variation added to the line and point distributions of (a); (c) Θ-spectrum of (a) (————) and (b) (————). (Reproduced from Renshaw and Ford, 1984, by permission of Kluwer Academic Publishers).

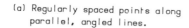

(a) Regularly spaced points along parallel, angled lines.

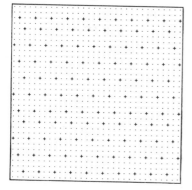

(b) Random variation added to the line and point distribution of (a).

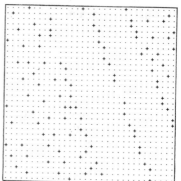

(c) θ-spectrum of (a) ———— and (b) ____ .

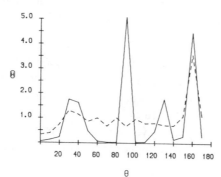

the spread of other species, such as the sand dune sedge *Carex arenaria* (L.). Bell and Tomlinson (1980) suggest that line-process systems might be appropriate for plants that inhabit sand dunes since the latter often form mobile bands of suitable habitat. Routledge (1987), for example, makes a detailed study of the spike rush *Eleocharis palustris* (L.), and then simulates an appropriate linear rhizomatous growth model to examine whether it reflects a sufficiently opportune colonization strategy. By limiting the amount of wandering in the direction of growth, the plant might well be able to increase its chance of occupying less crowded ground and thereby increase its fitness. However, *Eleocharis palustris* can be classified as a true guerilla, since it drives through the terrain and rarely leaves even a single tiller behind to consolidate its gains. Why is this? Would it not be better to adopt a mixed strategy by branching after escaping the crowded seedbed instead of relentlessly pursuing a course of death in either deep water or overpowering shrubs? Further extensive simulation exercises could well provide the key to understanding.

11.2   Spatial branching models for canopy growth

The general complexity of biological branching structures ensures that they stand apart from models discussed elsewhere in this text. For as Renshaw (1985) states:

> Simulation of ecological processes often proceeds from a simplistic viewpoint in which organisms are assumed to develop in time and/or space according to idealized, mathematical rules. This assumption is fine if all that is required is a crude qualitative similarity between a biological process and a mathematical model, but to develop simulation models which accurately reproduce the large scale behavioural features of complex systems such as trees involves discarding mathematical niceties, no matter how computationally convenient they may be, and introducing as much biological detail as is conceivably possible. Only then may a 'stepping-down procedure' be used which separately examines each component in the model to see whether it may be omitted, or at least simplified, without affecting the resulting patterns.

Note that mathematical analysis of even the simplest type of random walk model that can in any way be related to crown or root development would be too complicated to be of practical use to us here. Even introducing biologically meaningful *bending* rules for branch or root development gives rise to intractable mathematics; yet this is the relatively easy part since introducing *branching* involves a higher order of mathematical intractability still!

11.2.1   *Honda's deterministic model*

The relation between canopy structure and yield as the former progresses through different stages of development is of particular importance. Greater understanding could well lead to increased timber production (see Ford, 1985). In Sitka spruce, crown branching structure is controlled by just three features: the number of buds produced; the distance from the apex to where they develop; and their rate of growth relative to the mainstem continuation. Ford's purpose is to determine what attributes of branch growth are required in order to produce efficient crowns, and to achieve this it is necessary to learn how the controlling processes of tree growth affect the overall structure. So the first step is to develop simulation models which can successfully mimic real canopy growth.

Fortunately, a wide variety of tree crown forms can arise from varying just a few branching rules. For example, Honda (1971) uses a deterministic model of crown morphology to show that complex patterns can result from the multiplicative effect of simple rules. Repeated bifurcation of branches is generated under the assumptions that:

(i) branches are straight;

(ii) a mother branch forks into two daughters at angles θ_1, θ_2 in the plane whose steepest gradient coincides with its own gradient;

(iii) each length of the two daughter branches is shortened in the ratio R_1, R_2, respectively, to that of the mother one;

(iv) branching is in concurrent generations; and

(v) the angles θ_1 and θ_2 alternate at each generation.

Figure 11.6 shows the change (front view) in a simulated tree when $\theta_2 = 20°$, $30°$, $45°$, $60°$ and $80°$, for fixed $\theta_1 = 0°$, $R_1 = 0.9$ and $R_2 = 0.7$. Increasing θ_1 whilst maintaining a constant branching angle $\theta_1 + \theta_2$ results in the diminution of the mainstem and a change in the upper part of the crown form from conic to flat. Moreover, the ratio of the smaller of the two angles to the branching angle appears to have a close relationship to how clearly the simulated tree exhibits its main axis. An example order of axiality suggested by Honda for real trees is

fir, larch > ginkgo, birch, beech > maple, azalea.

Simulations show that the degree of apical dominance in the tree form is related to the ratio between R_1 and R_2.

11.2.2   *Description of measurements*

Whilst Honda's model is an excellent guide to the rich diversity of tree and plant forms that can be generated by repeated use of just a few simple deterministic rules, it is not applicable to Sitka spruce. Not only is assumption (ii) not met in real tree structures, but also his four parameters cannot be taken as remaining constant throughout the tree. Moreover, as considerable natural variation exists within a single tree, never mind between trees, stochastic and not deterministic descriptions are necessary.

In an attempt to determine an accurate representation of stochastic

Figure 11.6. Front view of three-dimensional simulated deterministic trees with constant $\theta_1 = 0°$, $R_1 = 0.9$, $R_2 = 0.7$ and $\theta_2 = $ (a) 20°, (b) 30°, (c) 45°, (d) 60° and (e) 80°. (Plate IV of Honda, 1971, reproduced by permission of Academic Press.)

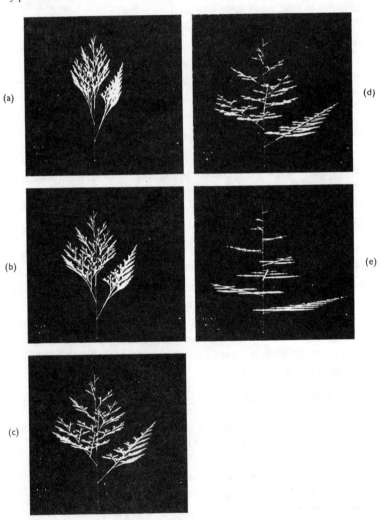

behaviour, Cochrane and Ford (1978) collected data from Sitka spruce trees growing in a 9 × 9 m square in the Rivox section of Greskine Forest, UK. Figure 11.7 shows the branching structure of a terminal mainstem, and illustrates the measurements that were made on the numbers, lengths and orientations of shoots within the developing canopy. Note that each current year's shoot has: (i) an apical bud which, in the next year, will continue to extend along the axis of growth of the parent shoot; (ii) a cluster of buds just below the apex which will form a whorl of branches; and, (iii) a more regularly distributed series of buds along its length which will develop as interwhorl branches.

The great advantage of working with Sitka spruce is that the various stages of shoot extension and bud formation comprise a distinct temporal sequence in the tree's development. Thus once each stage is understood, full canopy growth can be simulated through time. In 1973 a detailed analysis of five trees, selected as being representative of the whole population, was therefore

Figure 11.7. The branching structure of the terminal section of *Picea sitchensis* illustrating measurements made and the names used to describe them. (Reproduced from Cochrane and Ford, 1978, by permission of the British Ecological Society.)

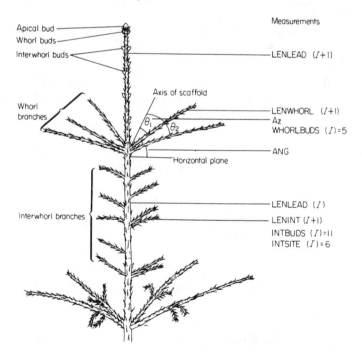

made by using the following measurements taken along the main stem of each tree:

(i) the number of whorl branches (WHORLBUDS);
(ii) the length of each whorl branch (LENWHORL);
(iii) the number of interwhorl branches (INTBUDS);
(iv) the length of each interwhorl branch (LENINT), and the distance of each interwhorl branch down the main stem from the whorl above it;
(v)/(vi) the angle of a branch to the horizontal (ANG) and vertical (AZ) planes.

Additional measurements were made in 1976 of the leader length (LENLEAD) together with (i), (iii), (v) and (vi) for 10 trees different from those used in 1973. Full details of the statistical models that were developed to represent the relationships between all these variables are contained in Cochrane and Ford (1978).

11.2.3   *Simulation of canopy structure*

The drift of Cochrane and Ford's work is as follows. First the growth mechanism is divided into three component processes.

(A) *Production*

The number of branches on the kth tree at growth level j is modelled by

$$n(k, j) = \mu + \alpha_{kj} + \delta_{kj}\varepsilon_{kj}$$

where μ is the population mean, α_{kj} is a Poisson random variable with a separate mean value for each tree (k), $\Pr(\delta_{kj} = +1) = \Pr(\delta_{kj} = -1) = \frac{1}{2}$, and ε_{kj} is a Poisson random variable with constant mean.

(B) *Dispersion*

The angle (azimuth) of the mth branch is modelled by

$$X(m) = (2\pi/n)m + \varepsilon_m$$

where ε_m is a Normal random variable with mean 0 and constant variance σ^2, and n is the number of buds.

(C) *Extension*

For constants a and b, the annual leader growth rate is assumed to be

$$R(I) = ab^{-I} \quad \text{for } 2 \leqslant I \leqslant i$$

where i is the age of the forest.

This decomposition of the growth mechanism into a set of elementary processes, and the determination of the statistical relationships between them, defines a series of mathematical rules which can then be assembled into a simulation model for canopy growth. Figure 11.8 shows such a simulated five-year-old Sitka spruce, and visual agreement with the structure of a real tree (Figure 11.9) is clearly excellent.

In their comparative study of real and simulated canopy growth, Cochrane and Ford (1978) found that a division could be made between components (A) to (C). Branch dispersion followed consistent rules from year to year, and whorl branches were always spatially regular around the stem irrespective of their numbers. This is a reflection of the regular distribution of branch initials which form on the apical bud. Interwhorl branches were absent immediately below the distal whorl and above the proximal one, and small aggregations of branches occurred in between. Angles between the vertical mainstem and the branch increased from the top of the tree towards the base of the crown, this increase being greater for interwhorl branches.

Not only does the simulation model provide an excellent geometrical description of tree growth, but it also provides considerable insight into the control system. The shape of simulated trees appears to be reasonably insensitive with respect to the choice of the underlying statistical distributions, only the choice of correct variances proves critical for the production of realistic patterns (see Figure 11.9).

In summary, development of this simulation model involves three clearly defined stages. The first is to obtain visually and statistically convincing realizations, the second is to determine the simplest set of rules which will still achieve such realizations, and the third is (ideally) to estimate parameter values which give rise to maximum stem-wood production. As stem-wood production is directly related to crown structure, clones could then be produced from trees which possess optimal properties. Of immediate benefit is the considerable increase in biological understanding which has been obtained through the determination of the relative importance of the component elementary processes that lead to the generation of realistic simulations.

11.3 Spatial branching models for structural root systems

Roots account for between 40% and 85% of net primary production in a wide range of ecosystems from grassland to forest. Typically, plant growth in non-agricultural conditions is limited more by soil-derived resources than by carbon dioxide or solar radiation. It therefore seems axiomatic that an understanding of the functioning of plants within natural

Figure 11.8. Simulated five-year old *Picea sitchensis* tree viewed in the *x–z* plane; parameters estimated from the data. (Reproduced by permission of Professor E. D. Ford.)

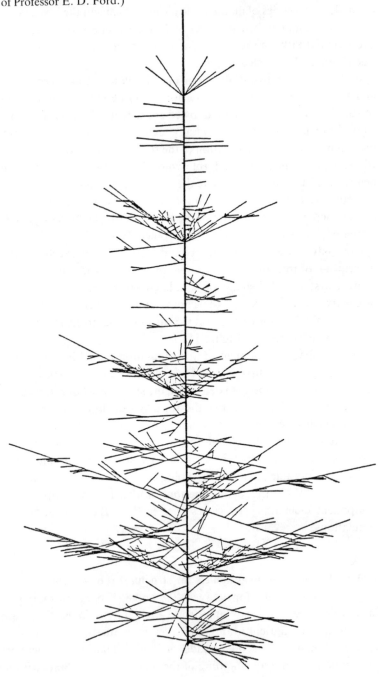

communities must demand an equal understanding of the behaviour of roots and root systems (Fitter, 1987).

Since the growth of a Sitka spruce canopy develops in a regular and systematic manner, its past history can be determined by inspecting its present structure. However, because root systems exist below ground level and have a far less well-defined structure, it is not possible to determine root development *through* time. Thus analysis of root systems involves determining spatial structure at a *fixed* time, with the objective of seeing whether the apparently haphazard nature of root growth is subject to a set of bending and branching rules that are broadly similar to those operating for canopy growth (Renshaw, 1985).

11.3.1   *Description of measurements*

Eight trees from Rivox were selected for study, and after they had been felled their structural root systems were exposed by hand during the summer of 1978. The spatial patterns of two of the root systems, viewed in the

Figure 11.9. Comparison of *Picea sitchensis* tree viewed in the x–z plane with simulated trees having variables with variances that are too high and too low. (Reproduced by permission of Professor E. D. Ford.)

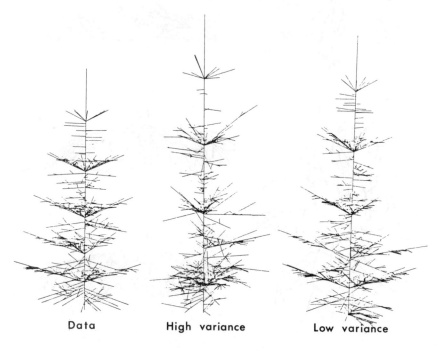

Data   High variance   Low variance

x–y, x–z and y–z planes, are shown in Figure 11.10. Here the elementary processes essentially comprise bending and forms of branching:

Figure 11.10. The spatial patterns of the measured structural root systems of two Sitka spruce: (a) x–y view of first root system; (b) x–z view of first root system; (c) y–z view of first root system; (d) x–y view of second root system; (e) x–z view of second root system; (f) y–z view of second root system. (Reproduced from Henderson, Ford and Renshaw, 1983, by permission of *Forestry*.)

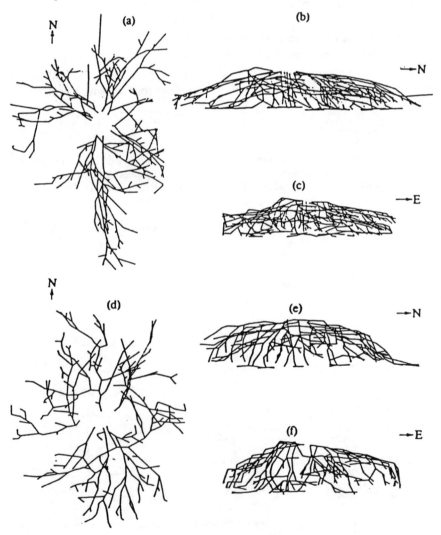

(i) *first-order* roots originate at the bole, roots initiated on them are termed *second-order*, and so on;

(ii) *lateral branching* points are where a new root is initiated and the parent root continues;

(iii) *forks* occur where a root splits into two equally sized roots which subtend equal angles to the original;

(iv) *proliferations* are where a root divides into a large number of much smaller roots;

(v) root *death* is said to occur when the diameter falls to 5 mm, smaller roots being insufficiently rigid to spring back to their original positions when displaced during excavation.

A statistical analysis of the measured structural root systems (Henderson, Ford, Renshaw and Deans, 1983) shows that the root branching process is far more systematic than was first apparent. First, in three of the root systems the horizontal directions of the first-order roots are significantly regularly spaced, and this property ensures that the soil region around the stem is evenly occupied. Second, lateral branching does not occur randomly, and whilst there are considerable differences between root systems in the mean length of root per lateral there are only small differences between root systems in the mean length of root per fork or bend. This suggests that bends, which are caused by fluctuations and obstructions in the soil environment, and forks, caused by damage to a growing root, are properties of the site and soil type, whilst lateral branching is determined by biological mechanisms and is therefore a property of the tree. Third, the angles and azimuths between laterals and their parents are typically large in comparison to the angle and azimuth changes at forks and bends, suggesting that there may be some form of inhibition operating which reduces the chance of two structural roots occupying and exploiting the same soil region. Fourth, changes in direction at bends and branching points tend to alternate clockwise then anticlockwise, so ensuring that individual root directions remain similar to their original directions. Thus root systems generally spread outwards away from the bole, with few root paths forming loops. This ability of root systems to effect an efficient spread through the soil is an important phenomenon, yet it is little understood. Do roots try to grow towards nutrients, so keeping away from existing roots which may have already depleted nutrients nearby?

11.3.2   *A temporal simulation model*

Not only are root patterns far more complex than canopy structures (compare Figures 11.9 and 11.10), but they also have the additional complication of developing within a heterogeneous bounded region, namely the ridges on which the trees are planted and the impenetrable mineral soil underneath. Moreover, whilst the canopy model could be based on distinct and observable annual growth increments, this is not possible for root systems and so their temporal development cannot be similarly reconstructed.

Any temporal model of root development must therefore involve an assumed, and thereby potentially inaccurate, growth sequence. However, interest in the consequences of using such an approach led Henderson and Renshaw (1980) to develop a fairly simple model based on the sequential production of straight-line segments between bends and branching points. Each of these segments has a random length and orientation relative to the previous step direction, and ends with the initiation of a number (n) of new segments, viz:

(a) $n = 0$ corresponds to root diameter falling below 5 mm (i.e. root death);

(b) $n = 1$ represents a bend;

(c) $n = 2, 3, \ldots$ corresponds to the production of new roots together with the continuing growth of the parent.

The following assumptions were found to fit the data well.

(1) Step lengths are independent and identically distributed gamma variables.

(2) If k segments are produced at a point then the orientation of the jth of these ($j = 1, \ldots, k$) is a (wrapped) Normal variable with constant variance but with mean depending on j and k. Successive bends tend to be in opposite directions.

(3) The number of new segments produced after any step is a random variable drawn from a distribution which depends on the number of branching points (r) between the stem and those segments. The probability of root death increases with r, as does the relative proportion of bends to branching points.

Henderson and Renshaw (1980) generate simulated root systems based on these rules; the length, angle and offspring distributions are estimated from the data. Comparison of observed and simulated systems shows that although the latter are generally of the correct size and complexity, the patterns produced are not sufficiently convincing. In particular, because branching points are not regularly spaced along the roots there are often long sequences of bends which can end a considerable distance away from the centre of the system, possibly in a cluster of branching points. This means that simulations under this model often produce much larger sparse regions, interspersed with very dense regions, than are observed in real root systems.

Thus the highly successful time-dependent simulation structure developed for canopy growth does not carry over to the rooting system.

11.3.3   *A fixed-time simulation model*

Henderson, Ford and Renshaw (1983) therefore discard the notion of time-development and construct instead a fixed-time empirical model. The total length of a root is determined first, and only then is the root broken up into a series of segments between bends and branching points. The simulation rules involved are fairly long and complicated, but the following brief summary gives the general picture.

(i) *Initial distribution of first-order roots*

The origins of the n first-order (in the Gravelius sense) roots have a roughly regular spacing around the root collar, their azimuths being $2\pi i/n + \varepsilon_i$ where ε_i is a Normal random variable with mean 0 and variance 0.11 radians². The initial angles of first-order roots are independently uniformly distributed between $-\pi/18$ and $5\pi/18$ radians.

(ii) *Root lengths*

Up to six orders of root are allowed, the root length distributions being uniform (order 1), gamma (orders 2, 3, 4), and exponential (orders 5, 6).

(iii) *Lateral branching*

The number of laterals produced by a root is a Poisson random variable with parameter proportional to root length; root origins are almost regularly spaced along the parent root with Normal deviations from perfect regularity.

(iv) *Root bending*

For all orders, the number of bends on a root is a Poisson random variable with parameter proportional to root length. Bends are uniformly dispersed along the root.

(v) *Root directions*

Roots change both angle and azimuth at bends and lateral branching points. For example, the azimuth change at bends is $\theta_B = \alpha|\varepsilon_B|$ where $|\varepsilon_B|$ denotes the magnitude of a Normal random variable with mean 0.21 radians and variance 0.085 radians²; α is $+1$ or -1 with probabilities p and $1 - p$, respectively, where $p = 0.38$ (0.62) if the previous azimuth direction is clockwise (anticlockwise).

(vi) *Branching angles and azimuths*

Azimuths (θ_L), for example, between parent roots and the initial directions of laterals are independent of both previous direction changes and positions in the soil. The model is $\theta_L = \gamma|\varepsilon_L|$ where $|\varepsilon_L|$ denotes the magnitude of a Normal random variable with mean 0.83 radians and variance 0.094 radians^2, and γ is $+1$ or -1 each with probability $\frac{1}{2}$.

(vii) *Soil boundaries*
Rules are constructed for root behaviour at the soil/atmosphere boundary, the sides of the parallel plough furrows on either side of the planting position, and the impenetrable soil substratum.

Fifty root systems were simulated and compared with the observed root systems, and the patterns produced were found to be visually convincing. Compare, for example, Figures 11.10 and 11.11 which show two real and two simulated systems, respectively. Visual agreement was backed up by the use of statistical descriptors such as the number of roots and the spatial distribution of root length. Both observed and simulated root systems include the same balance of densely and sparsely populated regions, with roots tending to spread outwards and few root paths forming loops.

A large number of empirical rules are incorporated into this simulation model: to see which rules are essential for the production of realistic branching structures simulations were performed under different conditions and the results were then compared with the excavated systems. For a full discussion of this study see Henderson, Ford and Renshaw (1983), but the main conclusions are as follows.

Direction changes at bends and forks must be small in order to ensure that the system spreads outwards from the stem, whilst lateral branches have to subtend large angles to their parents in order to exploit separate soil regions. It is also necessary for first-order roots to be almost regularly distributed around the stem, and for azimuth changes to have a tendency to be alternately clockwise then anticlockwise. Variances of the fitted distributions have the greatest influence on pattern, especially those which govern root directions. For example, azimuth variances that are too large result in many roots turning backwards towards the stem, so reducing outward spread; whilst angle variances that are too small lead to simulations in which fewer roots grow almost horizontally than are observed in the excavated systems. Note the similarity here with the aerial structure of Sitka spruce: simulated crown morphologies are unrealistic if the variances used in the simulations have either higher or lower values than those estimated from the data.

11.4   Discussion of the role of simulation
    The three processes we have presented in this chapter are far

removed from the relatively simple simulation models developed earlier, such as the simulated pure birth process (Section 2.1). For this we know both the deterministic solution

$$N(t) = \exp(\lambda t) \quad \text{(for } N(0) = 1)$$ (2.3)

Figure 11.11. Two simulated structural root systems: (a) x–y view of first simulation; (b) x–z view of first simulation; (c) y–z view of first simulation; (d) x–y view of second simulation; (e) x–z view of second simulation; (f) y–z view of second simulation. (Reproduced from Henderson, Ford and Renshaw, 1983, by permission of *Forestry*.)

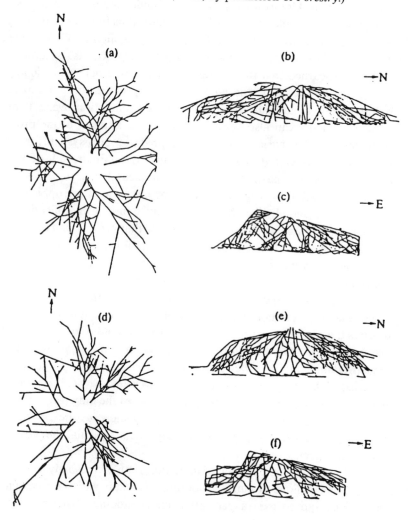

and the complete probability structure

$$p_i(t) = e^{-\lambda t}(1 - e^{-\lambda t})^{i-1} \quad (i = 1, 2, \ldots), \tag{2.9}$$

yet simulations still highlight important properties of the system. In particular they show quite clearly that it is the opening stages of a realization that determines the course of its future development (Figure 2.3).

The probability structure of this simple birth process can be determined fairly easily because individuals develop independently in a non-spatial environment. However, even introducing a straightforward interaction term can lead to substantial mathematical difficulties; the simple stochastic birth–death process, for example, becomes the far more mathematically involved logistic process. Unfortunately, biological understanding will not generally be enhanced by employing 'clever' mathematical techniques to develop large and complicated probability solutions. These are often totally opaque as far as revealing insight into the underlying stochastic process is concerned.

We have learned that the answer is to use both a deterministic approach and a simulation study: the latter not only provides insight through the nature of individual realizations, but repeated simulation over sets of given parameter values can also be used to derive an estimate of the probability structure itself. The basic Lotka–Volterra predator–prey system, for example, leads to a set of closed deterministic trajectories (Figure 6.1). Though we anticipated that stochastic trajectories might meander across this set, it took a simulated trajectory (Figure 6.3) to show just how catastrophic stochastic variation can actually be. Moreover, only a minor modification to the simulation program is needed to demonstrate that provision of cover for prey can lead to sustained stochastic cycles (Figure 6.13).

The philosophy behind the willow herb example is totally different. Here the emphasis lies not in the understanding of a process but in the ability of a particular statistical technique to identify both structure and scales of pattern in a spatial data set. Indeed, at the time of data collection it was by no means obvious that the below-ground invasion process could be discovered through an analysis of the above-ground shoot count. We have seen that the regularity inherent in a deterministic analysis throws up spurious wave patterns, whilst the process itself is too complicated to yield to a theoretical stochastic analysis. Simulation therefore provides the only way forward, and is especially useful here since it can be used to show just how variable root direction can be before the statistical analysis fails to signal the presence of a directional effect.

Ripley (1984) comments that, 'Simulation is an increasingly useful tool in stochastic modelling. It can help both statisticians and scientists envisage realizations and so gain a perceptual understanding of the process.' This remark is particularly pertinent in the spatial study of branching patterns, since the combination of branching and correlated step directions ensures that all but the simplest of models are mathematically intractable. Fortunately, with modern computing power the running of long and detailed simulation programs which accurately describe complex biological systems is no longer a problem. A hidden advantage in incorporating such complexity is that a considerable number of detailed biological questions must be answered in order to write the appropriate software, and it may be surprising to discover just how little is known about certain features of plant growth (Renshaw, 1985). Thus the procedure generates a source of important *academic* questions. Moreover, since stem-wood production is directly related to crown structure, and indirectly related to root structure through utilization of soil resource and resistance to windthrow, increased understanding of the branching mechanism is of potential *commercial* interest through the production of clones from trees which possess 'optimal properties'.

Statistical analysis of the Sitka spruce root systems shows that the underlying spatial branching process is more systematic than was first thought, though the degree of regularity is far less than that of the canopy structure. For both aerial and root structures the simulations appear to be fairly insensitive with respect to the chosen forms of the underlying distribution functions, yet sensitive to the correct choice of variances. If this was not so, then not only would it be virtually impossible to construct realistic simulations, but also a sensitivity analysis of the constituent parts of the branching structures would be impractical. Note the difference in approach though between the above- and below-ground simulation studies. Whilst it is possible to mimic the temporal growth of canopy structure, this cannot be done with structural roots because current data contain little information on time-development.

In spite of this restriction the root model provides a useful morphological description when knowledge at a fixed time is required. For instance, in many experiments estimates of total root length are important, but they are difficult to obtain because the excavation of complete structural root systems is often both laborious and expensive. Several authors (e.g. Bowen, 1964, and Savill, 1976) have described methods of excavating only part of the soil around a tree in order to obtain such estimates, and simulation is a very useful tool for comparing the effectiveness of the various methods proposed. To conclude, we shall illustrate how one such estimation technique was assessed by Henderson, Ford and Renshaw (1983).

Suppose the total length of a root system (y) is estimated from the known total length of roots (x) in a $90°$ segment. Then there exists an intuitive estimator of y, namely

$$y = 4x,$$

since one-quarter of the available soil volume has been excavated. Twenty root systems were simulated and their total root lengths determined together with the lengths of root in the 90° segment between the north and east directions. Surprisingly, the resulting plot of simulated root lengths did not lie close to the line $y = 4x$. Indeed, fitting a regression line produced an estimator with a far better fit, namely

$$y = 21.33 + 0.720x.$$

To compare the accuracy of these two estimation procedures twenty more root systems were then simulated and total root lengths were compared with expected values based on the root lengths in the 90° segment:

Estimator	Root mean square residual	Maximum residual magnitude
$y = 4x$	10.88 m	19.65 m
$y = 21.33 + 0.720x$	4.06 m	8.06 m

The intuitive estimator is clearly misleading, and a far better estimator is obtained by using a regression model based purely on simulated data. The line $y = 4x$ provides such a poor fit because both observed and simulated rooting patterns are rarely spread evenly around the stem, so any 90° segment tends to be either densely or sparsely occupied. Thus $4x$ either overestimates or underestimates the total root length, respectively.

Expensive though such simulation studies may be in terms of software development and lengthy computer runs, such costs pale into insignificance when compared with the financial rewards that can result from increased knowledge of tree structure.

REFERENCES

Aksland, M. (1975). A birth, death and migration process with immigration. *Advances in Applied Probability*, **7**, 44–60.

Allee, W.C., Emerson, A.E., Park, O., Park, T. & Schmidt, K.P. (1949). *Principles of Animal Ecology*. Philadelphia: W.B. Saunders.

Andel, J. van (1975). A study on the population dynamics of the perennial plant species *Chamaenerion angustifolium* (L.) Scop. *Oecologia*, **19**, 329–37.

Andrewartha, H.G. & Birch, L.C. (1954). *The Distribution and Abundance of Animals*. Chicago: University Press.

Armitage, P. (1952). The statistical theory of bacterial populations subject to mutation. *Journal of the Royal Statistical Society*, B, **14**, 1–33.

Aron, J.L. & May, R.M. (1982). The population dynamics of malaria. In *The Population Dynamics of Infectious Diseases: Theory and Applications*, ed. R.M. Anderson, pp. 139–79. London: Chapman and Hall.

Bailey, N.T.J. (1964). *The Elements of Stochastic Processes with Applications to the Natural Sciences*. New York: Wiley.

Bailey, N.T.J. (1967). The simulation of stochastic epidemics in two dimensions. *Proceedings of the Fifth Berkeley Symposium on Mathematical Statistics and Probability*, **4**, 237–57. Berkeley and Los Angeles: University of California Press.

Bailey, N.T.J. (1968). Stochastic birth, death and migration processes for spatially distributed populations. *Biometrika*, **55**, 189–98.

Bailey, N.T.J. (1975). *The Mathematical Theory of Infectious Diseases and its Applications*. 2nd edn. London: Griffin.

Bailey, V.A. (1931). The interaction between hosts and parasites. *Quarterly Journal of Mathematics (Oxford Series)*, **2**, 68–77.

Baltensweiler, W. (1971). The relevance of changes in the composition of larch budmoth populations for the dynamics of its numbers. In *Dynamics of Populations*, ed. P.J. den Boer & G.R. Gradwell, pp. 208–19. Wageningen: Centre for Agricultural Publishing.

Barnett, V.D. (1962). The Monte Carlo solution of a competing species problem. *Biometrics*, **18**, 76–103.

Bartholomew, D.J. (1967). *Stochastic Models for Social Processes*. 2nd edn. London: Wiley.

Bartlett, M.S. (1949). Some evolutionary stochastic processes. *Journal of the Royal Statistical Society*, B, **11**, 211–29.

Bartlett, M.S. (1957a). On theoretical models for competitive and predatory biological systems. *Biometrika*, **44**, 27–42.

Bartlett, M.S. (1957b). Measles periodicity and community size. *Journal of the Royal Statistical Society*, A, **120**, 48–70.

Bartlett, M.S. (1960). *Stochastic Population Models in Ecology and Epidemiology*. London: Methuen.

Bartlett, M.S. (1964). The spectral analysis of two-dimensional point processes. *Biometrika*, **51**, 299–311.

Bartlett, M.S. (1975). *The Statistical Analysis of Spatial Pattern*. London: Chapman and Hall.

Bartlett, M.S., Gower, J.C. & Leslie, P.H. (1960). A comparison of theoretical and empirical results for some stochastic population models. *Biometrika*, **47**, 1–11.

Beddington, J.R. (1974). Age distribution and the stability of simple discrete-time population models. *Journal of Theoretical Biology*, **47**, 65–74.

Beddington, J.R. & May, R.M. (1975). Time delays are not necessarily destabilizing. *Mathematical Biosciences*, **27**, 109–17.

Beddington, J.R., Free, C.A. & Lawton, J.H. (1975). Dynamic complexity in predator-prey models framed in difference equations. *Nature*, **255**, 58–60.

Bell, A.D. & Tomlinson, P.B. (1980). Adaptive architecture in rhizomatous plants. *Botanical Journal of the Linnean Society*, **80**, 125–60.

Birch, L.C. (1953). Experimental background to the study of the distribution and abundance of insects. III. The relations between innate capacity for increase and survival of different species of beetles living together on the same food. *Evolution*, **7**, 136–44.

Bowen, G.D. (1964). Root distribution of *Pinus radiata*. CSIRO Divisional Report 1/64.

Boyce, J.M. (1946). The influence of fecundity and egg mortality on the population growth of *Tribolium confusum* Duval. *Ecology*, **27**, 290–312.

Broadbent, S.R. & Kendall, D.G. (1953). The random walk of *Trichostrongylus retortaeformis*. *Biometrics*, **9**, 460–6.

Brownlee, J. (1911). The mathematical theory of random migration and epidemic distribution. *Proceedings of the Royal Society of Edinburgh*, **31**, 262–89.

Bulmer, M.G. (1974). A statistical analysis of the 10-year cycle in Canada. *Journal of Animal Ecology*, **43**, 701–18.

Bulmer, M.G. (1976). The theory of prey–predator oscillations. *Theoretical Population Biology*, **9**, 137–50.

Campbell, M.J. & Walker, A.M. (1977). A survey of statistical work on the Mackenzie River series of annual Canadian lynx trappings for the years 1821–1934 and a new analysis. *Journal of the Royal Statistical Society*, A, **140**, 411–31 (with Discussion, 448–68).

Carl, E.A. (1971). Population control in arctic ground squirrels. *Ecology*, **52**, 395–413.

Carlson, T. (1913). Über Geschwindigkeit und Grösse der Hefevermehrung in Würze. *Biochemische Zeitschrift*, **57**, 313–34.

Caswell, H. (1972). A simulation study of a time-lag population model. *Journal of Theoretical Biology*, **34**, 419–39.

Chandrasekhar, S. (1943). Stochastic problems in physics and astronomy. *Reviews of Modern Physics*, **15**, 1–89.

Chatfield, C. (1980). *The Analysis of Time Series: An Introduction*. 2nd edn. London: Chapman and Hall.

Chiang, C.L. (1968). *Introduction to Stochastic Processes in Biostatistics*. New York: Wiley.

Child, C.M. (1941). *Patterns and Problems of Development*. Chicago: University Press.

Cliff, A.D., Haggett, P., Ord, J.K. & Versey, G.R. (1981). *Spatial Diffusion: An Historical Geography of Epidemics in an Island Community*. Cambridge: University Press.

Cochrane, L.A. & Ford, E.D. (1978). Growth of a Sitka spruce plantation: analysis and stochastic description of the development of the branching structure. *Journal of Applied Ecology*, **15**, 227–44.

Cox, D.R. & Miller, H.D. (1965). *The Theory of Stochastic Processes*. London: Methuen.

Cox, D.R. & Smith, W.L. (1957). On the distribution of *Tribolium confusum* in a container. *Biometrika*, **44**, 328–35.

Crombie, A.C. (1945). On competition between different species of graminivorous insects. *Proceedings of the Royal Society of London*, B, **132**, 362–95.

Crowley, P.H. (1979). Predator-mediated coexistence: an equilibrium interpretation. *Journal of Theoretical Biology*, **80**, 129–44.

Cushing, J.M. (1977). *Integrodifferential Equations and Delay Models in Population Dynamics*. Lecture Notes in Biomathematics, ed. S. Levin, vol. 20. Berlin: Springer-Verlag.

Cvjetanović, B. (1982). The dynamics of bacterial infections. In *The Population Dynamics of*

Infectious Diseases: Theory and Application, ed. R.M. Anderson, pp. 38–66. London: Chapman and Hall.

Davidson, J. (1938a). On the ecology of the growth of the sheep population in South Australia. *Transactions of the Royal Society of South Australia*, **62**, 141–8.

Davidson, J. (1938b). On the growth of the sheep population in Tasmania. *Transactions of the Royal Society of South Australia*, **62**, 342–6.

Deevey, E.S. (1947). Life tables for natural populations of animals. *Quarterly Review of Biology*, **22**, 283–314.

Diggle, P.J. (1981). Binary mosaics and the spatial pattern of heather. *Biometrics*, **37**, 531–9.

Dubois, D.M. (1975). A model of patchiness for prey–predator plankton populations. *Ecological Modelling*, **1**, 67–80.

Dubois, D.M. & Monfort, G. (1978). Stochastic simulation of a space–time dependent predator–prey model. In *Compstat 78: Proceedings in Computational Statistics*, eds. L.C.A. Corsten and J. Hermans, pp. 384–90. Vienna: Physica-Verlag.

Elton, C. & Nicholson, M. (1942). The ten-year cycle in numbers of lynx in Canada. *Journal of Animal Ecology*, **11**, 215–44.

Elton, C.S. (1927). *Animal Ecology*. London: Sidgwick & Jackson.

Elton, C.A. (1958). *The Ecology of Invasions by Animals and Plants*. London: Methuen.

Faddy, M.J. & Slorach, I.H. (1980). Bounds on the velocity of spread of infection for a spatially connected epidemic process. *Journal of Applied Probability*, **17**, 839–45.

Feller, W. (1939). Die Grundlagen der Volterraschen Theorie des Kampfes ums Dasein in wahrscheinlichkeits theoretischen Behandlung. *Acta Biotheoretica*, **5**, 1–40.

Feller, W. (1966). *An Introduction to Probability Theory and its Applications*, 2nd edn, vol. 1. New York: Wiley.

Fisher, R.A. (1937). The wave of advance of advantageous genes. *Annals of Eugenics, London*, **7**, 355–69.

Fitter, A.H. (1987). An architectural approach to the comparative ecology of plant root systems. *The New Phytologist*, **106**, 61–77.

Ford, E.D. (1975). Competition and stand structure in some even-aged plant monocultures. *Journal of Ecology*, **63**, 311–33.

Ford, E.D. (1976). The canopy of a Scots pine forest: description of a surface of complex roughness. *Agricultural Meteorology*, **17**, 9–32.

Ford, E.D. (1985). Branching, crown structure and the control of timber production. Institute of Terrestrial Ecology Annual Report 1984: Cambridge.

Ford, E.D. & Renshaw, E. (1984). The interpretation of process from pattern using two-dimensional spectral analysis. *Vegetatio*, **56**, 113–23.

Fujii, K. (1967). Studies on interspecies competition between the azuki bean weevil *Callosobruchus chinensis* and the southern cowpea weevil *C. maculatus*. II. Competition under different environmental conditions. *Researches in Population Ecology*, **9**, 192–200.

Fujii, K. (1968). Studies on interspecies competition between the azuki bean weevil and the southern cowpea weevil. III. Some characteristics of strains of two species. *Researches in Population Ecology*, **10**, 87–98.

Gani, J. (1965). Stochastic models for bacteriophage. *Journal of Applied Probability*, **2**, 225–68.

Gause, G.F. (1932). Experimental studies on the struggle for existence. I. Mixed population of two species of yeast. *Journal of Experimental Biology*, **9**, 389–402.

Gause, G.F. (1934). *The Struggle for Existence*. Baltimore: Williams & Wilkins.

Gause, G.F. (1935). Experimental demonstration of Volterra's periodic oscillations in the numbers of animals. *Journal of Experimental Biology*, **12**, 44–8.

Gravelius, H. (1914). *Flusskunde*. Berlin: Goschen.

Gross, L.J. (1986). Ecology: an idiosyncratic overview. In Biomathematics, vol. 17, *Mathematical Ecology*, ed. T.G. Hallam & S.A. Levin, pp. 3–15. Berlin: Springer-Verlag.

Gurney, W.S.C. & Nisbet, R.M. (1978). Predator–prey fluctuations in patchy environments. *Journal of Animal Ecology*, **47**, 85–102.

Gurtin, M.E. & MacCamy, R.C. (1977). On the diffusion of biological populations. *Mathematical Biosciences*, **33**, 35–49.

Haldane, J.B.S. (1948). The theory of a cline. *Journal of Genetics*, **48**, 277–84.

Hallam, T.G. (1986). Community dynamics in a homogeneous environment. In Biomathematics, vol. 17, *Mathematical Ecology*, ed. T.G. Hallam & S.A. Levin, pp. 61–94. Berlin: Springer-Verlag.

Haskey, H.W. (1954). A general expression for the mean in a simple stochastic epidemic. *Biometrika*, **41**, 272–5.

Hassell, M.P. (1975). Density-dependence in single-species populations. *Journal of Animal Ecology*, **44**, 283–95.

Hassell, M.P., Lawton, J.H. & May, R.M. (1976). Patterns of dynamical behaviour in single-species populations. *Journal of Animal Ecology*, **45**, 471–86.

Helland, I.S. (1975). The condition for extinction with probability one in a birth, death and migration process. *Advances in Applied Probability*, **7**, 61–5.

Henderson, R. & Renshaw, E. (1980). Spatial stochastic models and computer simulation applied to the study of tree root systems. *Compstat 80*, 389–95.

Henderson, R., Ford, E.D. & Renshaw, E. (1983). Morphology of the structural root system of Sitka spruce. 2. Computer simulation of rooting patterns. *Forestry*, **56**, 137–53.

Henderson, R., Ford, E.D., Renshaw, E. & Deans, J.D. (1983). Morphology of the structural root system of Sitka spruce. 1. Analysis and quantitative description. *Forestry*, **56**, 121–35.

Hengeveld, R. (1989). *Dynamics of Biological Populations*. London: Chapman and Hall.

Hilborn, R. (1975). The effect of spatial heterogeneity on the persistence of predator–prey interactions. *Theoretical Population Biology*, **8**, 346–55.

Holgate, P. (1967). The size of elephant herds. *Mathematical Gazette*, **51**, 302–4.

Holling, C.S. (1965). The functional response of predators to prey density and its role in mimicry and population regulation. *Memoirs of the Entomological Society of Canada*, **45**, 3–60.

Honda, H. (1971). Description of the form of trees by the parameters of the tree-like body: effects of the branching angle and the branch length on the shape of the tree-like body. *Journal of Theoretical Biology*, **31**, 331–8.

Huffaker, C.B. (1958). Experimental studies on predation: dispersion factors and predator–prey interactions. *Hilgardia*, **27**, 343–83.

Huffaker, C.B., Shea, K.P. & Herman, S.G. (1963). Experimental studies on predation: complex dispersion and levels of food in an acarine predator–prey interaction. *Hilgardia*, **34**, 305–30.

Hutchinson, G.E. (1948). Circular causal systems in ecology. *Annals of the New York Academy of Sciences*, **50**, 221–46.

Hutchinson, G.E. (1978). *An Introduction to Population Ecology*. New Haven: Yale University Press.

Isham, V. (1988). Mathematical modelling of the transmission dynamics of HIV infection and AIDS: a review. *Journal of the Royal Statistical Society*, A, **151**, 5–30.

Jillson, D.A. (1980). Insect populations respond to fluctuating environments. *Nature*, **288**, 699–700.

Kelker, D. (1973). A random walk epidemic situation. *Journal of the American Statistical Association*, **68**, 821–3.

Kendall, D.G. (1948). On the role of variable generation time in the development of a stochastic birth process. *Biometrika*, **35**, 316–30.

Kendall, D.G. (1949). Stochastic processes and population growth. *Journal of the Royal Statistical Society*, B, **11**, 230–64.

Kendall, D.G. (1950). An artificial realization of a simple 'birth-and-death' process. *Journal of the Royal Statistical Society*, B, **12**, 116–19.

Kendall, D.G. (1956). Deterministic and stochastic epidemics in closed populations. *Proceedings of the Third Berkeley Symposium on Mathematical Statistics and Probability*, **4**, 149–65. Berkeley and Los Angeles: University of California Press.

Kendall, D.G. (1965). Mathematical models of the spread of infection. In *Mathematics and Computer Science in Biology and Medicine*, pp. 213–25. London: HMSO.

Kermack, W.O. & McKendrick, A.G. (1927). Contributions to the mathematical theory of epidemics. *Proceedings of the Royal Society of London*, A, 115, 700–21.

Keyfitz, N. (1977). *Introduction to the Mathematics of Population (with Revisions)*. Reading, Massachusetts: Addison-Wesley.

Kimura, M. (1953). 'Stepping stone' model of population. *Annual Report of the National Institute of Genetics, Japan*, 3, 62–3.

Kingman, J.F.C. (1969). Markov population processes. *Journal of Applied Probability*, 6, 1–18.

Kolmogorov, A.N. (1936). Sulla teoria di Volterra della lotta per l'esistenza. *Giornale dell' Istituto Italiano degli Attuari*, 7, 74–80.

Kraak, W.K., Rinkel, G.L. & Hoogenheide, J. (1940). Oecologische bewerking van de Europese ringgegevens van der Kievit (*Vanellus vanellus* L.). *Ardea*, 29, 151–7.

Krebs, C.J. (1985). *Ecology: The Experimental Analysis of Distribution and Abundance*. 3rd edn. New York: Harper & Row.

Krebs, C.J., Keller, B.L. & Tamarin, R.H. (1969). *Microtus* population biology: Demographic changes in fluctuating populations of *M. ochrogaster* and *M. pennsylvanicus* in southern Indiana. *Ecology*, 50, 587–607.

Leslie, P.H. (1958). A stochastic model for studying the properties of certain biological systems by numerical methods. *Biometrika*, 45, 16–31.

Leslie, P.H. & Gower, J.C. (1958). The properties of a stochastic model for two competing species. *Biometrika*, 45, 316–30.

Leslie, P.H. & Gower, J.C. (1960). The properties of a stochastic model for the predator–prey type of interaction between two species. *Biometrika*, 47, 219–34.

Levin, S.A. (1974). Dispersion and population interactions. *American Naturalist*, 108, 207–28.

Levin, S.A. (1978). Population models and community structure in heterogeneous environments. In *Mathematical Association of America Study in Mathematical Biology, vol. II: Populations and Communities*, ed. S.A. Levin, 439–76. Washington: Mathematical Association of America.

Levin, S.A. (1986). Population models and community structure in heterogeneous environments. In *Biomathematics, vol. 17, Mathematical Ecology*, ed. T.A. Hallam & S.A. Levin, 295–320. Berlin: Springer-Verlag.

Levin, S.A. & Paine, R.T. (1974). Disturbance, patch formation, and community structure. *Proceedings of the National Academy of Sciences, USA*, 71, 2744–7.

Levin, S.A. & Paine, R.T. (1975). The role of disturbance in models of community structure. In *Ecosystem Analysis and Prediction*, ed. S.A. Levin, 56–67. Philadelphia: SIAM-SIMS Conference, 1974.

Lotka, A.J. (1925). *Elements of Physical Biology*. Baltimore: Williams and Wilkins.

Luckinbill, L.S. (1973). Co-existence in laboratory populations of *Paramecium aurelia* and its predator *Didinium nasutum, Journal of Ecology*, 54, 1320–7.

MacArthur, R.H. & Wilson, E.O. (1967). *The Theory of Island Biogeography*. Princeton: University Press.

McBratney, A.B. & Webster, R. (1981). Detection of ridge and furrow pattern by spectral analysis of crop yield. *International Statistical Review*, 49, 45–52.

MacDonald, G. (1957). *The Epidemiology and Control of Malaria*. Oxford: University Press.

MacDonald, N. (1978). *Time Lags in Biological Models*. Lecture Notes in Biomathematics, ed. S. Levin, vol. 27. Berlin: Springer-Verlag.

MacDonald, N. (1983). *Trees and Networks in Biological Models*. New York: Wiley.

Mansfield, E. & Hensley, C. (1960). The logistic process: tables of the stochastic epidemic curve and applications. *Journal of the Royal Statistical Society*, B, 22, 332–7.

May, R.M. (1971a). Stability in multispecies community models. *Mathematical Biosciences*, 12, 59–79.

May, R.M. (1971b). Stability in model ecosystems. *Proceedings of the Ecological Society of Australia*, **6**, 18–56.

May, R.M. (1974a). Ecosystem patterns in randomly fluctuating environments. *Progress in Theoretical Biology*, **3**, 1–50.

May, R.M. (1974b). *Stability and Complexity in Model Ecosystems*, 2nd edn. Princeton: University Press.

May, R.M. (1974c). Biological populations with non-overlapping generations: stable points, stable cycles, and chaos. *Science*, **186**, 645–7.

May, R.M. (1975). Biological populations obeying difference equations: stable points, stable cycles and chaos. *Journal of Theoretical Biology*, **51**, 511–24.

May, R.M. (1986). When two and two do not make four: nonlinear phenomena in ecology. *Proceedings of the Royal Society of London*, **B**, **228**, 241–66.

Maynard Smith, J. (1974). *Models in Ecology*. Cambridge: University Press.

Mercer, W.B. & Hall, A.D. (1911). The experimental error of field trials. *Journal of Agricultural Science*, **4**, 107–32.

Mimura, M. & Murray, J.D. (1978). On a diffusive prey–predator model which exhibits patchiness. *Journal of Theoretical Biology*, **75**, 249–62.

Mollison, D. (1972). The rate of spatial propagation of simple epidemics. *Proceedings of the Sixth Berkeley Symposium on Mathematical Statistics and Probability*, **3**, 579–614.

Mollison, D. (1977). Spatial contact models for ecological and epidemic spread (with Discussion). *Journal of the Royal Statistical Society*, **B**, **39**, 283–326.

Moran, P.A.P. (1953a). The statistical analysis of the Canadian lynx cycle. I. Structure and prediction. *Australian Journal of Zoology*, **1**, 163–73.

Moran, P.A.P. (1953b). The statistical analysis of the Canadian lynx cycle. II. Synchronization and meteorology. *Australian Journal of Zoology*, **1**, 291–8.

Morgan, B.J.T. (1976). Stochastic models of grouping changes. *Advances in Applied Probability*, **8**, 30–57.

Morgan, B.J.T. (1984). *Elements of Simulation*. London: Chapman and Hall.

Morgan, B.J.T. & Hinde, J.P. (1976). On an approximation made when analysing stochastic processes. *Journal of Applied Probability*, **13**, 672–83.

Morgan, B.J.T. & Watts, S.A. (1980). On modelling microbial infections. *Biometrics*, **36**, 317–21.

Murray, J.D. (1974). *Asymptotic Analysis*. Oxford: Clarendon Press.

Myerscough, P.J. (1980). Biological Flora of the British Isles. *Epilobium angustifolium* L. (*Chamaenerion angustifolium* (L.) Scop.). *Journal of Ecology*, **68**, 1047–74.

Neyman, J., Park, T. & Scott, E.L. (1956). Struggle for existence. The *Tribolium* model. *Proceedings of the Third Berkeley Symposium on Mathematical Statistics and Probability*, **3**, 41–79. Berkeley and Los Angeles: University of California Press.

Nicholson, A.J. (1954). An outline of the dynamics of animal populations. *Australian Journal of Zoology*, **2**, 9–65.

Nicholson, A.J. (1957). The self-adjustment of populations to change. *Cold Spring Harbour Symposium on Quantitative Biology*, **22**, 153–73.

Nisbet, R.M. & Gurney, W.S.C. (1982). *Modelling Fluctuating Populations*. New York: Wiley.

Okubo, A. (1980). *Diffusion and Ecological Problems: Mathematical Models*. Berlin: Springer-Verlag.

Park, T. (1954). Experimental studies on interspecies competition. II. Temperature, humidity, and competition in two species of *Tribolium*. *Physiological Zoology*, **27**, 177–238.

Park, T., Leslie, P.H. & Mertz, D.B. (1964). Genetic strains and competition in populations of *Tribolium*. *Physiological Zoology*, **37**, 97–162.

Patil, V.T. (1957). The consistency and adequacy of the Poisson–Markov model for density fluctuations. *Biometrika*, **44**, 43–56.

Pearl, R. (1927). The growth of populations. *Quarterly Review of Biology*, **2**, 532–48.

Pearl, R. (1930). *Introduction of Medical Biometry and Statistics*. Philadelphia: Saunders.

Pearl, R. (1932). The influence of density of population upon egg production in *Drosophila melanogaster*. *Journal of Experimental Zoology*, **63**, 57–84.

Pearl, R. & Reed, L.J. (1920). On the rate of growth of the population of the United States since 1790 and its mathematical representation. *Proceedings of the National Academy of Sciences*, **6**, 275–88.

Peitgen, H.-O. & Richter, P.H. (1986). *The Beauty of Fractals: Images of Complex Dynamical Systems*. Berlin: Springer-Verlag.

Piaggio, H.T.H. (1962). *An Elementary Treatise on Differential Equations and Their Applications*. London: Bell and Sons.

Pielou, E.C. (1974). *Population and Community Ecology: Principles and Methods*. New York: Gordon and Breach.

Pielou, E.C. (1977). *Mathematical Ecology*. New York: Wiley.

Pimental, D., Nagel, W.P. & Madden, J.L. (1963). Space–time structure of the environment and the survival of parasite–host systems. *American Naturalist*, **97**, 141–67.

Platt, T. & Denman, K.L. (1975). Spectral analysis in ecology. *Annual Review of Ecology and Systematics*, **6**, 189–210.

Poole, R.W. (1977). Periodic, pseudoperiodic and chaotic population fluctuations. *Ecology*, **58**, 210–13.

Pratt, D.M. (1943). Analysis of population development in *Daphnia* at different temperatures. *Biological Bulletin*, **85**, 116–40.

Puri, P.S. (1968). Interconnected birth and death processes. *Journal of Applied Probability*, **5**, 334–49.

Rajendram, S. & Jayewickreme, S.H. (1951). Malaria in Ceylon. Part I. The control and prevention of epidemic malaria by the residual spraying of homes with DDT. *Indian Journal of Malariology*, **5**, 1–73.

Reid, C. (1899). *The Origin of the British Flora*. London: Dulau.

Renshaw, E. (1972). Birth, death and migration processes. *Biometrika*, **59**, 49–60.

Renshaw, E. (1973a). Interconnected population processes. *Journal of Applied Probability*, **10**, 1–14.

Renshaw, E. (1973b). The effect of migration between two developing populations. *Proceedings of the 39th Session of the International Statistical Institute*, **2**, 294–8.

Renshaw, E. (1974). Stepping-stone models for population growth. *Journal of Applied Probability*, **11**, 16–31.

Renshaw, E. (1977). Velocities of propagation for stepping-stone models of population growth. *Journal of Applied Probability*, **14**, 591–7.

Renshaw, E. (1980). The spatial distribution of *Tribolium confusum*. *Journal of Applied Probability*, **17**, 895–911.

Renshaw, E. (1982). The development of a spatial predator–prey process on interconnected sites. *Journal of Theoretical Biology*, **94**, 355–65.

Renshaw, E. (1984). Competition experiments for light in a plant monoculture: an analysis based on two-dimensional spectra. *Biometrics*, **40**, 717–28.

Renshaw, E. (1985). Computer simulation of Sitka spruce: spatial branching models for canopy growth and root structure. *IMA Journal of Mathematics Applied in Medicine and Biology*, **2**, 183–200.

Renshaw, E. (1986). A survey of stepping-stone models in population dynamics. *Advances in Applied Probability*, **18**, 581–627.

Renshaw, E. & Ford, E.D. (1983). The interpretation of process from pattern using two-dimensional spectral analysis: methods and problems of interpretation. *Applied Statistics*, **32**, 51–63.

Renshaw, E. & Ford, E.D. (1984). The description of spatial pattern using two-dimensional spectral analysis. *Vegetatio*, **56**, 75–85.

Rescigno, A. & Richardson, I.W. (1967). The struggle for life. I. Two species. *Bulletin of Mathematical Biophysics*, **29**, 377–88.

Rich, E.R. (1956). Egg cannibalism and fecundity in *Tribolium*. *Ecology*, **37**, 109–210.

Ricker, W.E. (1954). Stock and recruitment. *Journal of the Fisheries Research Board of Canada*, **11**, 559–623.

Ripley, B.D. (1984). Present position and potential developments: some personal views. Statistics in the natural sciences. *Journal of the Royal Statistical Society*, A, **147**, 340–8.

Ripley, B.D. (1987). *Stochastic Simulation*. New York: Wiley.

Rothschild, Lord (1953). A new method of measuring the activity of spermatozoa. *Journal of Experimental Biology*, **30**, 178–99.

Roughgarden, J. (1975). A simple model for population dynamics in stochastic environments. *American Naturalist*, **109**, 713–36.

Routledge, R.D. (1987). Rhizome architecture for dispersal in *Eleocharis palustris*. *Canadian Journal of Botany*, **65**, 1218–23.

Sang, J.H. (1950). Population growth in *Drosophila* cultures. *Biological Reviews*, **25**, 188–219.

Savill, P.S. (1976). The effects of draining and ploughing of surface water gleys on rooting and windthrow of Sitka spruce in Northern Ireland. *Forestry*, **49**, 133–41.

Schwerdtfeger, F. (1935). Studien über den Massenwechsel einiger Forstschädlinge. *Zeitschrift für Forst-u. Jagdwesen*, **67**, 15–38.

Scudo, F.M. (1971). Vito Volterra and theoretical ecology. *Theoretical Population Biology*, **2**, 1–23.

Seton, E.T. (1912). *The Arctic Prairies*. London: Constable.

Sherman, B. (1956). The limiting distribution of Brownian motion on a finite interval with instantaneous return. *Westinghouse Research Laboratory*, Scientific Paper 60-94698-3-P3.

Shiga, T. (1985). Mathematical results on the stepping stone model in population genetics. In *Population Genetics and Molecular Evolution*, ed. T. Ohta & K. Aoki. Tokyo: Japan Scientific Societies Press.

Shreve, R.L. (1966). Statistical law of stream numbers. *Journal of Geology*, **74**, 17–37.

Siegart, A.J.F. (1949). On the approach to statistical equilibrium. *Physical Review*, **76**, 1708–14.

Simberloff, D.S. (1976). Experimental zoogeography of islands: effects of island size. *Ecology*, **57**, 629–48.

Skellam, J.G. (1951). Random dispersal in theoretical populations. *Biometrika*, **38**, 196–218.

Skellam, J.G. (1967). Seasonal periodicity in theoretical population ecology. *Proceedings of the Fifth Berkeley Symposium on Mathematical Statistics and Probability*, **4**, 179–205.

Smith, R.H. & Mead, R. (1980). The dynamics of discrete-time stochastic models of population growth. *Journal of Theoretical Biology*, **86**, 607–27.

Soper, H.E. (1929). Interpretation of periodicity in disease-prevalence. *Journal of the Royal Statistical Society*, **92**, 34–73.

Strahler, A.N. (1952). Hypometric (area-altitude) analysis of erosional topography. *Bulletin of the Geological Society of America*, **63**, 1117–42.

Takashima, M. (1957). Note on evolutionary processes. *Bulletin of Mathematical Statistics*, **7**, 18–24.

Tamarin, R.H. (1977). Dispersal in island and mainland voles. *Ecology*, **58**, 1044–54.

Tanner, J.T. (1975). The stability and the intrinsic growth rates of prey and predator populations. *Ecology*, **56**, 855–67.

Tansley, A.G. (1939). *The British Islands and their Vegetation*. Cambridge: University Press.

Taylor, C.E. & Sokal, R.R. (1976). Oscillations in housefly population sizes due to time lags. *Ecology*, **57**, 1060–7.

Turing, A.M. (1952). The chemical basis of morphogenesis. *Philosophical Transactions of the Royal Society of London*, B, **237**, 37–72.

Ulbrich, J. (1930). *Die Bisamratte*. Dresden: Heinrich.

Usher, M.B. & Williamson, M.H. (1970). A deterministic matrix model for handling the birth, death and migration processes of spatially distributed populations. *Biometrics*, **26**, 1–12.

Utida, S. (1957). Cyclic fluctuations of population density intrinsic to the host–parasite system. *Ecology*, **38**, 442–9.

Utida, S. (1967). Damped oscillation of population density at equilibrium. *Researches in Population Ecology*, **9**, 1–9.

Verhulst, P.F. (1838). Notice sur la loi que la population suit dans son accroissement. *Corr. Math. et Phys. publ. par A. Quetelet*, T.X (also numbered T.II of the third series), 113–21.

Volterra, V. (1926). Fluctuations in the abundance of a species considered mathematically. *Nature*, **118**, 558–60.

Volterra, V. (1931). *Leçons sur la Théorie Mathématique de la Lutte pour la Vie*. Paris: Gauthier-Villars.

Watson, G.N. (1952). *Theory of Bessel Functions*. Cambridge: University Press.

Whittle, P. (1955). The outcome of a stochastic epidemic – a note on Bailey's paper. *Biometrika*, **42**, 116–22.

Whittle, P. (1957). On the use of the normal approximation in the treatment of stochastic processes. *Journal of the Royal Statistical Society*, B, **19**, 268–81.

Whittle, P. (1968). Equilibrium distributions for an open migration process. *Journal of Applied Probability*, **5**, 567–71.

Wildbolz, Th. & Baggiolini, M. (1959). Über das Mass der Ausbreitung des Apfelwicklers während der Einblageperiode. *Mitteilungen der Schweizerischen Entomologischen Gesellschaft*, **32**, 241–57.

Williams, E.J. (1961). The distribution of larvae of randomly moving insects. *Australian Journal of Biological Sciences*, **14**, 598–604.

Williams, T. & Bjerknes, R. (1972). Stochastic model for abnormal clone spread through epithelial basal layer. *Nature*, **236**, 19–21.

Williamson, M.H. (1975). The biological interpretation of time series analysis. *Bulletin of the Institute of Mathematics and its Applications*, **11**, 67–69.

Yule, G.U. (1925). A mathematical theory of evolution, based on the conclusions of Dr. J.C. Willis, F.R.S. *Philosophical Transactions of the Royal Society of London*, B, **213**, 21–87.

Zeigler, B.P. (1977). Persistence and patchiness of predator–prey systems induced by discrete event population exchange mechanisms. *Journal of Theoretical Biology*, **67**, 687–713.

AUTHOR INDEX

SUBJECT INDEX

Abundance, 205
Aerial shoots, 366
African elephants, 51
Apical dominance, 369
Autocorrelation
  definition of, 187, 190
  example of, 185, 186, 188
  in a fluctuating environment, 238–40,
    249–53, 255, 256
  for predator–prey process, 185–9, 203,
    214
Autocovariance
  bias in, 187, 239
  definition of, 187
  example of, 187, 188
  in a fluctuating environment, 238–40,
    248–50, 252
  relation to autocorrelation, 187
Aztecs, 325

BASIC, 23
Biocoenosis, 89
Birth process, 16–27, 142, 328
  assumptions in, 16
  deterministic, 17, 381
  first passage time, 24–7, 39
  simulation of, 20–3, 26–7, 381, 382
  stochastic population size, 17–20, 382
Birth–death process, 15, 33–41, 107, 280,
    299, 382
  assumptions in, 2, 33, 46
  deterministic, 2, 33, 34
  estimated growth rate, 33, 34
  extinction in, 2, 36, 37, 39, 40, 41, 66
  simulation of, 3, 38, 39, 281
  stochastic population size, 34–6
Branching architecture, 13, 14, 360, 361,
    368–84
  deterministic model for, 369, 370
  examples of, 360
  ordering for: Gravelius method, 360, 361,
    379; laws for, 361; Strahler method,
    360, 361
  Sitka spruce canopy, 361, 368–75, 383;
    branching description of, 371;

branching rules for, 369, 370, 372, 373;
    measurements for, 371, 372; simulation
    of, 361, 369, 372, 374, 375, 379
  Sitka spruce rooting system, 361, 373,
    375–81, 383; measurements for, 375–7;
    production rules, 375, 378–80;
    simulation models, 361, 384, (fixed-
    time) 379–81; (temporal) 377–9;
    statistical analysis of, 377, 383, 384
Brownian motion, 264–6, 272, 291, 294, 304
Business cycles, 89

Calandra oryzae, 55, 135, 136
Callosobruchus
  chinensis, 115, 116, 172
  maculatus, 113, 115, 116
Calluna vulgaris, 363
Carex arenaria, 368
Cannibalism, 167, 234
Carnivores, 167
Carrying capacity, definition of, 46
Chaotic data analysis, 114–17
  chaotic cycles, 114, 115
  stable cycles, 114–16
Chaotic models, 4, 5, 100–14
  chaotic cycles, 4, 5, 101, 102, 104–10, 112,
    113
  general, 105–7
  simple deterministic, 100–5
  simulation of, 107–10
  stable cycles, 4, 5, 101–3, 105, 106,
    108–11, 113
Chemostat, 100
Cicadas, 100
Climatic variation, 12
Coefficient of variation
  for birth process, 20, 24
  for birth–death process, 35
  for predator–prey process, 182–5
  for simple epidemic, 329
Coexistence, 137, 144, 148, 149, 173
Coherence time, 250–3
Colloidal particles, 44
Compartmental systems, 345